Natural Computing Series

For further volumes:
www.springer.com/series/4190

Julian F. Miller
Editor

Cartesian Genetic Programming

Springer

Editor
Dr. Julian F. Miller
Dept. of Electronics
The University of York
Heslington, YO10 5DD
UK
jfm7@ohm.york.ac.uk

Series Editors
G. Rozenberg (Managing Editor)
rozenber@liacs.nl

Th. Bäck, J.N. Kok, H.P. Spaink
Leiden Center for Natural Computing
Leiden University
Niels Bohrweg 1
2333 CA Leiden
The Netherlands

A.E. Eiben
Vrije Universiteit Amsterdam
Amsterdam
The Netherlands

ISSN 1619-7127 Natural Computing Series
ISBN 978-3-642-17309-7 e-ISBN 978-3-642-17310-3
DOI 10.1007/978-3-642-17310-3
Springer Heidelberg Dordrecht London New York

Library of Congress Control Number: 2011938670

ACM Computing Classification (1998): F.2, I.2, J.2, J.5

Cover design: KünkelLopka GmbH, Heidelberg

Printed on acid-free paper

Springer is part of Springer Science+Business Media (www.springer.com)

My heartfelt thanks to Peter Thomson for confirming the importance of warm water for creative thought.

Preface

This is the first book on Cartesian genetic programming (CGP). CGP is a form of automatic evolution of computer programs and other computational structures using ideas inspired by Darwin's theory of evolution by natural selection.

I still remember vividly, in 1993, when I first heard the term 'genetic algorithms' from a friend while I was working at Napier University in Edinburgh. I had been working at the time in the area of digital logic synthesis and optimization. I and some of my colleagues were designing and implementing new algorithms that would allow digital circuits to be built efficiently from a specification. It often took many months, even years, to design and implement these algorithms. My friend got to hear about this and explained the idea that computational problems could be solved using an algorithm that was inspired by Darwinian evolution. It was a Friday, and after he explained the basic idea he presented me with a copy of Lawrence David Davis's 'Handbook of Genetic Algorithms'. I asked him how long he thought it would take me to write a computer program using genetic algorithms to try to solve the optimization problem I had been working on. He said, 'Oh, a couple of hours, I would think'. That weekend, I devoured the book and wrote my program. By Monday, I had a working program and it was producing excellent results, better even than the purpose-built algorithm I and my colleagues had designed and published the previous year. When I met up with my friend again, the following week, I was excited, but also a little shamefaced, as he had said it should take a few hours, but I had spent the best part of the weekend working on the program! I explained, however, that the method was working well and giving better results than my previous algorithm. He then revealed that he had been joking about it requiring a couple of hours' work, and he too became very excited. From that moment to this, I have been fascinated by evolutionary algorithms and their power to solve problems efficiently, often in unexpected and extraordinary ways.

The contributors to this book include all the leading exponents of the subject. Although the volume is primarily aimed at postgraduates, researchers and academics, it is hoped that it may be useful to undergraduates who wish to learn about one of the leading techniques in genetic programming. For those completely new to evolutionary computation and genetic programming (GP), the introduction should give a sufficient, albeit brief, background. Five well-known representations used in GP are discussed. However, in reality there are many more that have been published in the research literature. Following up the many books and research references in these chapters would be an excellent place to start gaining a broader and more detailed understanding of the ever fascinating topic of evolutionary computing.

Chapter 2 explains the classic form of CGP. Three examples are given of how CGP genotypes could be used to represent computational structures in different domains. The three domains are digital circuits, mathematical equations and artworks. A detailed explanation of how CGP genotypes are decoded is given. The chapter also gives precise pseudocode algorithms for determining which nodes are active, for efficiently decoding a genotype and for calculating fitness, and a suitable evolutionary algorithm. Discussion is given of a suitable method of mutating genotypes, and there is also a brief discussion of recombination. A further section is included that gives advice about choosing parameter settings when running a CGP evolutionary algorithm. It is hoped that this will enable students and researchers to write their own programs so that they can evolve solutions to problems they are interested in. There is also a discussion of the important issue of genetic redundancy and how this leads to fitness neutrality. It is these properties that allow the evolutionary strategy to be a highly effective search method. CGP has largely been used for evolving programs without iteration or recursion. However, this is not a fundamental limitation of CGP. The chapter ends with a brief discussion of this topic. Given that it is hitherto relatively unexplored, it would form an interesting topic for further research.

One of the most powerful methods of problem solving is divide-and-conquer. This is where we solve a harder problem by breaking it down into a number of smaller, easier subproblems. This problem-solving methodology has long been utilized in forms of genetic programming and it is referred to by several names. Examples are automatically defined functions, automatically defined macros, module acquisition and program teams. Such methods are important in making GP systems *scale* up to larger or harder problem instances. Chapter 3 discusses problem decomposition in CGP, and how subfunctions can be automatically evolved and acquired. In CGP, these subfunctions are called *modules*. Two methods are described: one is called embedded CGP, and the other is modular CGP. The chapter includes details of experiments on many benchmark problems. It shows that CGP is a highly efficient and competitive technique when compared with other GP approaches. In addition, a number of methods of genetic recombination are discussed in this chapter including the use of *multiple* chromosomes in CGP.

In most forms of genetic programming, the genotype is composed of a *single* chromosome. Genetic recombination, or crossover, is carried out at a *gene* level. However, in biology, recombination also happens at the higher level of whole chromosomes. Humans have 23 pairs of chromosomes, while the kingfisher bird has 66 pairs, and the goldfish 51. In sexual recombination in humans, a child inherits each of the 23 chromosomes (which get duplicated) from *either* one of the parents and many inherited genetic traits are attributable to one parent or the other. The clearest example of this is in the inheritance of sex. If a sperm cell containing an X chromosome fuses with an ova (which always contains an X chromosome), then a female offspring is created, while if a sperm containing a Y chromosome fuses, it produces a male child. How does this relate to CGP? Well, many computational problems do not have a single output. A good example of this is provided by electronic circuits; for instance Chap. 3 considers an arithmetic logic circuit with 17 outputs. One has a choice to try to evolve a genotype with a single chromosome with 17 outputs, or one could evolve a genotype consisting of 17 chromosomes (each with a single output). In the latter case, it turns out that one can define an optimal form of sexual recombination for this case, where the offspring inherits the best chromosome from each of its parents.

Human programmers rarely, if ever, write programs that while being executed, change themselves. For one thing, such programs would be fiendishly difficult to debug. However, this does not mean that the automatic evolution of such programs cannot cope with such a problem. Chapter 4 takes this idea forward and discusses a relatively new form of CGP, called self-modifying CGP (SMCGP). In SMCGP, an evolved program (a phenotype) can actually modify itself to produce a new phenotype. In this way, a single evolved genotype can be *iterated* to produce a possibly infinite series of phenotypes, each of which is a program. This means that instead of evolving a program that solves one instance of a problem, one can attempt to evolve solutions to a possibly infinite sequence of problems. It turns out that in certain cases it can be proved mathematically that an evolved solution solves all instances, that is to say it is a general solution.

CGP developed from a method for evolving digital circuits, so it is not surprising that ever since researchers have been enhancing and modifying the technique and applying it to more challenging problems in electronic circuit design. Chapter 5 brings the reader up to date on this research. Although it has been known for some time that evolutionary algorithms are capable of producing innovative new designs that traditional synthesis methods were not capable of, there were always intractable limitations on the size of such evolved circuits. This issue is discussed in the chapter. The chapter also discusses the design of polymorphic circuits (where logic gates can assume more than one function depending on an additional physical variable). This is a good application area for CGP, since conventional circuit synthesis tools cannot handle such devices. Interestingly, the chapter describes how a specially designed chip, a hardware implementation of CGP, was used to speed up the evaluation of the design quality (fitness).

Another good area for application of CGP is in the design of specialized complex arithmetic circuits such as multiple-constant-multiplier circuits. These are circuits that multiply their single input by a number of constants. They are very useful in implementing low power finite-impulse response filter circuits. Finding optimal circuits of this type is known to be NP-complete. The chapter shows that CGP can be used to find designs that improve on those found by the state-of-the-art heuristic algorithm

For some time now, we have all been enjoying the benefits of Moore's famous law on the exponentially increasing transistor densities in integrated circuits. However, these densities have increased to the point that devices are approaching atomic sizes. This is causing increasing numbers of failures and lower production yields. Using models of transistors which incorporate random intrinsic variability together with a specially designed representation of CGP combined with a multi-objective algorithm is shown to be capable of designing novel CMOS circuit topologies.

Finally, Chap. 5 discusses two hardware implementations of CGP. The first uses embedded CGP (see Chap. 3) to design classifiers for electromyographic signals. Such methods are essential for the control of prosthetic hands. Results show that this approach can produce hardware prosthesis control classifiers that achieve accuracies close to that of state-of-the-art classification algorithms. In addition, CGP-evolved hardware classifiers offer compactness, fast computation and self-adaptability. The second hardware application uses CGP to implement application-specific caches. Not only does this approach significantly improve cache performance, but also the evolved mapping functions generalize very well. The technique also achieves faster execution times and consumes less energy than conventional cache architectures.

Chapter 6 presents three applications in which CGP can automatically generate novel image-processing algorithms whose performance is comparable to, or even exceeds, that of the best known conventional solutions. The first application deals with the automatic design of low-level image filters that are comparable in quality to conventional filters but with a lower implementation cost. In the second, CGP is used to design more advanced image operators such as dilation/erosion filters, using relatively complex elementary functions (sine, square root, etc.) Such designs may be useful in advanced image-processing software tools. The third application uses evolved CGP graphs to define transformations on medical images. It is shown that CGP image transformations can significantly improve classification accuracy for a number of predefined image features.

Chapter 7 describes a complete CGP design system implemented in hardware (on a field-programmable gate array). This has advantages over a CGP system running on a general processor in that it can achieve a significant speed-up for many applications and could be used in small low-power adaptive embedded systems.

It is well known that in evolutionary computation the evaluation of the quality of evolved solutions can be very time-consuming. In this regard, CGP is no exception. Chapter 8 describes how this problem can be significantly alleviated by exploiting the parallel processing capabilities of the graphics processing units (GPUs) that are found on modern graphics cards. GPUs are fast, highly parallel devices. GPUs are built for rapid processing of 3D graphics; however, it turns out that modern GPUs can also be programmed for more general-purpose computation. It is only recently that the genetic programming community has begun to start exploiting GPUs to accelerate the evaluation of encoded genetic programs. Indeed, CGP was the first GP technique implemented on GPUs that was suitable for a wide range of applications. The chapter describes how such an implementation has resulted in very significant performance increases.

Chapter 9 describes how CGP can be used to design an interesting new kind of artificial neural network (ANN), called the CGP Developmental Network (CGPDN). A neuron in the CGPDN is represented by seven CGP chromosomes encoding different desirable aspects of biological neurons (i.e. neuron body or soma, dendrites, axons, synapses, and dendritic and axonal branches). The CGPDN neuron is developmental in nature and when the programs encoded in the evolved CGP chromosomes are executed, they build a dynamic network in which neurons can replicate, die and change. Also, the dendritic and axonal branches can change their morphology and efficiency while at the same time solving a computational problem. It is shown that CGPDN is capable of learning in two well-known problems in artificial intelligence. The long-term vision of this work is to evolve programs that encode a general learning *capability*, rather than encoding a specific learned response (as in conventional ANNs).

Chapter 10 describes how CGP can be used in the creative arts. The chapter gives a brief overview of a field known as evolutionary art. This is where evolutionary algorithms or genetic programming is used to produce works of visual art. It is argued that some aspects of CGP prove to be highly advantageous in obtaining *contextual focus*. This is where there is simultaneously a drive towards a precise goal and an exploration of wider possibilities. It is argued that contextual focus is essential for building systems that mimic human creativity. Appropriately, the techniques described are used to generate creative portraits of Charles Darwin.

Evolutionary algorithms have long been applied to various problems in medicine as they provide not only good performance but are also highly flexible and adaptable. The final chapter in this book gives an overview of the types of medical problems that may be addressed and illustrates this by considering in detail a number of published case examples. It also presents case studies of how CGP can be utilized in the diagnosis of three medical conditions: breast cancer, Parkinson's disease and Alzheimer's disease. It gives advice on the common pitfalls, benefits and rewards of medical applications and, particularly, the tricky problem of obtaining patient data.

The articles in this book give a good indication of the wide applicability of CGP. The underlying reason for this is primarily because CGP is a general technique for evolving solutions to all sorts of computational problems. However, there are a number of applications of CGP that have been studied that have not been represented in this book. Some have been carried out with or by research colleagues, others by students. Examples include applications to bioinformatics, helicopter control, artificial neural networks, function optimization, artificial life, robot control, object recognition, sensor failure recovery, fault-tolerant circuits and the solution of differential equations. I offer my apologies to the many authors of such work for the lack of inclusion of this work in this volume. Perhaps, in the future a new volume on CGP will be produced.

It is important to realize that CGP is not a fixed technique; that is to say, it is constantly under development. It began with the classic technique described in Chap. 2, which is still widely used, but it developed into embedded or modular CGP, described in Chap. 3, and still more recently into the self-modifying form described in Chap. 4. No doubt this process will continue. However it develops, one thing is clear; it will become more efficient and more widely applicable. The pioneer of GP, John Koza, pointed out that the power of GP is proportional to the availability of computational power. Thus we are likely to see even more impressive results coming from GP (and CGP) in the future.

My very great thanks to my friend Peter Thomson, who, while having a warm shower, came up with a genetic representation that would be suitable for evolving circuits. This inexorably led me to Cartesian genetic programming.

The idea for this book was formed sometime in May 2009. I would like to thank Jerry Liu and James Walker for their help with final corrections. I would like especially to thank Ronan Nugent, senior editor for Computer Science at Springer-Verlag, for his encouragement and his great patience, as my own pledged deadlines slipped. Thanks, Ronan, for keeping faith. I would like also to thank my colleagues and friends who are respected CGP users and researchers and have contributed to this book. May you continue to spread the faith in CGP and retain enough scepticism to keep improving and applying it!

York, UK

March 2011 *Julian Miller*

Contents

List of Contributors

Wolfgang Banzhaf
Department of Computer Science, Memorial University of Newfoundland,
St. John's, NL, A1B 3X5, Canada, banzhaf@cs.mun.ca

Steve DiPaola
School of Interactive Arts and Technology, Simon Fraser University,
Surrey, BC, V3T 0A3, Canada, sdipaola@sfu.ca

Simon L. Harding
Istituto Dalle Molle di Studi sull'Intelligenza Artificiale (IDSIA)
Galleria 2, 6928 Manno-Lugano, Switzerland,
slh@evolutioninmaterio.com

Paul Kaufmann
Department of Computer Science, University of Paderborn,
33098 Paderborn, Germany, paul.kaufmann@upb.de

Gul Muhammad Khan
Electrical Engineering Department, NWFP, University of Engineering Technology,
Peshawar, Pakistan, gk502@nwfpuet.edu.pk

Taras Kowaliw
Faculty of Information Technology, Monash University,
Clayton, VIC 3800, Australia, taras@kowaliw.ca

Julian F. Miller
Department of Electronics, University of York,
York, YO10 5DD, UK, jfm7@ohm.york.ac.uk

Marco Platzner
Department of Computer Science, University of Paderborn,
33098 Paderborn, Germany, platzner@upb.de

Lukas Sekanina
Faculty of Information Technology, Brno University of Technology,
612 66 Brno, Czech Republic, sekanina@fit.vutbr.cz

Stephen L. Smith
Department of Electronics, University of York,
York, YO10 5DD, UK, sls5@ohm.york.ac.uk

Nathan Sorenson
School of Interactive Arts and Technology, Simon Fraser University,
Surrey, BC, V3T 0A3, Canada, nds6@sfu.ca

Zdenek Vasicek
Faculty of Information Technology, Brno University of Technology,
612 66 Brno, Czech Republic, vasicek@fit.vutbr.cz

James Alfred Walker
Department of Electronics, University of York,
York, YO10 5DD, UK, jaw500@ohm.york.ac.uk

Acronyms

ADF	automatically defined function
ADM	automatically defined macro
ANN	artificial neural network
ARL	adaptive representation through learning
BNF	Backus–Naur form
CAD	computer-aided detection or computer-aided design
CC	category classifier
CDM	category detection module
CE	computational effort
CGP	Cartesian genetic programming or Cartesian genetic program
CGPCN	CGP computational neuron
CGPDN	CGP developmental network
CLE	configurable logic element
CLB	configurable logic block
DAG	directed acyclic graph
DCGP	developmental Cartesian genetic programming
ECGP	embedded Cartesian genetic programming
EHW	evolvable hardware
EMG	electromyographic
EP	evolutionary programming
ES	evolutionary strategy
FIR	finite impulse response
FPGA	field-programmable gate array
GA	genetic algorithm
GE	grammatical evolution
GP	genetic programming
GPU	graphics processing unit
IOB	I/O block
IQR	interquartile range
IRCGP	implicit-context-representation Cartesian genetic programming

LGP	linear genetic programming
LUT	look-up table
MA	module acquisition
MAD	median absolute deviation
MCGP	modular Cartesian genetic programming
MC-CGP	multi-chromosome Cartesian genetic programming
MC-ECGP	multi-chromosome embedded Cartesian genetic programming
MCM	multiple-constant multiplier
ME	median number of evaluations
MOCGP	multi-objective Cartesian genetic programming
MOEA	multi-objective evolutionary algorithm
PDGP	parallel distributed genetic programming
PIB	programmable interconnect block
PSNR	peak signal-to-noise ratio
ROC	receiver operating characteristic
ROI	region of interest
SAAN	subgraph active-active node
SM	self-modification
SMCGP	self-modifying Cartesian genetic programming
TSP	travelling salesman problem
VRC	virtual reconfigurable circuit

Chapter 1
Introduction to Evolutionary Computation and Genetic Programming

Julian F. Miller

1.1 Evolutionary Computation

1.1.1 Origins

Evolutionary computation is the study of non-deterministic search algorithms that are based on aspects of Darwin's theory of evolution by natural selection [10]. The principal originators of evolutionary algorithms were John Holland, Ingo Rechenberg, Hans-Paul Schwefel and Lawrence Fogel. Holland proposed *genetic algorithms* and wrote about them in his 1975 book [19]. He emphasized the role of genetic recombination (often called 'crossover'). Ingo Rechenberg and Hans-Paul Schwefel worked on the optimization of physical shapes in fluids and, after trying a variety of classical optimization techniques, discovered that altering physical variables in a random manner (ensuring small modifications were more frequent than larger ones) proved to be a very effective technique. This gave rise to a form of evolutionary algorithm that they termed an *evolutionary strategy* [38, 40]. Lawrence Fogel investigated evolving finite state machines to predict symbol strings of symbols generated by Markov processes and non-stationary time series [15]. However, as is so often the case in science, various scientists considered or suggested search algorithms inspired by Darwinian evolution much earlier. David Fogel, Lawrence Fogel's son, offers a detailed account of such early pioneers in his book on the history of evolutionary computation [14]. However, it is interesting to note that the idea of artificial evolution was suggested by one of the founders of computer science, Alan Turing, in 1948. Turing wrote an essay while working on the construction of an electronic computer called the Automatic Computing Engine (ACE) at the National Physical Laboratory in the UK. His employer happened to be Sir Charles Darwin, the grandson of Charles Darwin, the author of 'On the Origin of Species'. Sir Charles dismissed the article as a "schoolboy essay"! It has since been recognized that in the article Turing not only proposed artificial neural networks but the field of artificial intelligence itself [44].

J.F. Miller (ed.), *Cartesian Genetic Programming*, Natural Computing Series,
DOI 10.1007/978-3-642-17310-3_1, © Springer-Verlag Berlin Heidelberg 2011

1.1.2 Illustrating Evolutionary Computation: The Travelling Salesman Problem

Solving many computational problems can be formulated as a problem of trying to find a string of symbols that lead to a solution. For instance, a well-known problem in computer science is called the travelling salesman problem (TSP). In this problem one wishes to find the shortest route that will visit a collection of cities. The route begins and ends on the same city and must visit every other city exactly once. Assuming that the cities are labelled with the symbols $C_1, C_2, C_3, \ldots, C_n$, then any solution is just a permutation of cities.

The way evolutionary algorithms would proceed on this problem is broadly as set out in Procedure 1.1. Firstly, it is traditional to refer to the string of symbols as a chromosome (or, sometimes, a genotype). The evolutionary algorithm starts by generating a number of chromosomes using randomness; this is called the initial population (step 1). Thus, after this step we would have a number of random permutations of routes (each one visiting each city once). The next step is to evaluate each chromosome in the population. This is referred to as determining the *fitness* of the members of the population (step 2). In the case of the TSP, the fitness of a chromosome is typically the distance covered by the route defined in the chromosome. Thus, in this case one is trying to minimize the fitness. The next step is to select the members of the population that will go forward to create the new population of potential solutions (step 3). Often these chromosomes are called 'parents' as they are used to create new chromosomes (often called 'children'). Generally, children are generated using two operations. The first is called *recombination* or *crossover*. A child is usually generated from two parents, by selecting genes (in this case cities) from each. For example, consider the two parent strings

$$c_2, c_9, c_1, c_8, c_5, c_7, c_3, c_6, c_4, c_{10} \text{ and } C_3, C_8, C_7, C_4, C_6, C_9, C_{10}, C_1, C_5, C_2.$$

Let us suppose that the child is created by taking the first five genes from one parent and the second five from the other. The child chromosome would look like this:

$$c_2, c_9, c_1, c_8, c_5, C_9, C_{10}, C_1, C_5, C_2.$$

However, when we inspect this we have an immediate problem. The child is not a permutation of the ten cities. In such cases, a *repair* procedure needs to be applied that takes the invalid chromosome and rearranges it in some way so that it becomes a valid permutation. In fact, there are a number of ways this can be done, each with advantages and disadvantages [13, 28]. We discuss briefly here an alternative method of recombination of permutation-based chromosomes which does not require a repair procedure. It was proposed by Anderson and Ashlock and is called 'merging crossover' (or MOX) [2]. It works as follows [28]. First, the two parent permutations are randomly merged (i.e. selecting with probability 0.5 a city from either parent) to create a list of twice the size. For example, suppose we have six cities and the two parents are

$$p_1 = c_3, c_1, c_4, c_6, c_5, c_2 \text{ and } p_2 = C_6, C_4, C_2, C_1, C_5, C_3.$$

After random merging we might obtain

$$c_3, C_6, C_4, c_1, c_4, c_6, C_2, C_1, c_5, C_5, C_3, c_2.$$

After this, we split the extended list in such a way that the first occurrence (reading left to right) of each value in the merged list gives the ordering of cities in the first child, and the second occurrence does so in the second child. Thus we obtain the two child chromosomes

$$d_1 = c_3, C_6, C_4, c_1, C_2, c_5 \text{ and } d_2 = c_4, c_6, C_1, C_5, C_3, c_2.$$

In many other computational problems, solution strings may not be permutational in nature (i.e. strings of binary digits or floating-point numbers). The next step in the generic evolutionary algorithm (step 6) is to mutate some parents and offspring. Mutation means making a random alteration to the chromosome. If the chromosome is a permutation, we should make random changes in such a way that we preserve the permutational nature of the chromosome. One simple way of doing this would be to select two cities at random in the chromosome and swap their positions. For instance, taking d_1 above, and swapping the first city with the third, we obtain the mutated version

$$\tilde{d_1} = C_4, C_6, c_3, c_1, C_2, c_5.$$

Appropriate mutations depend on the nature of the chromosome representation. For instance, with binary chromosomes one usually defines a mutation to be a single inversion of a binary gene. Step 7 of the evolutionary algorithm forms the new population from some combination of parents, offspring and their mutated counterparts. In the optional step 8, some parents are promoted to the next generation without change. This is referred to as *elitism*; as we shall see later in this book, this is very often used in Cartesian genetic programming.

The purpose behind the recombination step, 5, is to combine elements of each parent into a new potential solution. The idea behind it is that as evolution progresses, partial solutions (sometimes referred to as *building blocks*) can be formed in chromosomes which, when recombined, are closer to the desired solution. It has the potential to produce chromosomes that are genetically intermediate between the parent chromosomes. Mutation is often thought of as a local random search operator. It makes a small change, thus moving a chromosome to another that is not very distant (in genotype space) from the parent. As we will see in Chap. 2, the degree of genetic change that occurs through the application of a mutation operator is strongly dependent on what the chromosome represents. In Cartesian genetic programming

Procedure 1.1 Evolutionary algorithm

```
1: Generate initial population of size p. Set number of generations, g = 0
2: repeat
3:     Calculate the fitness of each member of the population
4:     Select a number of parents according to quality
5:     Recombine some, if not all, parents to create offspring chromosomes
6:     Mutate some parents and offspring
7:     Form new population from mutated parents and offspring
8:     Optional: promote a number of unaltered parents from step 4 to the new population
9:     Increment the number of generations g ← g + 1
10: until (g equals number of generations required) OR (fitness is acceptable)
```

it turns out that a single gene change can lead to a huge change in the underlying program encoded in the chromosome (the phenotype).

1.2 Genetic Programming

The automatic evolution of computer programs is known as genetic programming (GP). The origins of GP (and evolutionary computation) go back to the origins of evolutionary algorithms [14]. As early as 1958, Friedberg devised an algorithm that could evaluate the quality of a computer program, make some random changes to it and then test it again to check for improvements, and so on [17, 18]. Smith utilized a form of GP in his PhD thesis in 1980 [42]. In 1981, Forsyth advocated the use of GP for artificial intelligence. He evolved Boolean expressions[1] for three prediction problems: (a) whether heart patients would survive treatment in a hospital, (b) to distinguish athletes good at sprinting from those good at longer distances and (c) to predict British soccer results [16]. In 1985, Cramer evolved sequential programs in the computer languages JB and TB [9]. The latter have the form of symbolic expression trees. He used a few assembler-like functions (coded as positive integers, with integer arguments). Unaware of Cramer's work, also in 1985, Schmidhüber experimented with GP in LISP, later reimplementing it in a form of PROLOG [12, 39]. However, GP started to become more widely known after the publication of John Koza's book in 1992 [20]. One of the obvious difficulties in evolving computer programs is caused by the fact that computer programs are highly constrained and must obey a specific grammar in order to be compiled.

[1] Forsyth used the primitive functions AND, OR, NOT, EQ, NEQ, the comparatives GT, LT, GE, and LE, and the arithmetic operations +, minus, $\times$ and $\div$.

1.2.1 GP Representation in LISP

LISP was invented by John McCarthy in 1958 and is one of the oldest high-level computer languages [25]. Even today, it is widely used by researchers in artificial intelligence. In LISP, all programs consist of S-expressions of lists of symbols enclosed in parentheses. For instance, calls to functions are written as a list with the function name first, followed by its arguments. For example, a function `f` that takes four arguments would be written in LISP as `(f arg1 arg2 arg3 arg4)`. All LISP programs can be written in the form of data structures known as trees. This simplifies the task of obtaining valid programs by applying genetic operations. In 1992, John Koza published a comprehensive work on evolution of computer programs in the form of LISP expressions [20]. The example in Fig. 1.1 shows the tree representation of the LISP expression `(SIGMA (SET-SV (* (% X J))))` (Koza [20], p. 470). This happens to be an evolved S-expression that computes a solution to the differential equation (1.1), where the initial value $y_{initial}$ is 2.718, corresponding to an initial value $x_{initial}$ of 1.0:

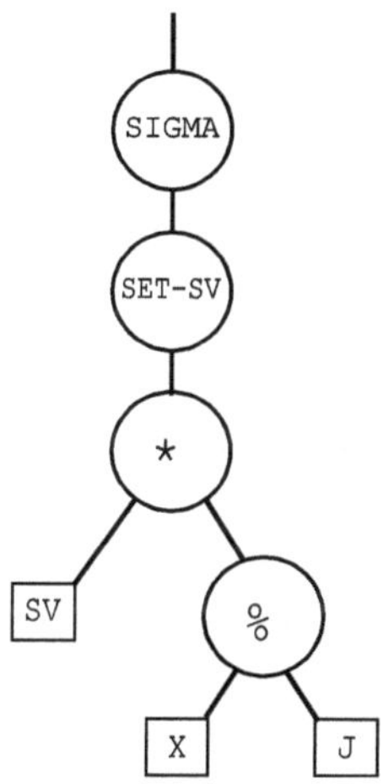

Fig. 1.1 Program tree that represents the S-expression `(SIGMA (SET-SV (* (% X J))))` [20].

$$\frac{dy}{dx} - y = 0. \tag{1.1}$$

The function `SIGMA` is a one-argument function that adds its argument to the value stored in a register called `SV`. It is defined to increment an indexing variable `J` each time it does this. There is a maximum allowed value for `J` of 15 (i.e. the maximum number of iterative summations is 15). The function `%` is a two-argument function that implements protected division. It is protected, as it is defined to return 1 if the denominator is close to zero. The function `SET-SV` assigns its single argument to the register. `SV` and `J` are defined to have initial values of one. The program sums the successive arguments $x, x^2/2, x^3/3!, \ldots$, which approximates a solution to (1.1), $e^x - 1$.

LISP programs (and programs in general) are highly constrained and obey precisely defined syntax. Ensuring that one can generate random trees is a first step. As we saw, evolutionary algorithms generally use the operations of recombination and mutation to generate new candidate solutions. Recombination in tree-based GP involves exchanging subtrees between parent chromosomes. Mutation involves substituting a subtree by a randomly generated one. Detailed accounts of how this can be accomplished are well known [20, 5, 36, 37]. It should be noted that the size of chromosomes in tree-based GP are variable, as recombination and mutation can create offspring of different sizes.

Human programmers solve problems by divide-and-conquer; that is to say, we break down algorithmic problems into smaller subprocedures that are encapsulated as functions. Such subprocedures are usually reused many times by the main program code. Koza enabled his LISP GP system to automatically construct and use such subprocedures. He called them *automatically defined functions* (ADFs) [21]. In his second book, he showed how they can be used in a GP system and the many computational advantages they bring. ADFs become particularly important in solving harder instances of problems; thus they are seen as a very important in bringing scalability to GP. They also make programs more comprehensible.

1.2.2 Linear or Machine Code Genetic Programming

In linear genetic programming (LGP), programs are a constrained linear set of operations and terminals (inputs). Such programs are fairly similar to programs written in machine code. Originally, Banzhaf evolved chromosomes in the form of bitstrings of length 225 [3, 4]. He defined terminals and functions using a five-bit code; thus his programs could encode up to 45 program instructions. The function set comprised the eight functions `PLUS, MINUS, TIMES, DIV, POW, ABS, MOD and IFEQ`, where `DIV` is protected division, `IFEQ(a,b) = 1` if `a` equals `b` and zero otherwise, and `MOD(a,b)= a mod b` (protected). Five-bit codes with the most significant bit equal to zero code for one of the functions (the eight-function list is duplicated to make 16 entries). The remaining five-bit codes denote terminals (inputs). Banzhaf evolved solutions to the problem of integer sequence prediction. The terminal set consisted of `(I0, I1, 0, 1, 2)`. Randomly altered or generated chromosome strings would not necessarily parse into compilable code and so a simple repair strategy needed to be implemented. A simple example showing the process of translating a chromosome bit string into compilable code (phenotype) is shown in Table 1.1.

Table 1.1 The genotype–phenotype mapping process in linear GP

111100111010011111011101111100011100...
Transcribes to
`0 PLUS I0 1 I0 I0 I1...`
After repair
`PLUS PLUS I0 1 I0 I0 I1...`
After parsing
`PLUS(PLUS(I0,1),I0)`
After editing
`function z1(I0,I1)`
`return PLUS(PLUS(I0,1),I0)`

Peter Nordin with Wolfgang Banzhaf, considerably extended this earlier approach to produce a genetic algorithm that manipulates machine code instructions [29, 31, 30]. This technique allowed the use of most program constructs: arrays, arithmetic operators, subroutines, if–then–else, iteration, recursion, strings and list functions. It also allowed incorporating any C function into the GP function set. They used the SPARC instruction set operating on SUN workstations. In this, processor instructions have a 32-bit word length. Crossover operated in such a way that only whole 32-bit instructions were swapped between individuals (thus maintaining validity). Mutation operated by selecting an instruction at random and checked whether it had an embedded constant (a) or whether it was an operation involving only registers (b). If it was case (a), a mutation was a bit-flip in the constant; if case (b), then either the instruction data source or the destination registers were mutated. The resulting system was shown to be up to 2000 times faster than a LISP implementation on some problems.

A more recent form of LGP uses chromosomes representing variable-length strings composed of simple statements in the C programming language [6, 7]. Individual instructions are encoded in a four-component vector. For instance, the C assignment `v[i]= v[j] + c;` would be coded as `(id(+), i, j, c)`. Employing very few possible instructions has proved to be effective in many problems. These are shown in Table 1.2.

Table 1.2 Instruction types in linear GP: op refers to allowed arithmetic operations (i.e. +, minus, *, /) and `cmp` is typically $>, \leq$

Instruction type	Function definition
Arithmetic operation 1	`v[i] = v[j] op v[k]`
Arithmetic operation 2	`v[i] = v[j] op c`
Conditional branch 1	`if (v[i] cmp v[k])`
Conditional branch 2	`if (v[i] cmp c))`
Function call	`v[i]=f([v[k])`

Crossover swaps only a whole number of instructions. Mutation is responsible for changing individual instructions by randomly replacing the instruction identifier, a variable or the constant (if it exists) by equivalents from valid ranges. Constants are modified within a user-defined standard deviation from the current value.

1.2.3 Grammar-Based Approaches

Compilers use grammars to define the legal expressions of a computer language. Thus a natural approach to genetic programming is to explicitly evolve chromosomes that obey a given grammar. This would mean that evolving all kinds of constrained structures or languages could be tackled by using the same generic approach but choosing a different grammar. A recent review of such approaches can be found in [26]. In this section, we will briefly discuss perhaps the most widely known and published grammatical approach, namely *grammatical evolution* (GE) [32, 33].

In GE, variable-length binary-string genomes are used grouped into codons of eight bits. The integer value defined by the codon is used via a mapping function to select an appropriate production rule from a grammar defined using the Backus–Naur form (BNF). BNF grammars consist of *terminals* which are items that can appear in the language (i.e. +, *, x, sin, 3.14) and *non-terminals*, which can be expanded into one or more terminals and non-terminals. A grammar can be represented by the tuple {N, T, P, S}, where N is the set of non-terminals, T is a set of terminals, P is a set of production rules that maps the elements of N to T, and S is a start symbol that is a member of N. When there are a number of productions that can be applied, the choice is delimited with the OR symbol, '|'. An example is given in Table 1.3

Table 1.3 Example BNF form

Non-terminals			expr, op, pre-op	
Terminals			sin, +, -, /, *, x, 1.0, (,)	
Start symbol			<expr>	
Production rules (1)	<expr>::		=<expr><op><expr>	(0)
		|	(<expr><op><expr>)	(1)
		|	<pre-op>(<expr>)	(2)
		|	<var>	(3)
(2)	<op>::=		+	(0)
		|	-	(1)
		|	/	(2)
		|	*	(3)
(3)	<pre-op>::=		sin	(0)
(4)	<var>::=		x	(0)
		|	1.0	(1)

Table 1.4 An example genotype in GE. Here, the eight-bit binary codons have all been packed as integers for convenience

220	240	220	203	101	53	202	203	102	55	223	202	243	134	35	202	203	140	39	202	203	102

The mapping process of mapping the genotype shown in Table 1.4 is carried out as follows. The leftmost non-terminal is selected and the symbol noted (i.e. expr, op, pre-op). We denote the codon by C, and the number of production rules for a given expression is N. The rule to apply is $R = C \bmod N$. Then the symbol is rewritten according to this rule. The decoding process continues until no rules can be applied. Note than the genotype is assumed to be circular (it wraps around) so that the first codon follows the last codon. Table 1.5 shows the complete decoding process for the genotype shown in Table 1.4.

Table 1.5 Decoding the example genotype in Table 1.4. Given an expression, with a number of production rules N and a codon C, a rewriting rule is chosen according to $R = C \bmod N$. This is applied to build the next expression

Expression	*C*	*N*	*R*	Rule
<expr>	220	4	0	<expr>::=<expr><op><expr>
<expr><op><expr>	240	4	0	<expr>::=<expr><op><expr>
<expr><op><expr><op><expr>	220	4	0	<expr>::=<expr><op><expr>
<expr><op><expr><op><expr><op><expr>	203	4	3	<expr>::=<var>
<var><op><expr><op><expr><op><expr>	101	2	1	<var>::=1.0
1.0<op><expr><op><expr><op><expr>	53	4	1	<op>::= -
1.0-<expr><op><expr><op><expr>	202	4	2	<expr>::=<pre-op>(<expr>)
1.0-<pre-op>(<expr>)<op><expr><op><expr>				<pre-op>::=sin
1.0-sin(<expr>)<op><expr><op><expr>	203	4	3	<expr>::=<var>
1.0-sin(<var>)<op><expr><op><expr>	102	2	0	<var>::=x
1.0-sin(x)<op><expr><op><expr>	55	4	3	<op>::= *
1.0-sin(x)*<expr><op><expr>	223	4	3	<expr>::=<var>
1.0-sin(x)*<var><op><expr>	202	2	0	<var>::=x
1.0-sin(x)*x<op><expr>	243	4	3	<op>::= *
1.0-sin(x)*x*<expr>	134	4	2	<expr>::=<pre-op>(<expr>)
1.0-sin(x)*x*<pre-op>(<expr>)				<pre-op>::=sin
1.0-sin(x)*x*sin(<expr>)	35	4	3	<expr>::=<var>
1.0-sin(x)*x*sin(<var>)	202	2	0	<var>::=x
1.0-sin(x)*x*sin(x)				

Since, in GE, genotypes are binary strings, no special crossover or mutation operators are required. The genotype-to-phenotype mapping process will always generate syntactically correct individuals. In addition to the standard genetic operators of mutation (point) and crossover, a codon duplication operator is also used. Duplication involves randomly selecting a number of codons to duplicate, and the starting position of the first codon in this set. The duplicated codons are placed at the end of the chromosome. So the genotype is of variable length.

1.2.4 PushGP

A stack-based computer language called Push has been developed by Lee Spector [43]. His GP system using Push (called PushGP) allows many advanced GP features, such as multiple data types, automatically defined subroutines and control structures. Push was also defined by Spector to support a self-adaptive form of evolutionary computation called *autoconstructive evolution*. An autoconstructive evolution system is an evolutionary computation system that adaptively constructs its own mechanisms of reproduction and diversification as it runs. That is to say, methods of recombination and mutation can be evolved in the system rather than, as is more usual, imposed from the start.

In stack-based computer languages,[2] arguments are passed to instructions via global data stacks. This contrasts with argument-passing techniques based on registers. In stack-based argument passing the programmer first specifies (or computes) arguments that are pushed onto the stack, and then executes an instruction using those arguments. For example, consider adding 2 and 7. This would be written in postfix notation as 2 7 +, and this code specifies that 2 and then 7 should be pushed onto the stack, and then the + instruction should be executed. The + instruction removes the top two elements of the stack, adds them together and pushes the result back onto the stack. If extra arguments are present, they are ignored automatically as each instruction takes only the arguments that it needs from the top of the stack. A stack instruction containing too few arguments would normally be signalled as a run-time error and require program termination; however in Push, an instruction with insufficient arguments is simply ignored (treated as a NOOP).

Push handles multiple data types by providing a stack for each type. There is a stack for integers, a stack for floating-point numbers, a stack for Boolean values, a stack for data types (called TYPE), a stack for program code (CODE) and others. Each instruction takes the inputs that it requires from appropriate stacks and pushes outputs onto appropriate stacks. Using the CODE stack allows Push to handle recursion and subprocedures. The CODE stack also allows evolved programs to push themselves or parts of themselves onto the CODE stack. It is this mechanism that allows programs to define new genetic operators (i.e. recombination and mutation) to create their own offspring.

A Push language reference can be found in [43]. [3]

---

[2] The language Forth is a well-known example.

[3] Source code is available for research versions of the Push interpreter and PushGP from Lee Spector's website.

1.2.5 Cartesian Graph-Based GP

Unlike trees, where there is always a unique path between any pair of nodes, graphs allow more than one path between any pair of nodes. If we assume all nodes carry out some computational function, representing functions in the form of graphs is more compact than trees since they allow the reuse of previously calculated subgraphs. Graphs are also attractive representations, since they are widely used in many areas of computer science and engineering [1, 11, 8]. Indeed, neural networks are graphs.

It appears that the first person to evolve graph-based encodings using a Cartesian grid was Sushil Louis in 1990 [22, 23]. In a technical report, Louis described a binary genotype that encodes a network of digital logic gates, in which gates in each column can be connected to the gates in the previous column. Figure 1.2 shows an image extracted from Louis's 1993 PhD thesis [24].

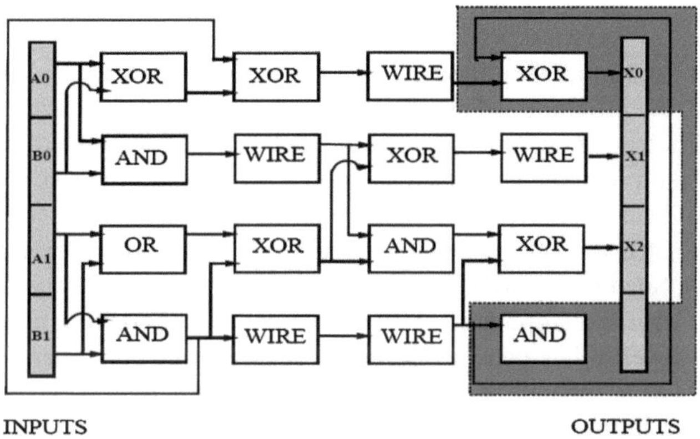

Fig. 1.2 Sushil Louis's evolved 2-bit adder. Image extracted from [24].

Independently, and inspired by neural networks, Poli proposed a graph-based form of GP called parallel distributed GP (PDGP) [34, 35]. PDGP, in principle, allows the evolution of standard tree-like programs, logic networks, neural networks, recurrent transition networks and finite state automata. Some of these are made possible by associating labels with the edges in the program graph. In addition to the usual function and terminal sets, PDGP requires the definition of a set of links that determine how nodes are connected. The labels on the links depend on what is to be evolved. For example, in neural networks, the link labels are numerical constants for the neural-network weights. An example of PDGP is given in Fig. 1.3.

In Fig. 1.3, the programs output is defined to be at coordinates (0,0). Connections point upwards and are allowed only between nodes belonging to adjacent rows. Like Louis, Poli included an identity function so that nodes in non-adjacent rows can still

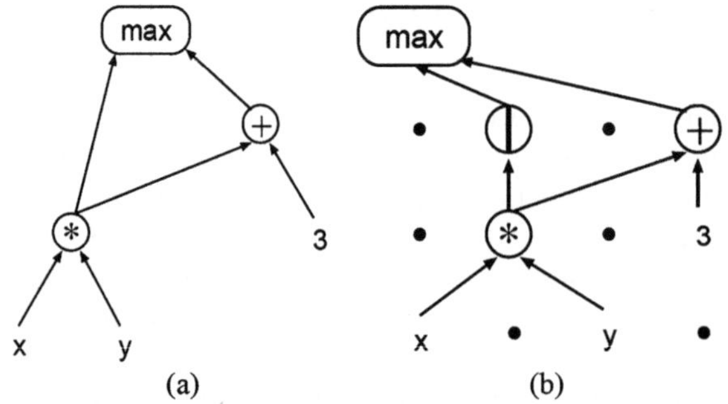

Fig. 1.3 (a) Graph-based representation of the expression max(x*y, 3+x*y), (b) Grid-based representation of the graph in (a). Image extracted from [35].

connect with each other (see the pass-through node in the figure). When PDGP is implemented, the program is represented as an array with the same topology as that of the grid. Each node contains a function label and the horizontal displacement of the nodes in the previous layer used as arguments for the function. The horizontal displacement is an offset from the position of the calling node. Functions or terminals are associated to every node in the grid even if they are not referenced in the program path. Such nodes are inactive (introns).

The basic crossover operator of PDGP is called subgraph active–active node (SAAN) crossover. It is a generalization for graphs of the crossover generally used in tree-based GP to recombine trees. SAAN crossover is defined below, and an example is shown in Fig. 1.4.

1. A random active node is selected in each parent (this is the crossover point),
2. A subgraph including all the active nodes which are used to compute the output value of the crossover point in the first parent is extracted,
3. The subgraph is inserted into the second parent to generate the offspring (if the x coordinate of the insertion node in the second parent is not compatible with the width of the subgraph, the subgraph is wrapped around).

Poli used two forms of mutation in PDGP. A *global* mutation inserts a randomly generated subgraph into an existing program. A *link* mutation changes a random connection in a graph by firstly selecting a random function node, then selecting a random input link of such a node and, finally, altering the offset associated with the link.

As we will see in Chap. 2, Cartesian GP (CGP) also encodes directed graphs; however, the genotype is just a one-dimensional string of integers. Also, CGP genetic operators operate directly on the chromosome, while in PDGP they act directly on the graph. Furthermore, CGP has almost always used mutation (in particular, Poli's link mutation) as its main search operation; however, as we will see later in

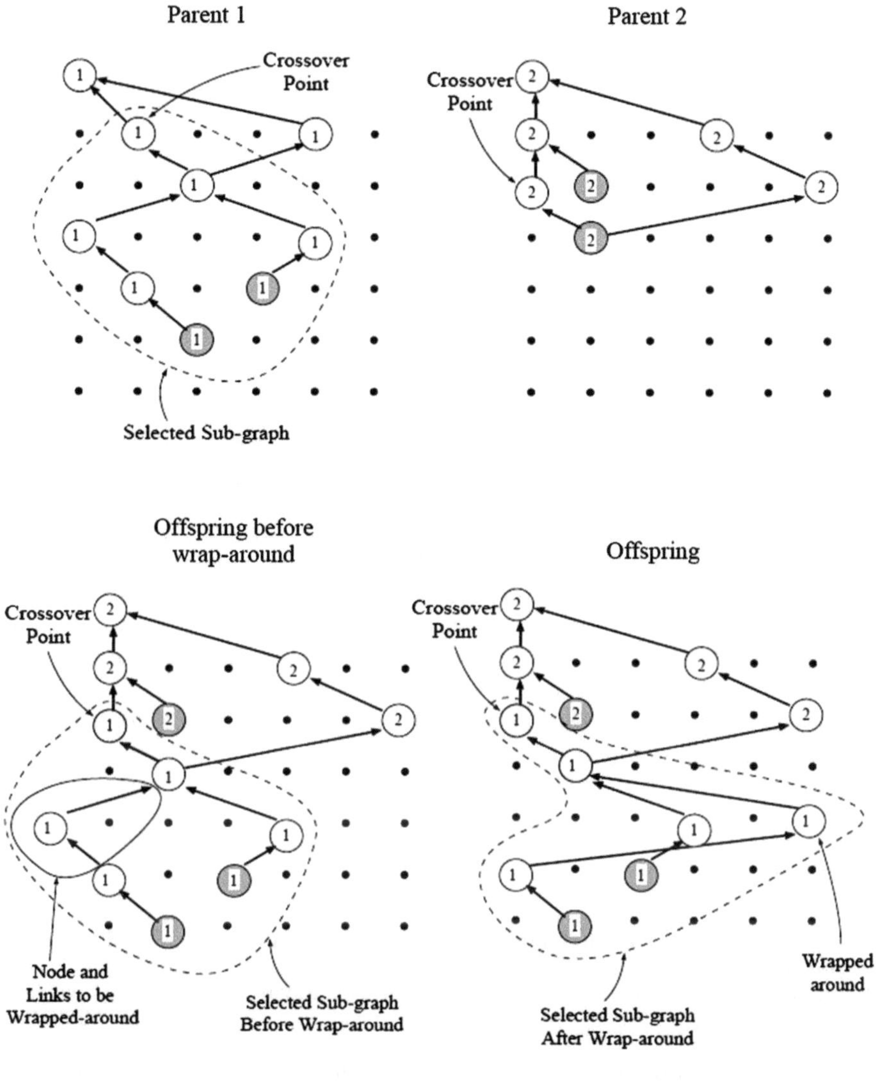

Fig. 1.4 An example of subgraph active–active node (SAAN) crossover [35].

this book, a number of crossover methods have been investigated for CGP. Rather than offsets, CGP uses absolute addresses to determine where nodes obtain their input data from. Graph connectivity is also controlled by a parameter called *levels-back* and nodes can obtain inputs from any of the previous nodes within the range of columns defined by this parameter. In addition, CGP evolutionary algorithms tend to be a form of evolutionary strategy using small populations and elitism.

1.2.6 Bloat

When evolutionary algorithms are applied to many representations of programs, a phenomenon called *bloat* happens. This is where, as the generations proceed, the chromosomes become larger and larger without any increase in fitness. Such programs generally have large sections of code that contain inefficient or redundant subexpressions. This can be a handicap, as it can mean that processing such bloated programs is time-consuming. Eventually, an evolved program could even exceed the memory capacity of the host computer. In addition, the evolved solutions can be very hard to understand and are very inelegant. There are many theories about the causes of bloat and many proposed practical remedies [36, 37, 41]. It is worth noting that Cartesian GP cannot suffer from genotype growth, as the genotype is of fixed size; in addition, it also appears not to suffer from phenotypic growth [27]. Indeed, it will be seen that program sizes remain small even when very large genotype lengths are allowed (see Sect. 2.7).

References

1. Aho, A.V., Ullman, J.D., Hopcroft, J.E.: Data Structures and Algorithms. Addison–Wesley (1983)
2. Anderson, P.G., Ashlock, D.A.: Advances in Ordered Greed. In: C.H. Dagli (ed.) Intelligent Engineering Systems Through Artificial Neural Networks, vol. 14, pp. 223–228. ASME Press (2004)
3. Banzhaf, W.: Genetic programming for pedestrians. In: S. Forrest (ed.) Proc. International Conference on Genetic Algorithms, p. 628. Morgan Kaufmann (1993)
4. Banzhaf, W.: Genetic Programming for pedestrians. Tech. Rep. 93-03, Mitsubishi Electric Research Labs, Cambridge, MA (1993)
5. Banzhaf, W., Nordin, P., Keller, R.E., Francone, F.D.: Genetic Programming: An Introduction. Morgan Kaufmann (1999)
6. Brameier, M., Banzhaf, W.: A Comparison of Linear Genetic Programming and Neural Networks in Medical Data Mining. IEEE Transactions on Evolutionary Computation **5**(1), 17–26 (2001)
7. Brameier, M.F., Banzhaf, W.: Linear Genetic Programming. Springer (2006)
8. Chartrand, G., Lesniak, L., Zhang, P.: Graphs and Digraphs, fifth edn. Chapman and Hall (2010)
9. Cramer, N.L.: A Representation for the Adaptive Generation of Simple Sequential Programs. In: J.J. Grefenstette (ed.) Proc. International Conference on Genetic Algorithms and their Applications. Carnegie Mellon University, USA (1985)

10. Darwin, C.: On the Origin of Species by Means of Natural Selection, or the Preservation of Favoured Races in the Struggle for Life. John Murray (1859)
11. Deo, N.: Graph Theory with Applications to Engineering and Computer Science. Prentice-Hall (2004)
12. Dickmanns, D., Schmidhüber, J., Winklhofer, A.: Der genetische Algorithmus: Eine Implementierung in Prolog. Fortgeschrittenenpraktikum, Institut für Informatik, Technische Universität München (1987)
13. Eiben, A.E., Smith, J.E.: Introduction to Evolutionary Computing. Springer (2007)
14. Fogel, D.B.: Evolutionary Computation: The Fossil Record. Wiley-IEEE Press (1998)
15. Fogel, L.J., Owens, A.J., Walsh, M.J.: Artificial Intelligence through Simulated Evolution. John Wiley (1966)
16. Forsyth, R.: BEAGLE A Darwinian Approach to Pattern Recognition. Kybernetes **10**(3), 159–166 (1981)
17. Friedberg, R.: A learning machine: Part I. IBM Journal of Research and Development **2**, 2–13 (1958)
18. Friedberg, R., Dunham, B., North, J.: A learning machine: Part II. IBM Journal of Research and Development **3**, 282–287 (1959)
19. Holland, J.H.: Adaptation in natural and artificial systems: An introductory analysis with applications to biology, control, and artificial intelligence. University of Michigan Press, Ann Arbor, MI (1975)
20. Koza, J.R.: Genetic Programming: On the Programming of Computers by Natural Selection. MIT Press, Cambridge, Massachusetts, USA (1992)
21. Koza, J.R.: Genetic Programming II: Automatic Discovery of Reusable Programs. MIT Press, Cambridge, Massachusetts (1994)
22. Louis, S., Rawlins, G.J.E.: Using Genetic Algorithms to Design Structures. Tech. Rep. 326, Department of Computer Science, Indiana University (1990)
23. Louis, S., Rawlins, G.J.E.: Designer Genetic Algorithms: Genetic Algorithms in Structure Design. In: Proc. International Conference on Genetic Algorithms, pp. 53–60. Morgan Kauffmann (1991)
24. Louis, S.J.: Genetic algorithms as a computational tool for design. Ph.D. thesis, Department of Computer Science, Indiana University (1993)
25. McCarthy, J.: Recursive functions of symbolic expressions and their computation by machine, part I. Communications of the ACM, 184–195 (1960)
26. McKay, R., Hoai, N., Whigham, P., Shan, Y., O'Neill, M.: Grammar-based Genetic Programming: a survey. Genetic Programming and Evolvable Machines **11**(3), 365–396 (2010)
27. Miller, J.F., Smith, S.L.: Redundancy and Computational Efficiency in Cartesian Genetic Programming. IEEE Transactions on Evolutionary Computation **10**(2), 167–174 (2006)
28. Mumford, C.L.: New Order-Based Crossovers for the Graph Coloring Problem. In: T. Runarsson, H.G. Beyer, E. Burke, J. Merelo-Guervós, L. Whitley, X. Yao (eds.) Parallel Problem Solving from Nature - PPSN IX, *LNCS*, vol. 4193, pp. 880–889 (2006)
29. Nordin, P.: A Compiling Genetic Programming System that Directly Manipulates the Machine Code. In: K.E. Kinnear (ed.) Advances in Genetic Programming, pp. 311–331. MIT Press (1994)
30. Nordin, P.: Evolutionary program induction of binary machine code and its applications. Ph.D. thesis, Department of Computer Science, University of Dortmund, Germany (1997)
31. Nordin, P., Banzhaf, W.: Evolving Turing-Complete Programs for a Register Machine with Self-modifying Code. In: Proc. International Conference on Genetic Algorithms, pp. 318–327. Morgan Kaufmann Publishers Inc., San Francisco, CA, USA (1995)
32. O'Neill, M., Ryan, C.: Grammatical Evolution. IEEE Transactions on Evolutionary Computation **5**(4), 349–358 (2001)
33. O'Neill, M., Ryan, C.: Grammatical Evolution: Evolutionary Automatic Programming in an Arbitrary Language. Springer (2003)
34. Poli, R.: Parallel distributed genetic programming. Tech. Rep. CSRP-96-15, School of Computer Science, University of Birmingham (1996)

35. Poli, R.: Evolution of Graph-Like Programs with Parallel Distributed Genetic Programming. In: E. Goodman (ed.) Proc. International Conference on Genetic Algorithms, pp. 346–353. Morgan Kaufmann (1997)
36. Poli, R., Langdon, W.B.: Foundations of Genetic Programming. Springer (2002)
37. Poli, R., Langdon, W.B., McPhee, N.F.: A field guide to genetic programming. Published via http://lulu.com and freely available at http://www.gp-field-guide.org.uk (2008)
38. Rechenberg, I.: Evolutionsstrategie - Optimierung technischer Systeme nach Prinzipien der biologischen Evolution. Ph.D. thesis, Technical University of Berlin, Germany (1971)
39. Schmidhüber, J.: Evolutionary principles in self-referential learning. Diploma thesis, Institut für Informatik, Technical University of München (1987)
40. Schwefel, H.P.: Numerische Optimierung von Computer-Modellen. Ph.D. thesis, Technical University of Berlin (1974)
41. Silva, S., Costa, E.: Dynamic limits for bloat control in genetic programming and a review of past and current bloat theories. Genetic Programming and Evolvable Machines **10**, 141–179 (2009)
42. Smith, S.F.: A Learning System Based on Genetic Adaptive Algorithms. Ph.D. thesis, University of Pittsburgh (1980)
43. Spector, L., Robinson, A.: Genetic Programming and Autoconstructive Evolution with the Push Programming Language. Genetic Programming and Evolvable Machines **3**, 7–40 (2002)
44. Turing, A.: Intelligent Machinery. In: D. Ince (ed.) Collected Works of A. M. Turing: Mechanical Intelligence. Elsevier Science (1992)

Chapter 2
Cartesian Genetic Programming

Julian F. Miller

2.1 Origins of CGP

Cartesian genetic programming grew from a method of evolving digital circuits developed by Miller et al. in 1997 [8]. However the term 'Cartesian genetic programming' first appeared in 1999 [5] and was proposed as a general form of genetic programming in 2000 [7]. It is called 'Cartesian' because it represents a program using a two-dimensional grid of nodes (see Sect. 2.2).

2.2 General Form of CGP

In CGP, programs are represented in the form of directed acyclic graphs. These graphs are represented as a two-dimensional grid of computational nodes. The genes that make up the genotype in CGP are integers that represent where a node gets its data, what operations the node performs on the data and where the output data required by the user is to be obtained. When the genotype is decoded, some nodes may be ignored. This happens when node outputs are not used in the calculation of output data. When this happens, we refer to the nodes and their genes as 'non-coding'. We call the program that results from the decoding of a genotype a phenotype. The genotype in CGP has a fixed length. However, the size of the phenotype (in terms of the number of computational nodes) can be anything from zero nodes to the number of nodes defined in the genotype. A phenotype would have zero nodes if all the program outputs were directly connected to program inputs. A phenotype would have the same number of nodes as defined in the genotype when every node in the graph was required. The genotype–phenotype mapping used in CGP is one of its defining characteristics.

The types of computational node functions used in CGP are decided by the user and are listed in a function look-up table. In CGP, each node in the directed graph

J.F. Miller (ed.), *Cartesian Genetic Programming*, Natural Computing Series,
DOI 10.1007/978-3-642-17310-3_2, © Springer-Verlag Berlin Heidelberg 2011

represents a particular function and is encoded by a number of genes. One gene is the address of the computational node function in the function look-up table. We call this a *function gene*. The remaining node genes say where the node gets its data from. These genes represent addresses in a data structure (typically an array). We call these *connection genes*. Nodes take their inputs in a feed-forward manner from either the output of nodes in a previous column or from a program input (which is sometimes called a terminal). The number of connection genes a node has is chosen to be the maximum number of inputs (often called the arity) that any function in the function look-up table has. The program data inputs are given the absolute data addresses 0 to n_i minus 1 where n_i is the number of program inputs. The data outputs of nodes in the genotype are given addresses sequentially, column by column, starting from n_i to $n_i + L_n - 1$, where L_n is the user-determined upper bound of the number of nodes. The general form of a Cartesian genetic program is shown in Fig. 2.1.

If the problem requires n_o program outputs, then n_o integers are added to the end of the genotype. In general, there may be a number of output genes (O_i) which specify where the program outputs are taken from. Each of these is an address of a node where the program output data is taken from. Nodes in columns cannot be connected to each other. The graph is directed and feed-forward; this means that a node may only have its inputs connected to either input data or the output of a node in a previous column. The structure of the genotype is seen below the schematic in Fig. 2.1. All node function genes f_i are integer addresses in a look-up table of functions. All connection genes C_{ij} are data addresses and are integers taking values between 0 and the address of the node at the bottom of the previous column of nodes.

CGP has three parameters that are chosen by the user. These are the *number of columns*, the *number of rows* and *levels-back*. These are denoted by n_c, n_r and l, respectively. The product of the first two parameters determine the maximum number of computational nodes allowed: $L_n = n_c n_r$. The parameter l controls the connectivity of the graph encoded. Levels-back constrains which columns a node can get its inputs from. If $l = 1$, a node can get its inputs only from a node in the column on its immediate left or from a primary input. If $l = 2$, a node can have its inputs connected to the outputs of any nodes in the immediate left two columns of nodes or a primary input. If one wishes to allow nodes to connect to any nodes on their left, then $l = n_c$. Varying these parameters can result in various kinds of graph topologies. Choosing the number of columns to be small and the number of rows to be large results in graphs that are tall and thin. Choosing the number of columns to be large and the number of rows to be small results in short, wide graphs. Choosing levels-back to be one produces highly layered graphs in which calculations are carried out column by column. An important special case of these three parameters occurs when the number of rows is chosen to be one and levels-back is set to be the number of columns. In this case the genotype can represent any bounded directed graph where the upper bound is determined by the number of columns.

As we saw briefly in Chap. 1, one of the benefits of a graph-based representation of a program is that graphs, by definition, allow the implicit reuse of nodes, as a

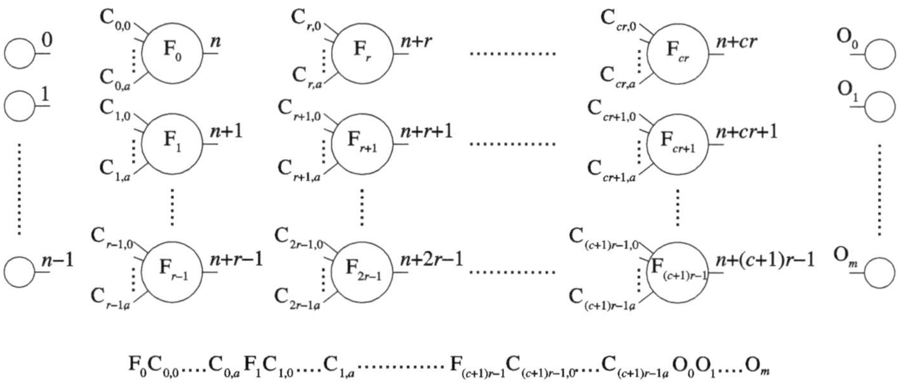

Fig. 2.1 General form of CGP. It is a grid of nodes whose functions are chosen from a set of primitive functions. The grid has n_c columns and n_r rows. The number of program inputs is n_i and the number of program outputs is n_o. Each node is assumed to take as many inputs as the maximum function arity a. Every data input and node output is labeled consecutively (starting at 0), which gives it a unique data address which specifies where the input data or node output value can be accessed (shown in the figure on the outputs of inputs and nodes).

node can be connected to the output of any previous node in the graph. In addition, CGP can represent programs having an arbitrary number of outputs. In Sect. 2.7, we will discuss the advantages of non-coding genes. This gives CGP a number of advantages over tree-based GP representations.

2.3 Allelic Constraints

In the previous section, we discussed the integer-based CGP genotype representation. The values that genes can take (i.e. alleles) are highly constrained in CGP. When genotypes are initialized or mutated, these constraints should be obeyed.

First of all, the alleles of function genes f_i must take valid address values in the look-up table of primitive functions. Let n_f represent the number of allowed primitive node functions. Then f_i must obey

$$0 \leq f_i \leq n_f. \tag{2.1}$$

Consider a node in column j. The values taken by the connection genes C_{ij} of all nodes in column j are as follows. If $j \geq l$,

$$n_i + (j-l)n_r \leq C_{ij} \leq n_i + jn_r. \tag{2.2}$$

If $j < l$,

$$0 \leq C_{ij} \leq n_i + jn_r. \tag{2.3}$$

Output genes O_i can connect to any node or input:

$$0 \leq O_i < n_i + L_n. \tag{2.4}$$

2.4 Examples

CGP can represent many different kinds of computational structures. In this section, we discuss three examples of this. The first example is where a CGP genotype encodes a digital circuit. In the second example, a CGP genotype represents a set of mathematical equations. In the third example, a CGP genotype represents a picture. In the first example, the type of data input is a single bit; in the second, it is real numbers. In the art example, it is unsigned eight-bit numbers.

2.4.1 A Digital Circuit

The evolved genotype shown in Fig. 2.2 arose using CGP genotype parameters $n_c = 10$, $n_r = 1$ and $l = 10$. It represents a digital combinational circuit called a two-bit parallel multiplier. It multiplies two two-bit numbers together, so it requires four inputs and four outputs. There are four primitive functions in the function set (logic gates). Let the first and second inputs to these gates be a and b. Then the four functions (with the function gene in parentheses) are AND(a,b)(0), AND(a,NOT(b))(1), XOR(a,b)(2) and OR(a,b)(3). One can see in Fig. 2.2 that two nodes (with labels 6 and 10) are not used, since no circuit output requires them. These are non-coding nodes. They are shown in grey.

Figure 2.2 shows a CGP genotype and the corresponding phenotype.

2.4.2 Mathematical Equations

Suppose that the functions of nodes can be chosen from the arithmetic operations plus, minus, multiply and divide. We have allocated the function genes as follows. Plus is represented by the function gene being equal to zero, minus is represented

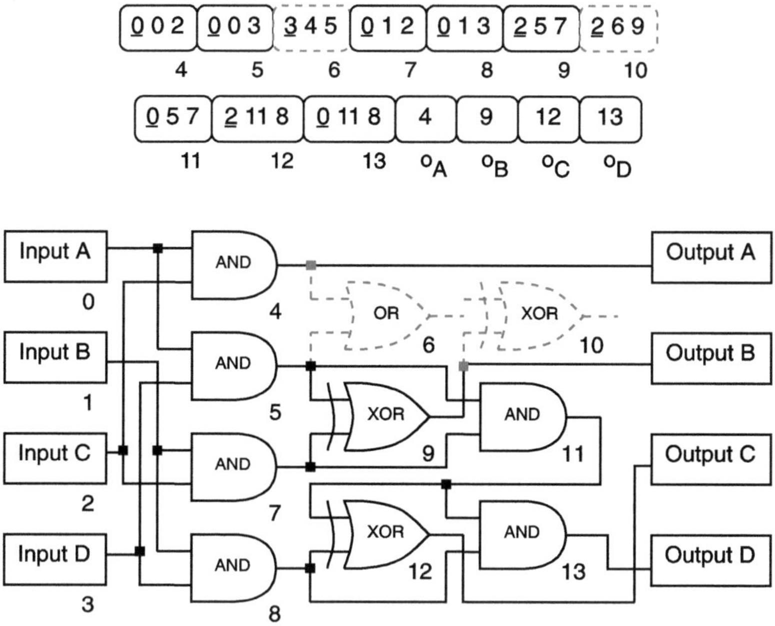

Fig. 2.2 A CGP genotype and corresponding phenotype for a two-bit multiplier circuit. The underlined genes in the genotype encode the function of each node. The function look-up table is AND (0), AND with one input inverted (1), XOR (2) and OR (3). The addresses are shown underneath each program input and node in the genotype and phenotype. The inactive areas of the genotype and phenotype are shown in grey dashes (nodes 6 and 10).

by one, multiply by two and divide by three. Let us suppose that our program has two real-valued inputs, which symbolically we denote by x_0 and x_1. Let us suppose that we need four program outputs, which we denote O_A, O_B, O_C and O_D. We have chosen the number of columns n_c to be three and the number of rows n_r to be two. In this example, assume that levels-back, l is two. An example genotype and a schematic of the phenotype are shown in Fig. 2.3. The phenotype is the following set of equations:

$$
\begin{aligned}
O_A &= x_0 + x_1 \\
O_B &= x_0 * x_1 \\
O_C &= \frac{x_0 * x_1}{x_0{}^2 - x_1} \\
O_D &= x_0{}^2.
\end{aligned}
\tag{2.5}
$$

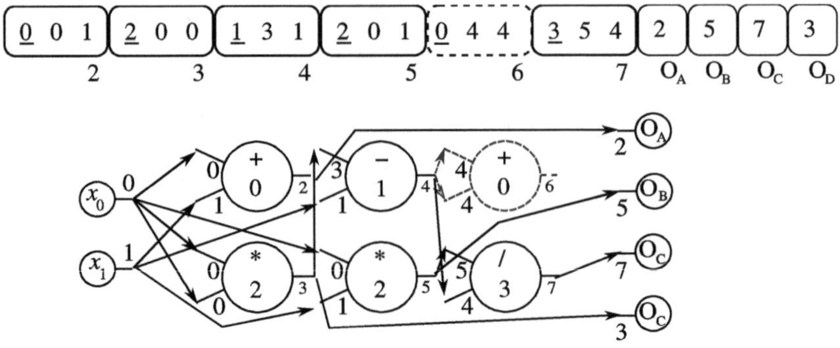

Fig. 2.3 A CGP genotype and corresponding schematic phenotype for a set of four mathematical equations. The underlined genes in the genotype encode the function of each node. The function look-up table is add (0), subtract (1), multiply (2) and divide (3). The addresses are shown underneath each program input and node in the genotype and phenotype. The inactive areas of the genotype and phenotype are shown in grey dashes (node 6).

2.4.3 Art

A simple way to generate pictures using CGP is to allow the integer pixel Cartesian coordinates to be the inputs to a CGP genotype. Three outputs can then be allowed which will be used to determine the red, green and blue components of the pixel's colour. The CGP outputs have to be mapped in some way so that they only take values between 0 and 255, so that valid pixel colours are defined. A CGP program is executed for all the pixel coordinates defining a two-dimensional region. In this way, a picture will be produced. In Fig. 2.4, a genotype is shown with a corresponding schematic of the phenotype. The set of function genes and corresponding node functions are shown in Table 2.1. In a later chapter, ways of developing art using CGP will be considered in detail.

The functions in Table 2.1 have been carefully chosen so that they will return an integer value between 0 and 255 when the inputs (Cartesian coordinates) x and y are both between 0 and 255. The evolved genotype shown in Fig. 2.4 uses only four function genes: 5, 9, 6 and 13. We denote the outputs of nodes by g_i, where i is the output address of the node. The red, green and blue channels of the pixel values (denoted r, g, b) are given as below:

Table 2.1 Primitive function set used in art example

Function gene	Function definition
0	x
1	y
2	$\sqrt{x+y}$
3	$\sqrt{\mid x-y \mid}$
4	$255(\mid \sin(\frac{2\pi}{255}x) + \cos(\frac{2\pi}{255}y) \mid)/2$
5	$255(\mid \cos(\frac{2\pi}{255}x) + \sin(\frac{2\pi}{255}y) \mid)/2$
6	$255(\mid \cos(\frac{3\pi}{255}x) + \sin(\frac{2\pi}{255}y) \mid)/2$
7	$\exp(x+y) \pmod{256}$
8	$\mid \sinh(x+y) \mid \pmod{256}$
9	$\cosh(x+y) \pmod{256}$
10	$255 \mid \tanh(x+y) \mid$
11	$255(\mid \sin(\frac{\pi}{255}(x+y)) \mid$
12	$255(\mid \cos(\frac{\pi}{255}(x+y)) \mid$
13	$255(\mid \tan(\frac{\pi}{8*255}(x+y)) \mid$
14	$\sqrt{\frac{x^2+y^2}{2}}$
15	$\mid x \mid^y \pmod{256}$
16	$\mid x+y \mid \pmod{256}$
17	$\mid x-y \mid \pmod{256}$
18	$xy/255$
19	$\begin{cases} x & y=0 \\ x/y & y \neq 0 \end{cases}$

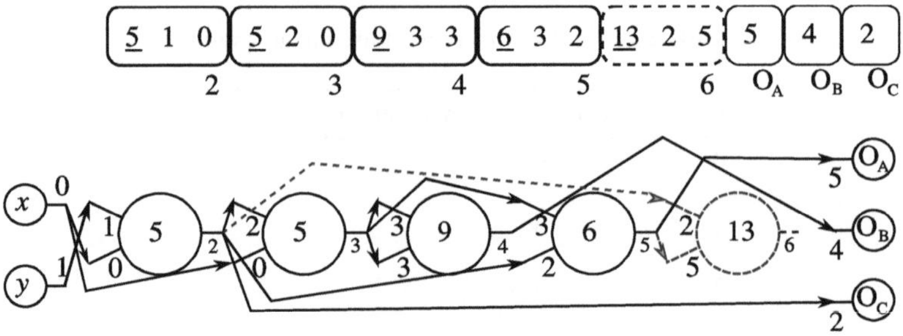

Fig. 2.4 A CGP genotype and corresponding schematic phenotype for a program that defines a picture. The underlined genes in the genotype encode the function of each node. The function look-up table is given in Table 2.1. The addresses are shown underneath each program input and node in the genotype and phenotype. The inactive areas of the genotype and phenotype are shown in grey dashes (node 6).

$$
\begin{aligned}
g_2 &= 255\left(\left|\cos\left(\frac{2\pi}{255}y\right) + \sin\left(\frac{2\pi}{255}x\right)\right|\right)\Big/2,\\
g_3 &= 255\left(\left|\cos\left(\frac{2\pi}{255}g_2\right) + \sin\left(\frac{2\pi}{255}x\right)\right|\right)\Big/2,\\
g_4 &= \cosh(2g_3) \pmod{256},\\
g_5 &= 255\left(\left|\cos\left(\frac{2\pi}{255}g_3\right) + \sin\left(\frac{2\pi}{255}g_2\right)\right|\right)\Big/2,\\
r &= g_5,\\
g &= g_4,\\
b &= g_2.
\end{aligned}
\tag{2.6}
$$

When these mathematical equations are executed for all 256^2 pixel locations, they produce the picture (the actual picture is in colour) shown in Fig. 2.5.

2.5 Decoding a CGP Genotype

So far, we have illustrated the genotype-decoding process in a diagrammatic way. However, the algorithmic process is recursive in nature and works from the output genes first. The process begins by looking at which nodes are 'activated' by being directly addressed by output genes. Then these nodes are examined to find out which nodes they in turn require. A detailed example of this decoding process is shown in Fig. 2.6.

It is important to observe that in the decoding process, non-coding nodes are not processed, so that having non-coding genes presents little computational overhead.

Fig. 2.5 Picture produced when the program encoded in the evolved genotype shown in Fig. 2.4 is executed. It arose in the sixth generation. The user decides which genotype will be the next parent.

There are various ways that this decoding process can be implemented. One way would be to do it recursively; another would be to determine which nodes are active (in a recursive way) and record them for future use, and only process those. Procedures 2.2 and 2.1 in the next section detail the latter. The possibility of improving efficiency by stripping out non-coding instructions prior to phenotype evaluation has also been suggested for Linear GP [1].

2.5.1 Algorithms for Decoding a CGP Genotype

In this section, we will present algorithmic procedures for decoding a CGP genotype. The algorithm has two main parts: the first is a procedure that determines how many nodes are used and the addresses of those nodes. This is shown in Procedure 2.1. The second presents the input data to the nodes and calculates the outputs for a single data input. We denote the maximum number of addresses in the CGP graph by $M = L_n + n_i$, the total number of genes in the genotype by L_g, the number of genes in a node by n_n, and the number of active or used nodes by n_u.

In the procedure, a number of arrays are mentioned. Firstly, it takes the CGP genotype stored in an array $G[L_g]$ as an argument. It passes back as an argument an array holding the addresses of the nodes in the genotype that are used. We denote this by *NP*. It also returns how many nodes are used. Internally, it uses a Boolean array holding whether any node is used, called $NU[M]$. This is initialized to *FALSE*. An array *NG* is used to store the node genes for any particular node. It also assumes that a function $Arity(F)$ returns the arity of any function in the function set.

Once we have the information about which nodes are used, we can efficiently decode the genotype *G* with some input data to find out what data the encoded pro-

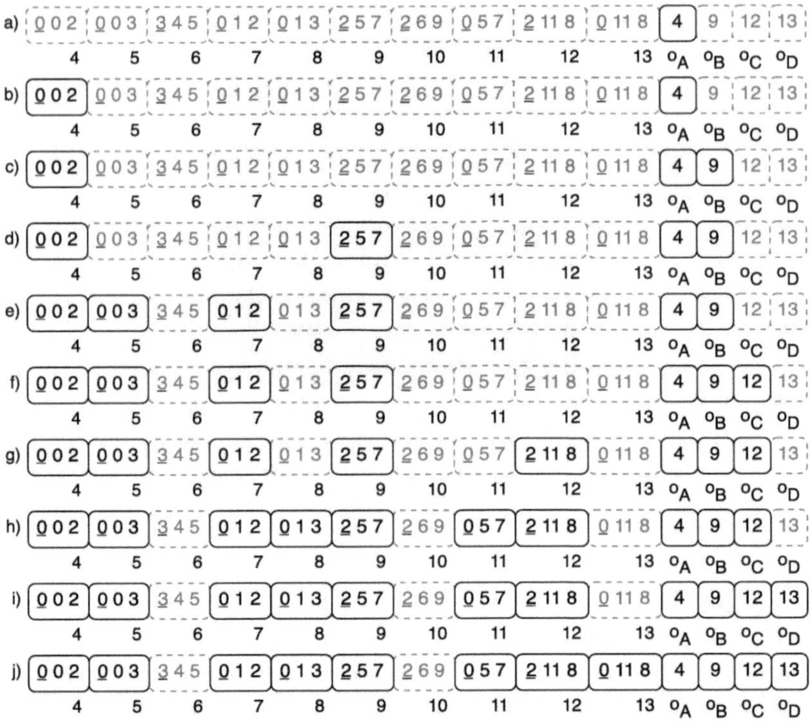

Fig. 2.6 The decoding procedure for a CGP genotype for the two-bit multiplier problem. (a) Output A (o_A) connects to the output of node 4; move to node 4. (b) Node 4 connects to the program inputs 0 and 2; therefore the output A is decoded. Move to output B. (c) Output B (o_B) connects to the output of node 9; move to node 9. (d) Node 9 connects to the output of nodes 5 and 7; move to nodes 5 and 7. (e) Nodes 5 and 7 connect to the program inputs 0, 3, 1 and 2; therefore output B is decoded. Move to output C. The procedure continues until output C (o_C) and output D (o_D) are decoded (steps (f) to (h) and steps (i) to (j) respectively). When all outputs are decoded, the genotype is fully decoded.

gram gives as an output. This procedure is shown in Procedure 2.2. It assumes that input data is stored in an array *DIN*, and the particular item of that data that is being used as input to the CGP genotype is *item*. The procedure returns the calculated output data in an array *O*. Internally, it uses two arrays: *o*, which stores the calculated outputs of used nodes and *in*, which stores the input data being presented to an individual node. The symbol *g* is the address in the genotype *G* of the first gene in a node. The symbol *n* is the address of a node in the array *NP*. The procedure assumes that a function *NF* implements the functions in the function look-up table.

The fitness function required for an evolution algorithm is given in Procedure 2.3. It is assumed that there is a procedure *EvaluateCGP* that, given the CGP cal-

Procedure 2.1 Determining which nodes need to be processed

```
NodesToProcess(G,NP) // return the number of nodes to process
for all i such that 0 ≤ i < M do
    NU[i] = FALSE
end for
for all i such that L_g − n_o ≤ i < L_g do
    NU[G[i]] ← TRUE
end for
for all i such that M − 1 ≥ i ≥ n_i do // Find active nodes
    if NU[i] ← TRUE then
        index ← n_n(i − n_i)
        for all j such that 0 ≤ j < n_n do // store node genes in NG
            NG[j] ← G[index + j]
        end for
        for all j such that 0 ≤ j < Arity(NG[n_n − 1]) do
            NU[NG[j]] ← TRUE
        end for
    end if
end for
n_u = 0
for all j such that n_i ≤ j < M do // store active node addresses in NP
    if NU[j] = TRUE then
        NP[n_u] ← j
        n_u ← n_u + 1
    end if
end for
return n_u
```

Procedure 2.2 Decoding CGP to get the output

```
DecodeCGP(G,DIN,O,n_u,NP,item)
for all i such that 0 ≤ i < n_i do
    o[i] ← DIN[item]
end for
for all j such that 0 ≤ j < n_u do
    n ← NP[j] − n_i
    g ← n_n n
    for all i such that 0 ≤ i < n_n − 1 do // store data needed by a node
        in[i] ← o[G[g + i]]
    end for
    f = G[g + n_n − 1] // get function gene of node
    o[n + n_i] = NF(in, f) // calculate output of node
end for
for all j such that 0 ≤ j < n_o do
    O[j] ← o[G[L_g − n_o + j]]
end for
```

culated outputs O and the desired program outputs $DOUT$, calculates the fitness of the genotype f_i for a single input data item. The procedure assumes that there are N_{fc} fitness cases that need to be considered. In digital-circuit evolution the usual number of fitness cases is $n_o 2^{n_i}$. Note, however, that Procedure 2.1 only needs to be executed once for a genotype, irrespective of the number of fitness cases.

Procedure 2.3 Calculating the fitness of a CGP genotype

```
FitnessCGP(G)
n_u ← NodesToProcess(G,NP)
fit ← 0
for all i such that 0 ≤ i < N_fc do
    DecodeCGP(G,DIN,O,n_u,NP,item)
    f_i = EvaluateCGP(O,DOUT,i)
    fit ← fit + f_i
end for
```

2.6 Evolution of CGP Genotypes

2.6.1 Mutation

The mutation operator used in CGP is a point mutation operator. In a point mutation, an allele at a randomly chosen gene location is changed to another valid random value (see Sect. 2.3). If a function gene is chosen for mutation, then a valid value is the address of any function in the function set, whereas if an input gene is chosen for mutation, then a valid value is the address of the output of any previous node in the genotype or of any program input. Also, a valid value for a program output gene is the address of the output of any node in the genotype or the address of a program input. The number of genes in the genotype that can be mutated in a single application of the mutation operator is defined by the user, and is normally a percentage of the total number of genes in the genotype. We refer to the latter as the *mutation rate*, and will use the symbol μ_r to represent it. Often one wants to refer to the actual number of gene sites that could be mutated in a genotype of a given length L_g. We give this quantity the symbol μ_g, so that $\mu_g = \mu_r L_g$. We will talk about suitable choices for the parameters μ_r and μ_g in Sect. 2.8.

An example of the point mutation operator is shown in Fig. 2.7, which highlights how a small change in the genotype can sometimes produce a large change in the phenotype.

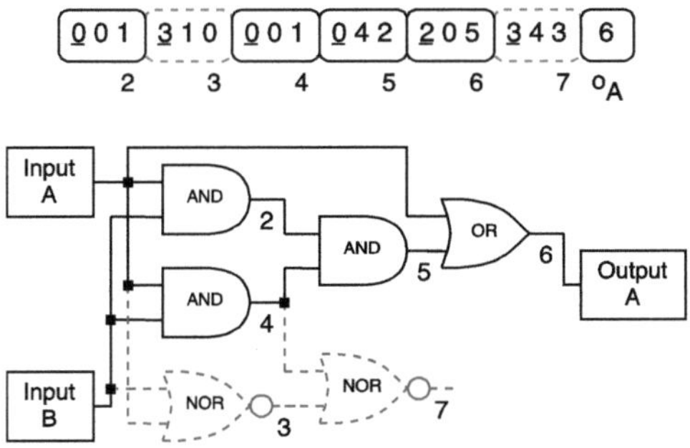

(a) Before mutation

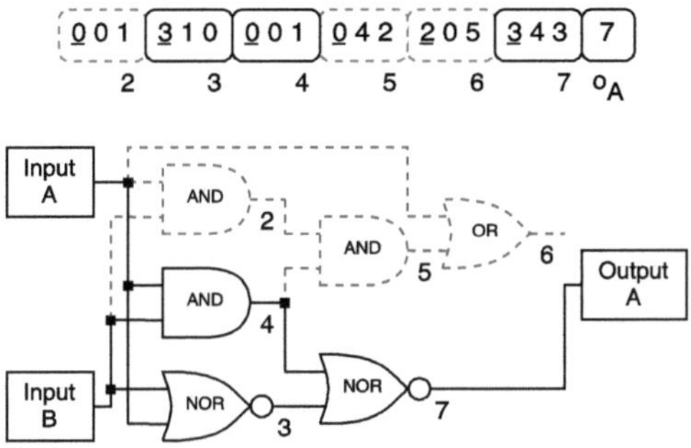

(b) After mutation

Fig. 2.7 An example of the point mutation operator before and after it is applied to a CGP genotype, and the corresponding phenotypes. A single point mutation occurs in the program output gene (o_A), changing the value from 6 to 7. This causes nodes 3 and 7 to become active, whilst making nodes 2, 5 and 6 inactive. The inactive areas are shown in grey dashes.

2.6.2 Recombination

Crossover operators have received relatively little attention in CGP. Originally, a one-point crossover operator was used in CGP (similar to the n-point crossover in genetic algorithms) but was found to be disruptive to the subgraphs within the chromosome, and had a detrimental affect on the performance of CGP [5]. Some work by Clegg et al. [2] has investigated crossover in CGP (and GP in general). Their approach uses a floating-point crossover operator, similar to that found in evolutionary programming, and also adds an extra layer of encoding to the genotype, in which all genes are encoded as a floating-point number in the range $[0,1]$. A larger population and tournament selection were also used instead of the $(1+4)$ evolutionary strategy normally used in CGP, to try and improve the population diversity.

The results of the new approach appear promising when applied to two symbolic regression problems, but further work is required on a range of problems in order to assess its advantages [2]. Crossover has also been found to be useful in an image-processing application as discussed in Sect. 6.4.3. Crossover operators (cone-based crossover) have been devised for digital-circuit evolution (see Sect. 3.6.2). In situations where a CGP genotype is divided into a collection of chromosomes, crossover can be very effective. Sect. 3.8 discusses how new genotypes created by selecting the best chromosomes from parents' genotypes can produce *super-individuals*. This allows difficult multiple-output problems to be solved.

2.6.3 Evolutionary Algorithm

A variant on a simple evolutionary algorithm known as a $1+\lambda$ evolutionary algorithm [9] is widely used for CGP. Usually λ is chosen to be 4. This has the form shown in Procedure 2.4.

Procedure 2.4 The $(1+4)$ evolutionary strategy

```
1:  for all i such that 0 ≤ i < 5 do
2:      Randomly generate individual i
3:  end for
4:  Select the fittest individual, which is promoted as the parent
5:  while a solution is not found or the generation limit is not reached do
6:      for all i such that 0 ≤ i < 4 do
7:          Mutate the parent to generate offspring i
8:      end for
9:      Generate the fittest individual using the following rules:
10:     if an offspring genotype has a better or equal fitness than the parent then
11:         Offspring genotype is chosen as fittest
12:     else
13:         The parent chromosome remains the fittest
14:     end if
15: end while
```

On line 10 of the procedure there is an extra condition that when offspring genotypes in the population have the same fitness as the parent and there is no offspring that is better than the parent, in that case an *offspring* is chosen as the new parent. This is a very important feature of the algorithm, which makes good use of redundancy in CGP genotypes. This is discussed in Sect. 2.7.

2.7 Genetic Redundancy in CGP Genotypes

We have already seen that in a CGP genotype there may be genes that are entirely inactive, having no influence on the phenotype and hence on the fitness. Such inactive genes therefore have a neutral effect on genotype fitness. This phenomenon is often referred to as neutrality. CGP genotypes are dominated by redundant genes. For instance, Miller and Smith showed that in genotypes having 4000 nodes, the percentage of inactive nodes is approximately 95%! [6].

The influence of neutrality in CGP has been investigated in detail [7, 6, 10, 13, 14] and has been shown to be extremely beneficial to the efficiency of the evolutionary process on a range of test problems. Just how important neutral drift can be to the effectiveness of CGP is illustrated in Fig. 2.8.

This shows the normalized best fitness value achieved in two sets of 100 independent evolutionary runs of 10 million generations [10]. The target of evolution was to evolve a correct three-bit digital parallel multiplier. In the first set of runs, an offspring could replace a parent when it had the same fitness as the parent and there was no other population member with a better fitness (as in line 10 of Procedure 2.4). In the figure, the final fitness values are indicated by diamond symbols. In the second set, a parent was replaced only when an offspring had a strictly better fitness value. These results are indicated by plus symbols. In the case where neutral drift was allowed, 27 correct multipliers were evolved. Also, many of the other circuits were very nearly correct. In the case where no neutral drift was allowed, there were no runs that produced a correct multiplier, and the average fitness values are considerably lower.

It is possible that by analyzing within an evolutionary algorithm whether mutational offspring are phenotypically different from parents, one may be able to reduce the number of fitness evaluations. Since large amounts of non-coding genes are helpful to evolution, it is more likely that mutations will occur only in non-coding sections of the genotype; such genotypes will have the same fitness as their parents and do not need to be evaluated. To accomplish this would require a slight change to the evolutionary algorithm in Procedure 2.4. One would keep a record of the nodes that need to be processed in the genotype that is promoted (i.e. array *NP* in Sect. 2.5.1). Then, if an offspring had exactly the same nodes that were active as in the parent, it would be assigned the parent's fitness. Whether in practice this happens sufficiently often to warrant the extra processing required has not been investigated.

2.8 Parameter Settings for CGP

To arrive at good parameters for CGP normally requires some experimentation on the problem being considered. However, some general advice can be given. A suitable mutation rate depends on the length of the genotype (in nodes). As a rule of thumb, one should use about 1% mutation if a maximum of 100 nodes are used (i.e. $n_c n_r = 100$). Let us assume that all primitive functions have two connection genes

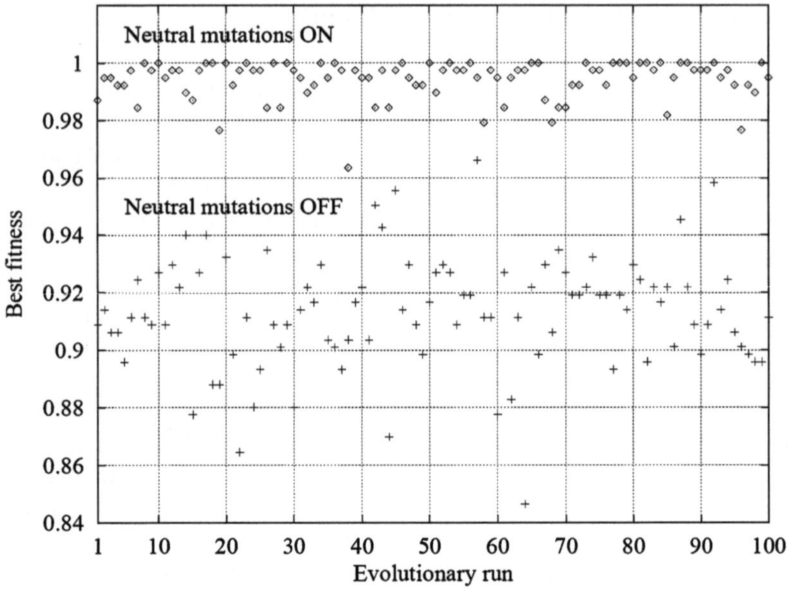

Fig. 2.8 The normalized best fitness value achieved in two sets of 100 independent evolutionary runs of 10 million generations. The target of evolution was to evolve a correct three-bit digital parallel multiplier. In one set of runs, neutral drift was allowed, and in the other, neutral drift was not allowed. The evolutionary algorithm was unable to produce a correct circuit in the second case.

and the problem has a single output. Then a genotype with a maximum of 100 nodes will require 301 genes. So 1% mutation would mean that up to three genes would be mutated in the parent to create an offspring. Experience shows that to achieve reasonably fast evolution one should arrange the mutation rate μ_r to be such that the number of gene locations chosen for mutation is a slowly growing function of the genotype length. For instance, in [6] it was found that $\mu_g = 90$ proved to be a good value when $L_g = 12,004$ (4000 two-input nodes and four outputs). This corresponds to $\mu_r = 0.75\%$. When $L_g = 154$ (50 two-input nodes and four outputs), a good value of μ_g was 6, which corresponds to $\mu_r = 4\%$. Even smaller genotypes usually require higher mutation rates still for fast evolution.

Generally speaking, when optimal mutation rates are used, larger genotypes require fewer fitness evaluations to achieve evolutionary success than do smaller genotypes. For instance, Miller and Smith found that the number of fitness evaluations (i.e. genotypes whose fitness is calculated) required to successfully evolve a two-bit multiplier circuit was lower for genotypes having 4000 nodes than for genotypes of smaller length [6]! The way to understand this is to think about the usefulness of neutral drift in the evolution of CGP genotypes. Larger CGP genotypes have a much larger percentage of non-coding genes than do smaller genotypes, so the potential for neutral drift is much larger. This is another illustration of the great importance of neutral drift in evolutionary algorithms for CGP.

So, we have seen that large genotypes lead to more efficient evolution; however, given a certain genotype length, what is the optimal number of columns n_c and number of rows n_r? The advice here is as follows. If there are no problems with implementing arbitrary directed graphs, then the recommended choice of these parameters is $n_r = 1$ with $l = n_c$. However, if for instance one is evolving circuits for implementation of evolved CGP genotypes on digital devices (with limited routing resources), it is often useful to choose $n_c = n_r$. It should be stressed that these recommendations are 'rules of thumb', as no detailed quantitative work on this aspect has been published.

CGP uses very small population sizes (5 in the case described in Sect. 2.6.3). So one should expect large numbers of generations to be used. Despite this, in numerous studies, it has been found that the average number of fitness evaluations required to solve many problems can be favourably compared with other forms of GP (see for instance [5, 12]).

2.9 Cyclic CGP

CGP has largely been used in an acyclic form, where graphs have no feedback. However, there is no fundamental reason for this. The representaion of graphs used in CGP is easily adapted to encode cyclic graphs. One merely needs to remove the restriction that alleles for a particular node have to take values less than the position (address) of the node. However, despite this, there has been little published work where this restriction has been removed. One exception is the recent work of Khan et al., who have encoded artificial neural networks in CGP [3]. They allowed feedback and used CGP to evolve recurrent neural networks. Other exceptions are the work of Walker et al., who applied CGP to the evolution of machine code of sequential and parallel programs on a MOVE processor [11] and Liu et al., who proposed a dual-layer CGP genotype representation in which one layer encoded processor instructions and the other loop control parameters [4]. Also Sect. 5.6.2.1 describes how cyclic analogue circuits can be encoded in CGP.

Clearly, such an investigations would extend the expressivity of programs (since feedback implies either recursion or iteration). It would also allow both synchronous and asynchronous circuits to be evolved. A full investigation of this topic remains for the future.

References

1. Brameier, M., Banzhaf, W.: A Comparison of Linear Genetic Programming and Neural Networks in Medical Data Mining. IEEE Transactions on Evolutionary Computation **5**(1), 17–26 (2001)

2. Clegg, J., Walker, J.A., Miller, J.F.: A New Crossover Technique for Cartesian Genetic Programming. In: Proc. Genetic and Evolutionary Computation Conference, pp. 1580–1587. ACM Press (2007)
3. Khan, M.M., Khan, G.M., Miller, J.F.: Efficient representation of Recurrent Neural Networks for Markovian/Non-Markovian Non-linear Control Problems. In: A.E. Hassanien, A. Abraham, F. Marcelloni, H. Hagras, M. Antonelli, T.P. Hong (eds.) Proc. International Conference on Intelligent Systems Design and Applications, pp. 615–620. IEEE (2010)
4. Liu, Y., Tempesti, G., Walker, J.A., Timmis, J., Tyrrell, A.M., Bremner, P.: A Self-Scaling Instruction Generator Using Cartesian Genetic Programming. In: Proc. European Conference on Genetic Programming, *LNCS*, vol. 6621, pp. 299–310. Springer (2011)
5. Miller, J.F.: An Empirical Study of the Efficiency of Learning Boolean Functions using a Cartesian Genetic Programming Approach. In: Proc. Genetic and Evolutionary Computation Conference, pp. 1135–1142. Morgan Kaufmann (1999)
6. Miller, J.F., Smith, S.L.: Redundancy and Computational Efficiency in Cartesian Genetic Programming. IEEE Transactions on Evolutionary Computation **10**(2), 167–174 (2006)
7. Miller, J.F., Thomson, P.: Cartesian Genetic Programming. In: Proc. European Conference on Genetic Programming, *LNCS*, vol. 1802, pp. 121–132. Springer (2000)
8. Miller, J.F., Thomson, P., Fogarty, T.C.: Designing Electronic Circuits Using Evolutionary Algorithms: Arithmetic Circuits: A Case Study. In: D. Quagliarella, J. Periaux, C. Poloni, G. Winter (eds.) Genetic Algorithms and Evolution Strategies in Engineering and Computer Science: Recent Advancements and Industrial Applications, pp. 105–131. Wiley (1998)
9. Rechenberg, I.: Evolutionsstrategie – Optimierung technischer Systeme nach Prinzipien der biologischen Evolution. Ph.D. thesis, Technical University of Berlin, Germany (1971)
10. Vassilev, V.K., Miller, J.F.: The Advantages of Landscape Neutrality in Digital Circuit Evolution. In: Proc. International Conference on Evolvable Systems, *LNCS*, vol. 1801, pp. 252–263. Springer (2000)
11. Walker, J.A., Liu, Y., Tempesti, G., Tyrrell, A.M.: Automatic Code Generation on a MOVE Processor Using Cartesian Genetic Programming. In: Proc. International Conference on Evolvable Systems: From Biology to Hardware, *LNCS*, vol. 6274, pp. 238–249. Springer (2010)
12. Walker, J.A., Miller, J.F.: Automatic Acquisition, Evolution and Re-use of Modules in Cartesian Genetic Programming. IEEE Transactions on Evolutionary Computation **12**, 397–417 (2008)
13. Yu, T., Miller, J.F.: Neutrality and the evolvability of Boolean function landscape. In: Proc. European Conference on Genetic Programming, *LNCS*, vol. 2038, pp. 204–217. Springer (2001)
14. Yu, T., Miller, J.F.: Finding Needles in Haystacks is not Hard with Neutrality. In: Proc. European Conference on Genetic Programming, *LNCS*, vol. 2278, pp. 13–25. Springer (2002)

Chapter 3
Problem Decomposition in Cartesian Genetic Programming

James Alfred Walker, Julian F. Miller, Paul Kaufmann and Marco Platzner

3.1 Introduction

Modular structures are commonplace in conventional human design principles. Nature is also abundant with modular structures and processes. Divide-and-conquer [8] is a top-down design paradigm with the ability to recursively decompose a problem into many subproblems, solve each of the subproblems and then recombine the solutions to the subproblems to solve the entire problem. Also, if a subproblem is repeated, it only needs to be solved once, as the solution can be reused for the duplicate problems.

Modularity is the degree to which an entity can be represented in terms of smaller functional blocks. These smaller functional blocks are known as modules. Modularity is neither a process nor an ability; it is simply a method of describing how an entity is constructed. For example, if a problem can be decomposed into a small number of subproblems, then the problem is described as being highly modular. One of the main advantages of modularity is that it allows the simplification of complex problems into small functional blocks, which can be independently solved. This also allows the solutions to be written more compactly and efficiently.

On examining the types of modularity in GP, it is possible to classify it into three categories: functional reuse, structural reuse and compartmentalization.

Functional reuse occurs when the output of a function is constant but it is reused as input to other functions. This type of modularity occurs naturally in representations that are based on graphs, such as CGP, and it has been shown that exploiting this type of modularity improves performance [15, 17]. However, this form of modularity does not exist in tree-based GP.

Structural reuse can be seen as an extension of functional reuse, and occurs when a small group of components can be grouped together to form a functional unit, which can be reused elsewhere in the genotype, taking *different* inputs. This is by far the most common form of modularity in tree-based GP and has been exploited in a number of methods. The most popular of these is automatically defined func-

J.F. Miller (ed.), *Cartesian Genetic Programming*, Natural Computing Series,
DOI 10.1007/978-3-642-17310-3_3, © Springer-Verlag Berlin Heidelberg 2011

tions (ADFs) [12], which co-evolve reusable programs with a main GP program. An alternative to ADFs is module acquisition (MA) [1], which introduces two operators, 'compress' and 'expand', which dynamically construct and destroy reusable functions from sections of the genotype. MA is less influenced by user knowledge than ADFs are, allowing evolution to decide on things such as the number of functions, the number of inputs per function and what primitives the functions consist of. Another alternative to ADFs is adaptive representation through learning (ARL) [22, 21, 23], see which shares more similarities with MA than with ADFs, except that it uses heuristics to form population statistics to determine useful areas of the program to construct into reusable functions, whereas MA just picks pieces of the program at random. All three methods have been shown to improve the performance of GP by exploiting modularity, when compared with the non-modular approaches. However, all approaches have their drawbacks. ADFs require user-defined knowledge for defining the parameters of each ADF; MA can produce too many modules in the module library, making it harder to identify good modules; and the heuristics used by ARL have been shown to be not as effective as defining reusable functions at random.

It has also been shown that CGP without any form of structural reuse is capable of performing better than GP with ADFs on a number of problems [15, 17]. This implies that if CGP could exploit a form of structural reuse, then further performance improvements might be possible. This chapter describes a possible approach to exploiting modularity through structural reuse in CGP in Sect. 3.2, which is mainly based on the MA approach, but also takes some inspiration from ADFs and ARL. This new approach is called *embedded CGP* (ECGP), and is capable of dynamically acquiring, evolving, and reusing modules to exploit modularity. Alternative approaches to acquiring modules within ECGP are discussed in Sect. 3.6. An enhancement to ECGP that allows the use of nested modules (which is also possible with ADFs) to see if further performance improvements are possible is described in Sect. 3.7. This is referred to as *modular CGP* (MCGP). Finally, an approach that uses the concept of multiple chromosomes in order to allow CGP and ECGP to exploit modularity through compartmentalization is described in Sect. 3.8.

3.2 Embedded Cartesian Genetic Programming (ECGP)

ECGP is an enhancement of the CGP representation described in Chap. 2 [28, 29, 30, 31], which also incorporates ideas from module acquisition [1]. A module in ECGP is a small CGP program, which can be used as a function by the main CGP program. In ECGP, modules are created and destroyed during the evolutionary process, through the use of the compress and expand operators, in a similar manner to module acquisition. However, ECGP extends the idea of automatic acquisition and reuse of modules by also incorporating a collection of module mutation operators that allow the evolution of modules. ECGP implements, in addition, an implicit selection pressure towards the formation of useful modules (i.e. ones that are associated with fitness improvement). In so doing, it brings to CGP a form of automatically defined function [12].

3.2.1 Genotype Representation

The CGP representation has to be slightly modified in order to allow the automatic acquisition and reuse of modules in ECGP. Each gene in an ECGP genotype is now represented using a *pair* of integers, rather than just a single integer as in CGP. This is illustrated in Fig. 3.1.

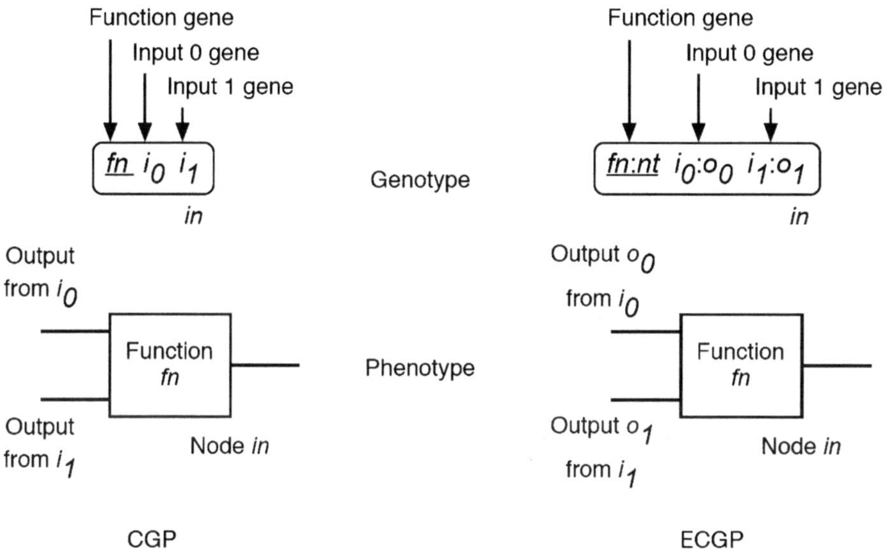

Fig. 3.1 Sections of CGP and ECGP genotypes encoding a single node and the corresponding phenotype for the node. In both cases, all of the genes are labelled. The separate components of each gene are also labelled as follows: function, *fn*; node type, (*nt*); node indexes that the node inputs are taken from, i_0, i_1; node outputs that the node inputs are taken from, o_0, o_1; index of this node, *in*.

One reason for this is that the modules are capable of having multiple outputs, and the CGP representation has hitherto only encoded nodes representing primitive functions with single outputs. However, each node encoded in the genotype still consists of a function gene and a number of input genes (dictated by the arity of the function represented by the node). For each node input encoded in the ECGP genotype, the first integer of the pair encodes the node or program input that the node input connects to (using its *node index*). This is the same principle that is used in the CGP representation. However, the second integer of the pair encodes the output of the node or program input (specified by the first integer of the pair) that the node input takes its value from (remember, nodes in ECGP can have multiple outputs).

Another reason for encoding each gene using a pair of integers is to allow the introduction of node types into the ECGP representation. Node types allow the identification of three different kinds of nodes encoded in the genotype:

- primitive functions (node type 0);
- modules containing an original section of the genotype (node type I);
- reused modules containing a replicated section of the genotype (node type II).

Different node types need to be identified in the genotype, as the evolutionary operators act differently on the nodes depending on their node type (this is explained further in Sect. 3.2.3). Node types are encoded in the genotype as the second integer of the pair which encodes the function gene of each node, whereas the first integer encodes the primitive function (as it would in CGP) or the module which the node represents (using values from a look-up table). An example of CGP and ECGP genotypes that encode the same phenotype is shown in Fig. 3.2 illustrates the differences between the representations.

In contrast to the fixed-length representation used in CGP, the ECGP *genotype* is a bounded variable-length representation (in terms of the number of nodes encoded in the genotype, and the number of genes used to encode each node). The number of nodes encoded in the genotype decreases when sections of the genotype are encapsulated into modules. Alternatively, the number of nodes encoded in the genotype increases when modules are expanded back into sections of the genotype. The number of genes used to encode the inputs of a node in the genotype can also vary in two main ways:

- when a node in the genotype represents a module and the number of inputs possessed by that module is mutated by the module mutation operators (thereby altering the number of genes required to encode the inputs of the module);
- when a module is introduced into the genotype by a compress or genotype point mutation operators (as the node representing the module requires extra genes to encode all of the module inputs).

The next section discusses the representation used to encode the modules in ECGP, and how the modules are stored once they have been created.

3.2.2 Modules

3.2.2.1 Module Representation

A module is represented as a bounded variable-length genotype, which has the same characteristics as an ECGP genotype, except that the number of nodes encoded in

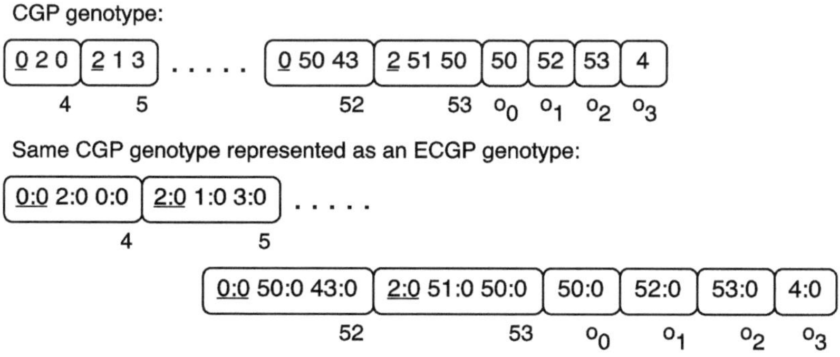

Fig. 3.2 Examples of CGP and ECGP genotypes encoding the same phenotype for a two-bit multiplier circuit (four inputs, four outputs). For each encoded ECGP node, the underlined pair of integers encodes the function and the node type; the remaining pairs of integers encode the node inputs. Every node encoded in this CGP genotype represents a single-output primitive function; therefore every node encoded in the ECGP genotype is of node type 0, and the second integer of each pair encoding the node inputs is always 0. The node index is underneath each node.

the module genotype remains fixed. However, the number of module outputs encoded in the module genotype can vary. The module genotype consists of a list of integers and is split into two parts: the module header and the module body. The module header contains four integers and stores information about the module. These four integers encode the module identifier, the number of module inputs, the number of nodes contained in the module, and the number of module outputs, respectively. The module body encodes the connections and functions of the nodes contained in the module, and the module outputs (similar to program outputs) in the same manner as in ECGP. An example of a module genotype showing the separate components is shown in Fig. 3.3.

The size of a module genotype is determined by the number of nodes and module outputs it encodes. The number of nodes encoded in the module genotype is bounded between a minimum limit of two (any fewer and it would be either an empty module or a primitive function) and a maximum limit, which is set by the user. Also, the number of module outputs encoded in the module genotype is bounded between a minimum limit of one (otherwise there would be no way to connect to the module and access its result) and a maximum of n module outputs, where n is equal to the number of nodes contained in the module (one module output per node). The number of module inputs a module is allowed to have is also restricted between a minimum of two and a maximum of $2n$ module inputs. However, the number of module inputs allowed does not affect the size of the module genotype, as the module inputs are not encoded in the module genotype.

In its current form, ECGP only allows modules to contain nodes representing primitive functions, rather than nodes representing other modules. This is to prevent

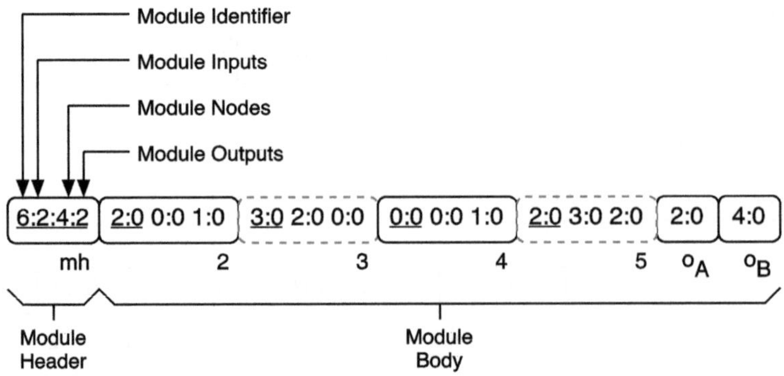

Fig. 3.3 The representation of a module genotype in ECGP. The first section of the module genotype is the module header (*mh*), where each of the genes is labelled. The remainder of the module genotype is the module body, which is represented in the same way as in ECGP.

the module point mutation operator from modifying existing module genotypes, and the compress operator from creating new module genotypes that contain other modules. This was necessary to prevent excessive nesting of modules that leads to extreme code growth through the reuse of modules, which leads to problems with 'stack overflow' and 'out of memory' errors. A method to prevent these implementation problems, while at the same time allowing modules within modules, will be discussed in Sect. 3.7.

Not all of the node outputs have to be connected in the module genotype. This leads to a form of redundancy in modules that is identical to that occurring in CGP. This can be seen in Fig. 3.4, an example of a half-adder circuit, where nodes 3 and 5 of the module genotype are not connected. The contents of a module are immune from the main genotype point mutation operator. However, the module itself is allowed to be mutated by the *module* mutation operators (which include a module point mutation operator and four structural mutation operators) described in Sect. 3.2.4.

3.2.2.2 Module Storage

Once a module is created, the module genotype is stored in a global *module list*, which is an extension of the primitive-function list and is shared by all individuals in the population. Any node in the genotype of a member of the population can be mutated to represent any module or primitive function present in either list (providing the rules for node type are obeyed; see Sect. 3.2.3.2 for more information on node types). The module list is dynamic and has no restrictions on its maximum size. However, when the fittest individual of the population is promoted to the

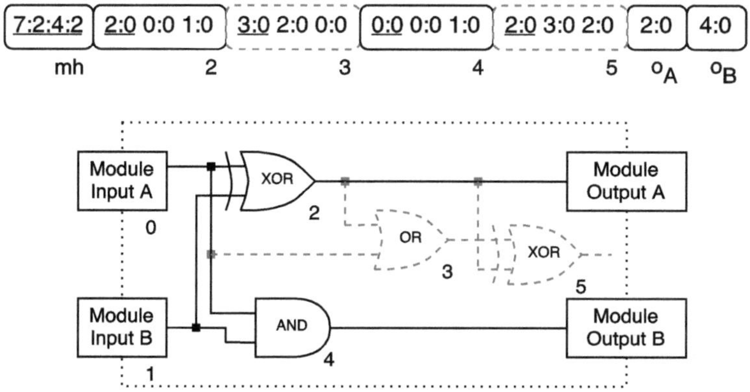

Fig. 3.4 The genotype and corresponding phenotype of an ECGP module constructing a half adder circuit. The first section of the genotype is the module header (*mh*). For each node, the underlined genes encode the function and the remaining genes encode the node inputs. The function look-up table is AND (0), XOR (2) and OR (3). The index labels are shown underneath each module input and node in the genotype and phenotype. The inactive areas of the genotype and phenotype are shown in grey dashes. The dotted box represents the edges of the module.

next generation (chosen in accordance with the evolutionary strategy used in Sect. 3.2.5), the module list is updated to include only those modules present in the fittest individual, thereby deleting all modules present in the module list that were found only in less fit individuals of the population. It was found that this creates a regulatory control of the module list and prevents excessive growth of the module list. An example of a module list is shown in Fig. 3.5.

3.2.3 Genotype Operators

ECGP extends CGP by allowing the use of dynamic acquisition, evolution and the reuse of modules. This is achieved through extra mutation operators, which are used in conjunction with the genotype point mutation operator of CGP.

3.2.3.1 Compress and Expand

The compress operator constructs modules by selecting two random points in the genotype (in accordance with the module size limits) and encapsulates all the nodes (of node type 0) between these two points into a new module. This is encoded into a module genotype of the form described earlier, in Sect. 3.2.2.1. If there are any

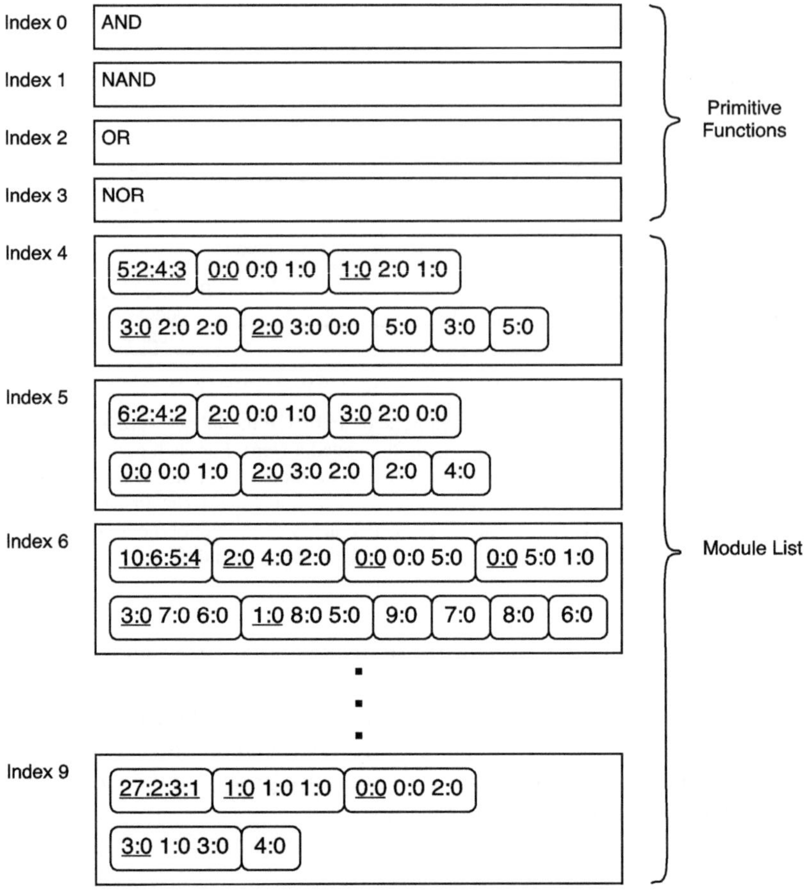

Fig. 3.5 The ECGP function set for a digital circuit problem, which comprises of a fixed length primitive function set, and a variable length module list. The module list currently contains six module genotypes. Therefore, the current ECGP function set contains 10 functions.

nodes of type I or type II between the two selected points, the compress operator does not perform any operation (this is because, at present, we do not allow modules within modules), and does not try to encapsulate a different section of genotype. An example of the compress operator creating a module in the genotype is shown in Fig. 3.6.

The number of module inputs that a module is initialized with is determined by the number of connections between the inputs of the nodes which are going to be encapsulated into a module and the outputs of any previous nodes or program inputs (terminals) which are in the genotype when the module is created. This also

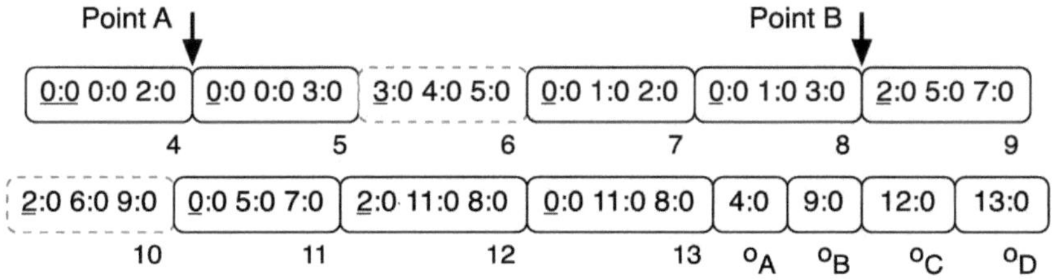

(a) A section of the genotype is chosen by randomly choosing points A and B.

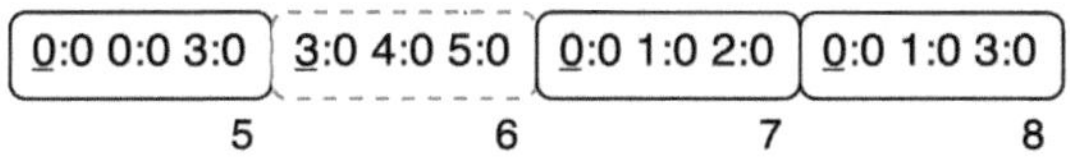

(b) The chosen section is removed from the genotype.

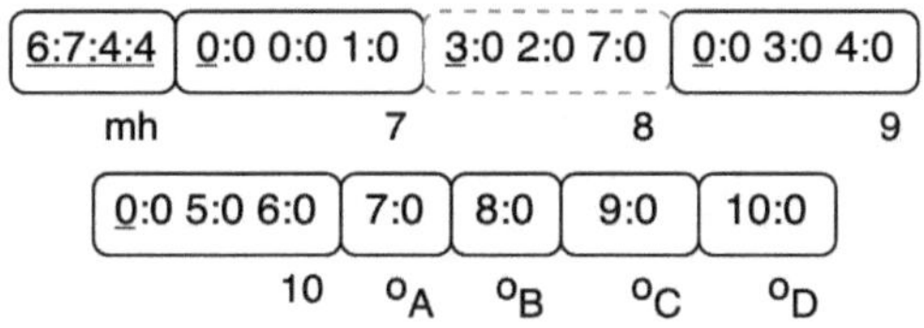

(c) The chosen section is converted into a module genotype by adding a module header (containing the module identifier (6) and the numbers of module inputs (7), nodes (4) and module outputs (4)), and also four module output genes. All nodes and their inputs are also relabelled.

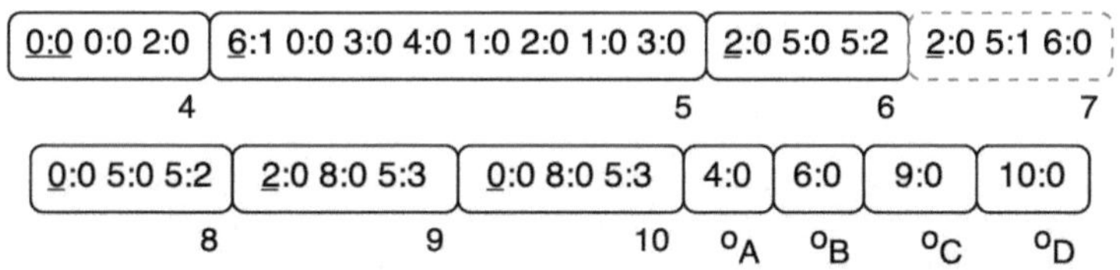

(d) A node of type I representing the new module is added to the genotype in place of the removed section. All nodes, node inputs and outputs occurring after the inserted node are relabelled.

Fig. 3.6 The four steps (a–d) of the compress operator.

includes connections from the inputs of nodes which are going to be encapsulated into the module that are currently inactive, as evolution of the module may make these nodes active at a later time. If there are repeated connections to the output of a previous node, each connection is assigned its own module input. This is thought to be less disruptive to the functionality of the module when it is evolved than if a single module input was assigned for all of the connections to the same node output. Likewise, the number of module outputs possessed by a module is determined by the number of connections between the inputs of the later nodes in the genotype (to the right of the right-hand module boundary) and the outputs of the nodes that are going to be encapsulated in the module, when it is created. Once again, this also includes connections from the inputs of later nodes in the genotype which are currently inactive, as the compress operator needs to preserve all connections in the genotype, in case nodes are activated by mutation at a later time. An example showing the assignments of module inputs and outputs is shown in Fig. 3.7.

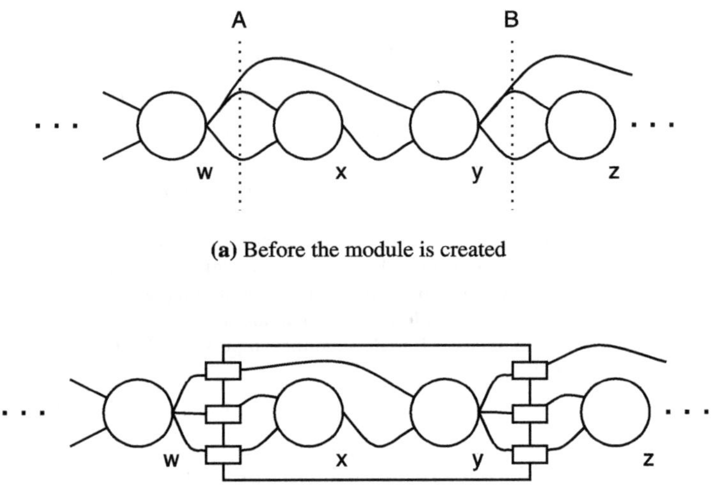

(a) Before the module is created

(b) After the module is created

Fig. 3.7 Assigning module inputs and module outputs during module creation between the two dashed lines A and B. In (a), three connections are made to the output of node *w* from nodes *x* and *y*, so three module inputs are created in (b). Likewise, three connections are made to the output of node *y*, two from node *z* and one from a node later in the genotype, in (a). Therefore, three module outputs are created in (b). The module is represented as a box in (b). The module inputs are the three boxes on the left edge of the module, and the module outputs are the three boxes on the right edge of the module.

Any module created by the compress operator is represented in the genotype as a type I node. The function gene of any type I node is immune from the genotype point mutation operator. A type I node can only be removed by the expand operator.

The expand operator reverses the effect of the compress operator by randomly selecting a type I node in the genotype of an individual, and replacing the type I node with the nodes contained in the module represented by the type I node. This operation can only be applied to type I nodes, as the nodes contained inside a module represented by a type I node were originally part of the genotype. An example of the expand operator is provided by the reverse of Fig. 3.6.

The inputs of all of the later nodes in the genotype of the individual are updated in the final stage of both the compress and the expand operators, so all of the connections remain intact. The compress and expand operators make only a structural change to the genotype and have no effect on genotype fitness, as the genotype before and after the action of these operators represents the same directed graph.

The expand operator has *twice* the probability of being applied to the genotype compared with the compress operator. This was shown to be a good choice through a trial-and-error experiment in early work. Therefore, modules represented by type I nodes are twice as likely to be destroyed, compared with a type I node representing a new module being created. This induces a pressure on the modules within a genotype to replicate in order to improve their chance of survival. The only way that modules can replicate in a genotype is through the action of the genotype point mutation operator, which changes the function gene of a node into a value that represents a module, thereby performing a module duplication. This also performs a transition from a module being represented by a type I node to a type II node, which reduces the probability of the module being removed from the genotype. This is discussed further in Sect. 3.2.3.2.

Since an evolutionary strategy that promotes the single fittest genotype to the next generation (described in Sect. 3.2.5) is used the only way that modules can survive from one generation to the next is by being present in this promoted genotype. We found that this, coupled with the effects of the probabilities of the operators introduces an implicit pressure for good modules to replicate quickly in the genotype in order to survive. Thus the modules within the genotype undergo a struggle for survival. This is in addition to the usual fitness-based selection that is applied to the genotypes.

3.2.3.2 Genotype Mutation

Modules can replicate within the genotype through the action of the genotype point mutation operator. This is identical to the point mutation operator used in CGP, with the exception that it can mutate the function of a node to represent any of the primitive functions or available modules in the module list.

If the function gene of a type 0 node (a node representing a primitive function) is mutated by the genotype point mutation operator to represent a module, it is classed as a type II node. The genotype point mutation operator treats type 0 and type II nodes the same way. Therefore, the genotype point mutation operator can also mutate the function gene of a type II node to represent any of the predefined functions (thereafter referred to as a type 0 node) or available modules in the module list (still referred to as a type II node). Whenever a node changes type (from type 0 to type II or vice versa), the mutated node keeps however many of the original node's inputs it requires, and 'randomly' generates any extra inputs it may need. This may also happen when a type II node representing a module *A* is mutated to represent a module *B*, when module *B* has a different number of module inputs from module *A*.

The genotype point mutation operator can also mutate the input genes of any type 0, type I or type II node. The genotype point mutation operator in ECGP differs only slightly from that used in CGP. In ECGP, every input gene has two integers associated with it, a node index and an output of the node corresponding to the node index. Both of these values are mutated at the same time, to ensure a valid connection, as not all nodes have the same number of outputs.

Type II nodes are also immune from the expand operator. This was introduced to avoid the excessive growth of the genotype that could occur when type 0 nodes were replicated to form type II nodes that, in turn, were expanded back into the genotype.

To summarize, the properties of type 0, I and II nodes are shown in Table 3.1, with the effect the genotype operators have on them. The reason for having type I and II nodes is to try and reduce excessive growth of the genotype, and to help induce a selection pressure on the modules. The modules have to replicate in the genotype (i.e. make the transition from being represented by type I to type II nodes) and at the same time be associated with a high-fitness genotype in order to survive. Once the module is represented by a type II node, it is harder for the module to be removed from the module list, as it has a lower probability that it will be removed from the genotype (i.e. it cannot be expanded). This is advantageous, as it allows good modules to stay in the module list. However, it is also disadvantageous, as it could possibly allow the evolution of the genotype to progress at a slower rate, as it is harder to remove modules which contribute towards convergence on local optima.

Table 3.1 The effect of the operators on each node type

Node type	Action of compress	Action of expand	Action of genotype point mutation
0	Encapsulates into a module	Immune	Mutates function or node genes
I	Immune	Replaces with module contents	Mutates node genes
II	Immune	Immune	Mutates function or node genes

3.2.4 Module Operators

The module genotypes contained in the module list can also be evolved through the action of five different module mutation operators: module point mutation, add-input, add-output, remove-input and remove-output. All of the module mutation operators must comply with the restrictions on the number of module inputs and the number of module outputs at all times.

3.2.4.1 Module Point Mutation

The module point mutation operator is a restricted version of the ECGP genotype point mutation operator, as it can still mutate the input and function genes of any node encoded in the module genotype. However, it is not allowed to introduce any type II nodes into the module genotype. Therefore, it can only mutate the function gene of a node to represent one of the predefined primitive functions. The module point mutation operator can also mutate which node outputs the module outputs are connected to. However, a module output can never be mutated to connect directly to a module input, as this would bypass any processing that the nodes in the module may perform on the data presented at the module inputs (i.e. it would implement a 'junk' module).

3.2.4.2 Add-input

The add-input operator allows greater connectivity to the nodes contained in a module by increasing the number of module inputs by one, each time the add-input operator is applied to a module (provided the constraints for the number of module inputs a module is allowed are obeyed). When the add-input operator is applied to a module, the gene representing the number of module inputs in the module header is incremented by one. An extra gene is also inserted into every node (type I and type II) representing the module in the genotype of an individual (to update the arity of the node). The extra gene is randomly assigned a value for the new module input (in accordance with the feed-forward manner of CGP). The effect the add-input operator has on a module genotype and the corresponding individual genotype is shown in Fig. 3.8.

3.2.4.3 Add-output

The add-output operator also allows greater connectivity to the nodes contained in a module, by increasing the number of module outputs by one, each time the add-output operator is applied to a module (providing the constraints for the number of module outputs a module can possess are obeyed). When the add-output operator is applied to a module, the gene representing the number of module outputs in the

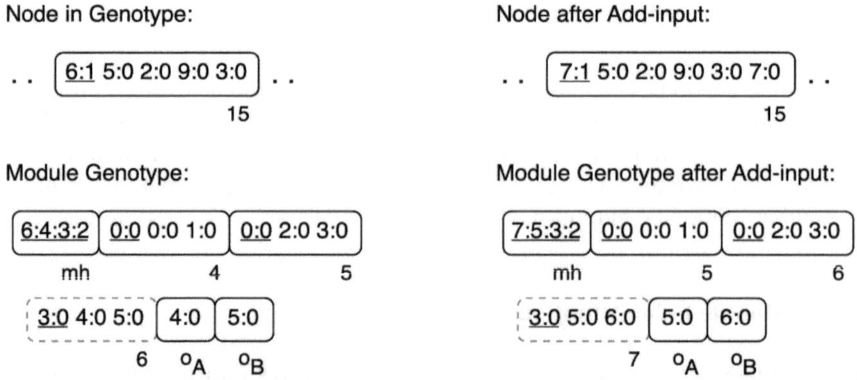

(a) A node representing module 6 in the genotype of an individual, and the corresponding module genotype for module 6, before the application of the add-input operator.

(b) After the add-input operator. The module input gene in the module header (mh) of the module genotype has been incremented by one and the module identifier gene has been changed to the next available number. The function gene of the node has been changed to reflect the new module identifier and an extra gene has been added for the new module input, whose value is randomly chosen.

Fig. 3.8 Before (a) and after (b) the application of the add-input operator.

module header is incremented by one. An extra gene is also added to the end of the module genotype. The extra gene is randomly assigned values for the node index and node output, which encode where the new module output is connected to. The effect the add-output operator has on a module genotype and an individual genotype is shown in Fig. 3.9.

3.2.4.4 Remove-input

The remove-input operator is the reverse of the add-input operator, and reduces the connectivity from the contents of a module by decreasing the number of module inputs by one each time the remove-input operator is applied to a module (once again making sure to obey the constraints imposed on a module regarding the number of module inputs). When the remove-input operator is applied to a module, the gene representing the number of module inputs in the module header is decremented by one, and the gene corresponding to the randomly chosen module input is removed from every node (type I and type II) representing the module in the genotype of the individual.

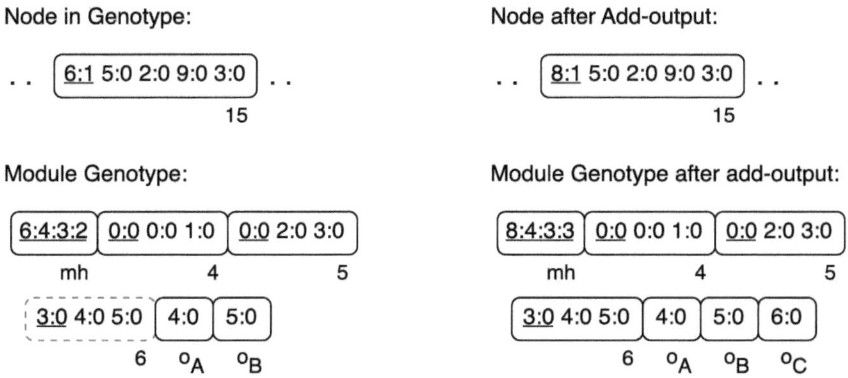

(a) A node representing module 6 in the genotype of an individual and the genotype for module 6 before the application of the add-output operator.

(b) After the add-output operator. The module output gene in the module header (mh) is incremented by one and an extra gene is added to the module genotype to encode the new module output, whose value is randomly chosen. The module identifier gene is also changed to the next available number. The function gene of the node is also updated to reflect the change in the module.

Fig. 3.9 Before (a) and after (b) the application of the add-output operator.

3.2.4.5 Remove-output

The remove-output operator is the reverse of the add-output operator, and reduces the connectivity to the contents of a module by decreasing the number of module outputs by one each time the remove-output operator is applied to a module (once again making sure to obey the constraints imposed on a module regarding the number of module outputs that it may have). When the remove-output operator is applied to a module, the gene representing the number of module outputs in the module header is decremented by one, and the gene corresponding to the randomly chosen module output is removed from the module genotype.

3.2.5 Evolutionary Strategy

ECGP uses the same evolutionary strategy as CGP, which is based on a $(\mu + \lambda)$ *evolutionary strategy* [24] with $\mu = 1$ and $\lambda = 4$, giving a population size of 5. The μ value indicates the number of individuals promoted to the next generation as parents and the λ value indicates the number of offspring generated from the promoted parents. The $(1+4)$ evolutionary strategy is defined in Procedure 3.1.

Procedure 3.1 The $(1+4)$ evolutionary strategy

```
1: for all i such that 0 ≤ i < 5 do
2:     Randomly generate individual i
3: end for
4: Select the fittest individual, which is promoted as the parent
5: while a solution is not found or the generation limit is not reached do
6:     for all i such that 0 ≤ i < 4 do
7:         Mutate the parent to generate offspring i
8:     end for
9:     Select the fittest individual using the following rules:
10:    if an offspring has a better or equal fitness than the parent then
11:        The offspring is promoted as the new parent
12:    else if many offspring have an equal fitness which is a better or equal fitness than the parent
       then
13:        A randomly selected offspring is promoted as the new parent
14:    else
15:        The parent is promoted
16:    end if
17: end while
```

In the rules for selecting the fittest individual (on lines 10–15 of Procedure 3.1), an offspring is always chosen over the parent when they have equal fitness, as the offspring is phenotypically identical to (in terms of fitness) but genetically different from the parent. This allows neutral exploration of the search space until a phenotypically better offspring is discovered. An example of the $(1+4)$ evolutionary strategy over three generations is shown in Fig. 3.10.

3.2.6 Benchmark Experiments

In the work described in this section, the ECGP approach and the original CGP approach were applied to a variety of benchmark problems, each of which varied in difficulty in different ways. A comparison is made between ECGP and CGP for each of the problems and, where possible, comparisons are also made with other GP techniques.

All of the experiments were run on a 2.2 GHz single processor desktop PC with 448 MB of memory. The time taken to complete 50 runs of each problem varied between a few seconds and a few hours, depending on the difficulty of the problem. ECGP took only fractionally longer to complete one thousand generations on any problem than CGP, showing that the computational time required for the overhead associated with module acquisition was quite small, and the computational time

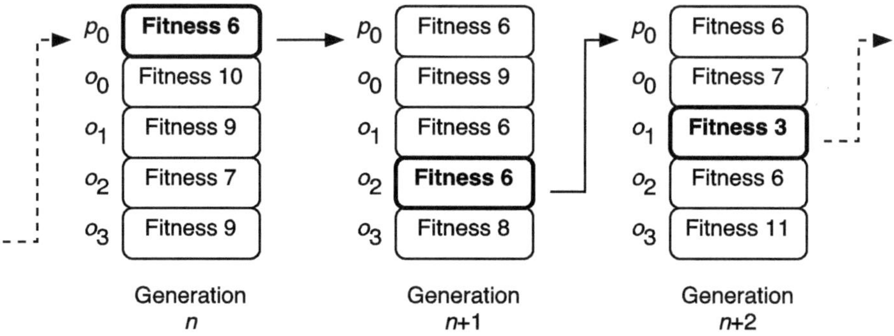

Fig. 3.10 An example of the $(1+4)$ evolutionary strategy over three generations. The individual promoted from each generation as the parent is shown in bold. In generation n, the existing parent, p_0, is promoted, whereas in generation $n+1$, offspring o_2 is randomly chosen for promotion instead of offspring o_1, and in generation $n+2$, offspring o_1 is promoted as the fittest.

taken for fitness evaluation (both for CGP and for ECGP) was by far the dominant factor.

3.2.6.1 Parameters

The parameters used for CGP and ECGP which were common to all of the experiments are shown in Table 3.2. The operator rates and probabilities were determined to be fairly optimal by means of an experimental optimization process in previous work [28, 29, 30].

In most GP experiments, the maximum number of generations allowed is set quite low (for example, 1000 generations), normally resulting in a success rate that is less than 100%. However, the aim of each experiment in this chapter was to achieve 100% success. Therefore, the maximum number of generations allowed for each independent run was set to 20 million. This allowed ample time for each independent run to find a solution (at which point the independent run terminated), whilst guaranteeing that each independent run always terminated after 20 million generations, if a solution was not found.

3.2.6.2 Suitable Maximum Module Size

In ECGP, the user sets a parameter, *ms*, for the maximum module size. This allows ECGP to construct modules of size n, where $2 \leq n \leq ms$. The main concern is how to determine a good maximum module size. To shed light on this issue, we applied

Table 3.2 The common parameters used for CGP and ECGP on all of the test problems. Parameters denoted by $^{\diamond}$ apply to ECGP only

Parameter	Value
Population size	5
Initial genotype size	100 nodes (300 genes)
Genotype point mutation rate	3% (9 genes)
Genotype point mutation probability	1
Compress/expand probability$^{\diamond}$	0.1/0.2
Module point mutation probability$^{\diamond}$	0.04
Add/remove input probability$^{\diamond}$	0.01/0.02
Add/remove output probability$^{\diamond}$	0.01/0.02
Module list initial contents$^{\diamond}$	Empty
Number of independent runs	50

ECGP to the even-4-parity problem (also used in Sect. 3.2.6.6), using the parameters in Table 3.2 and various maximum module sizes. The aim of the investigation was to see if any correlation exists between the maximum module size and the performance of ECGP. The results are shown in Fig. 3.11. It can be seen that there is no obvious correlation between the maximum module size and computational effort. However, there is a weak trend that implies that larger module sizes improve the performance of ECGP. One possible reason to explain this is that larger modules contain more inactive nodes, so that redundancy (the same type as in CGP) could be having an impact. Another possible reason is that having larger modules implies that they must be fewer in number. This is due to modules not being permitted within modules, thus implying that the existing modules are evolved for a longer period of time. All of these aspects will be investigated further in future work. For the moment, it appears that the best maximum module sizes for the even-4-parity problem are five, seven and 18. In this chapter, a maximum module size of five was used for most of the experiments, as this is the smallest of these values and allows faster evaluation of the evolved programs.

3.2.6.3 Computational Effort

In each experiment on digital circuits, the results for all independent runs were assessed using a statistic called the *computational effort*. This metric was introduced by Koza [11] as a measure of the computational effort required to solve a problem based on the data from all of the independent runs. The formula to calculate the computational effort is shown in (3.1). The notation is taken from [11] as follows: $N_s(i)$ is the number of successful independent runs by generation i; N_{total} is the total number of independent runs; $P(M,i)$ is the cumulative probability of success for an independent run with population size M producing a solution by generation i; $R(P(M,i),z)$ is the number of independent runs required to satisfy the success

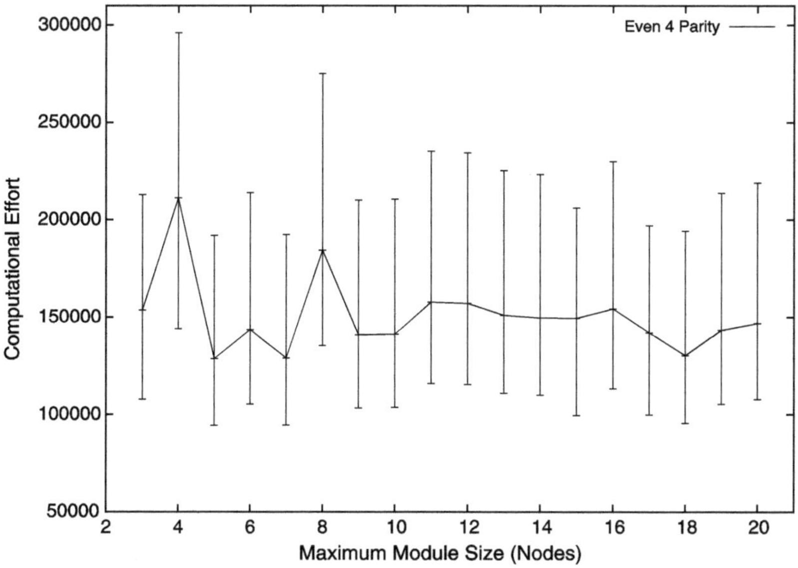

Fig. 3.11 A comparison between the maximum module size and the performance of ECGP in terms of computational effort for the even-4-parity problem. The error bars indicate the confidence interval for the true computational effort.

predicate by generation i with probability z; $I(M,i,z)$ is the number of individuals that need to be processed to produce a solution with probability z, using population size M, at generation i; CE is the minimum number of individuals to be processed with probability z, using population size M, and hence the minimum computational effort. In this chapter, we use $z = 0.99$.

$$\begin{aligned} P(M,i) &= \frac{N_s(i)}{N_{total}}, \\ R(P(M,i),z) &= \left\lceil \frac{\log(1-z)}{\log(1-P(M,i))} \right\rceil, \\ I(M,i,z) &= M \times R(P(M,i),z) \times (i+1), \\ CE &= \min_i I(M,i,z). \end{aligned} \tag{3.1}$$

The computational-effort statistic used here is a popular performance measure in the GP community. However, it is by no means perfect and has numerous inadequacies. Christensen and Oppacher [3] found that the ceiling operator in Eqn. 3.1 has a tendency to overestimate $R(z)$, whilst the min operator tends to underestimate the computational effort required. Furthermore, the underestimation increases for systems with a high number of generations, which is the case in the approach used in this chapter. Niehaus and Banzhaf [18] later found that the underestimation of the computational-effort statistic was inversely proportional to the number of runs used in the calculation, so for a small number of runs, the underestimation of the computational effort is very large. In the work described in this chapter only 50 independent runs are used (which is classed as a small number of runs) for each experiment, as this was the number of runs used in the work we were comparing the results with. Therefore, the computational-effort figures are likely to be underestimates of the theoretical value for computational effort and should be used only as a rough guide. However, Niehaus and Banzhaf [18] also found that as the probability of a run ending in failure increased, the computational effort deviated further from the theoretical value. In the work described in this chapter, every run continued until a solution was found, thereby producing a 100% success rate, which should improve the accuracy of the computational-effort results.

3.2.6.4 Confidence Interval

Walker et al. [32, 33] noted that computational effort is only a point statistic, with no confidence interval, so any comparisons made with other techniques are inconclusive. Instead they devised an approach for defining a 95% confidence interval for the true computational effort of a technique, using Wilson's method. The approach starts by defining formulae for the upper and lower bounds of the confidence interval for the number of successful runs given a probability z, which are shown in (3.2) and (3.3). The proportion of successes p is defined as $p = r/n$, where r is the number of successful runs and n is the total number of runs. The z_{norm} value was set to 1.96, as this was used in [32, 33].

$$upper(p,n) = \frac{2np + z_{norm}^2 + z_{norm}\sqrt{z_{norm}^2 + 4np(1-p)}}{2(n + z_{norm}^2)}, \tag{3.2}$$

$$lower(p,n) = \frac{2np + z_{norm}^2 - z_{norm}\sqrt{z_{norm}^2 + 4np(1-p)}}{2(n + z_{norm}^2)}. \tag{3.3}$$

Equations (3.2) and (3.3) can be used to define the upper and lower bounds of the confidence interval for the true computational effort at generation i, $\tau(i)$, by substituting $upper(p,n)$ and $lower(p,n)$ for $P(M,i)$ in (3.1) and dropping the ceiling operator from the $R(P(M,i),z)$ formula, as shown in the following equation:

$$MR(upper(p,n),z)(i+1) \leq \tau(i) \leq MR(lower(p,n),z)(i+1). \tag{3.4}$$

The confidence interval produced by (3.4) is always valid regardless of the probability of success or the number of runs [32, 33].

In order to find the confidence interval for the true minimum computational effort $\tau(j)$, the minimum generation j and the proportion of successes p at which the minimum computational effort occurs must be known. However, this is virtually always unknown when any form of evolutionary computation is used. Therefore, a good estimate for the minimum computational effort is required, in order to find these values. Koza's proposed approach for calculating the minimum computational effort (3.1) provides a good estimate for the minimum generation, and also the proportion of successes. Using this estimate for the minimum generation and the proportion of successes makes it possible to calculate the confidence interval for the estimated minimum computational effort [32, 33].

3.2.6.5 Non-parametric Statistics

Since there are still questions concerning the accuracy of the computational-effort statistic, a variety of other statistics have also been compiled. The experimental results in this chapter are positively skewed (i.e. not normally distributed), since the minimum number of evaluations is 1. Therefore parametric measures, such as the mean and standard deviation cannot be used, as they would not provide an accurate and meaningful representation of the data. Hence, a number of non-parametric statistics have been used. These are the median number of evaluations, the median absolute deviation (MAD) and the interquartile range (IQR). The MAD is a measure of the variability within a distribution, and is similar to the standard deviation, except that it is based on the median rather than the mean. The IQR measures the dispersion of the middle 50% of the distribution, and is the difference between the third and first quartiles.

The significance of the results in this chapter has also been assessed using the non-parametric Mann–Whitney U test [14] (also known as the Wilcoxon rank-sum test [34]), which assesses whether two independent samples come from the same distribution.

In order to allow others to compare their results with the figures presented in this chapter and conduct their own statistical tests, the CGP data sets collected from all runs is available from the CGP website.[1]

3.2.6.6 Even-Parity

The problem of evolving even-parity functions using GP with the primitive Boolean operations of AND, OR, NAND, and NOR has been shown to be very difficult and has been adopted by the GP research community as a good benchmark problem for testing the efficacy of new GP techniques [11]. It is particularly appropriate for testing module acquisition techniques, as even-parity functions are more compactly represented using XOR and XNOR functions. Also, smaller parity functions can help build larger parity functions. An example of this is shown in Fig. 3.12. Thus parity functions are naturally modular, and it is to be expected that they will be evolved more efficiently when such modules are provided. It is therefore of great

[1] http://www.cartesiangp.co.uk

interest to see whether modules that represent such functions are constructed automatically.

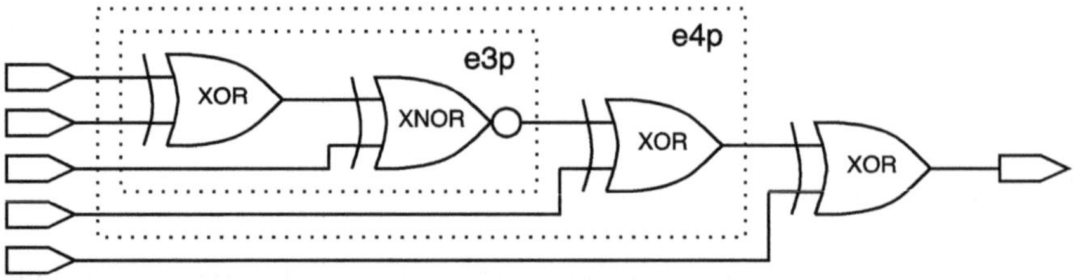

Fig. 3.12 An example of the even 5 parity circuit, compactly constructed from the XOR and XNOR functions. Some of the smaller parity functions, even-3-parity (e3p) and even-4-parity (e4p), which are used as building blocks in the circuit, are highlighted with dotted boxes.

The even-n-parity problem has n inputs and a single output, which produces a 1 if there is an even number of 1's in the inputs and 0 otherwise. In the work described in this section, CGP and ECGP were applied to the even-n-parity problem, where $n = 3,4,5,6,7,8$. The function set used when evolving the even-parity problem consisted of the Boolean functions AND, NAND, OR and NOR. The maximum module size used for ECGP was five. The computational effort figures for CGP and ECGP applied to the even-n-parity problem, where $n = 3,4,5,6,7,8$, are shown in Table 3.3.

Table 3.3 The computational effort (CE) figures for CGP and ECGP applied to even-parity problems of various sizes. Also included are the lower (CI_{lower}) and upper (CI_{upper}) bounds of the true computational-effort confidence interval

	CGP			ECGP		
Parity	CI_{lower}	CE	CI_{upper}	CI_{lower}	CE	CI_{upper}
three-bit	24,440	33,282	49,646	25,902	37,446	55,539
four-bit	106,546	151,683	210,235	148,045	201,602	300,724
five-bit	569,853	776,002	1,157,540	375,986	512,002	763,739
six-bit	2,235,463	3,044,162	4,540,890	718,836	978,882	1,460,171
seven-bit	8,409,124	11,451,202	17,081,434	1,412,762	1,923,842	2,869,741
eight-bit	22,902,612	31,187,842	46,522,022	2,960,877	4,032,002	6,014,423

In all 50 runs, both CGP and ECGP produced 100% successful solutions. As the complexity of the problem increases, it can be seen that ECGP performs significantly better than CGP, and the difference in performance between ECGP and CGP increases. This suggests that ECGP may perform even better than CGP on more complex problems. It appears that the performance increase found for ECGP is due to the discovery, preservation and reuse of partial solutions in the genotype, whereas CGP has to find each partial solution separately. For the low order parity functions, CGP performs better than ECGP. In this case, it is likely that the overhead of discovering useful modules and learning how to reuse the modules is responsible for the decrease in performance.

A closer examination of the solutions found by ECGP revealed that a high percentage of the nodes contained in the genotype represented modules. Also, the majority of the modules were reused repeatedly. Further inspection of the modules showed that a large number of them represented either an XOR or an XNOR function. Some of the evolved XOR and XNOR functions were represented in their most compact form, consisting of three Boolean functions. Other XOR and XNOR functions were represented in a much less efficient form, consisting of up to five Boolean functions. The majority of the modules also contained some inactive nodes, indicating that redundancy also occurs in the module genotype while the modules are being evolved. Figure 3.13 shows some examples of modules used in solutions evolved using ECGP to solve the even-parity problem.

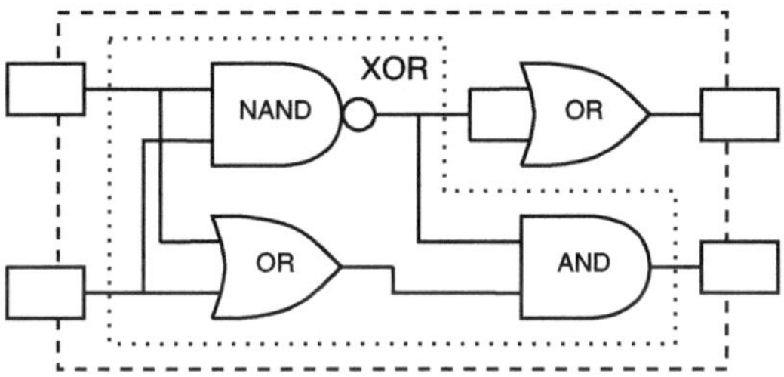

(a) XOR module

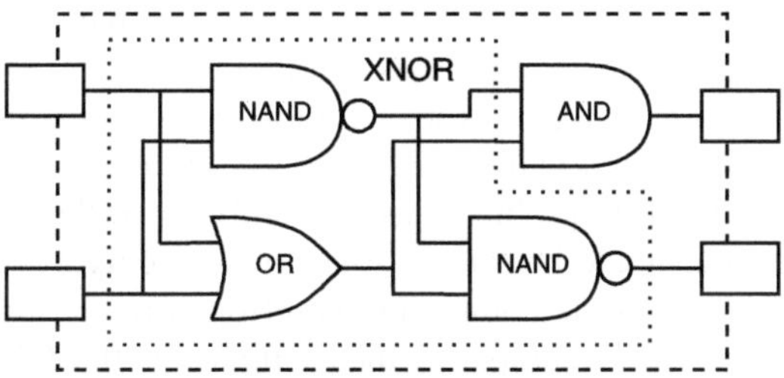

(b) XNOR module

Fig. 3.13 A selection of module phenotypes used in evolved solutions to the even-parity problem. The group of nodes in each module that build the XOR or XNOR function are surrounded by a dotted box. The XNOR module in part (b) also contains the XOR function from (a) if the other output is used instead. The dashed box represents the module.

Table 3.4 The computational effort figures for GP and EP, both with and without ADFs, applied to various-size even-parity problems. The computational-effort figures for GP and EP are taken from [12] and [2]. $^\diamond$ denotes computational-effort figures calculated with $z = 1.0$

Parity	GP	GP with ADFs	EP	EP with ADFs
three-bit	96,000	64,000	$^\diamond$28,500	63,000
four-bit	384,000	176,000	181,500	$^\diamond$118,500
five-bit	6,528,000	464,000	2,100,000	126,000
six-bit	70,176,000	1,344,000	–	$^\diamond$121,000
seven-bit	–	–	–	$^\diamond$169,000
eight-bit	–	–	–	321,000

Comparing the results of CGP and ECGP with those of GP shown in Table 3.4 for the even-n-parity problem, where $n = 3,4,5,6$, shows CGP and ECGP significantly outperforming GP. Moreover, the speed-up increases with problem complexity. However, GP has been reported to be unable to find any solutions on the higher-order parity functions [12]. Comparing ECGP on the one hand and GP with ADFs on the other, shows a similar trend for all of the even-parity problems tested. This is an interesting result, as GP allows a two-level hierarchy in the ADFs, thereby allowing the creation of sub-parity functions. ECGP, however, only allows a single-level hierarchy in its modules, where only the XOR or XNOR function can be created. Therefore, the most likely cause of the performance increase between CGP, ECGP and GP is the implicit reuse of nodes found in the directed-graph representation. In addition to the results in Table 3.4, GP with ADFs is capable of finding solutions to higher-order parity problems. However, different ADF parameter settings were used for the runs, which allowed the ADFs to have a larger number of arguments, therefore making an unfair comparison with the rest of the results. When a greater number of arguments are allowed in the ADFs, larger sub-parity functions can be created, causing an increase in performance [12]. This suggests that allowing a variable number of module inputs in ECGP in conjunction with larger module sizes could provide further performance increases in ECGP.

The computational-effort figures for evolutionary programming (EP) [2], with and without ADFs, also listed in Table 3.4, show a strange correlation, as EP with ADFs finds the even-6-parity problem easier to solve than the even-5-parity problem, when the opposite should be true. Also, some of the figures for EP were computed with $z = 1.0$, making the computational-effort formula invalid and comparisons more difficult. However, comparing the EP figures with CGP and ECGP shows that the results for CGP and ECGP scale much better with problem difficulty than do those for EP. When CGP and ECGP are compared against EP with ADFs, EP with ADFs is shown to perform much better than CGP and ECGP on all but the smallest parity problem. The ADFs in EP also use a multi-level hierarchy, like those found in GP. However, they seem to have a larger effect on the performance of EP than on that of GP, which could possibly be attributed to some of the extra mutation operators used in EP. As previously mentioned, ECGP uses only a single-level hierarchy, which in this experiment could construct only either the XOR or XNOR function, whereas EP, like GP, could construct various sub-parity functions. This again strongly suggests that if a multi-level hierarchy is allowed in ECGP, a significant improvement in performance could be possible.

As a further investigation to see if allowing larger modules capable of encoding smaller parity functions would improve performance, the following experiments were carried out. ECGP was rerun on the even-n-parity problem, where $n = 4, 5$, with a maximum module size of 20 nodes and a genotype size of 400 nodes (keeping the same ratio of maximum module size to genotype size as in the previous experiments). The extra resources provided a significant performance boost in the case of ECGP. The computational effort for the even-4-parity was 32,641, a speed-up of 6.18 compared with the original ECGP result for the same problem, which is also significantly better than that for EP with ADFs. For the even-5-parity, the results were not as impressive but still comparable to the result for EP with ADFs, as the computational-effort figure for ECGP was 130,081. This is still an improvement on the original results for ECGP by 3.94 times. However, all of the computational-effort figures for CGP and ECGP are heavily dependent on the number of nodes encoded in the genotype [16]. This can cause vast fluctuations in the performance of both techniques, thereby making fair comparisons with other techniques very difficult, especially non-graph-based approaches, where it is difficult to quantify the number of resources used.

The improvement in performance of ECGP in the further experiments owes its origin to the fact that modules are allowed to construct multiple XOR or XNOR functions inside a single module, thereby forming larger sub-parity functions than in the previous run of ECGP. However, the limit on module resources and the single-level hierarchy still remain, which explains why the speed-up decreases between the even-4 and even-5-parity results, and why ECGP still performs worse than EP with ADFs for the latter. From these results, it is possible to say that allowing modules to use other modules inside themselves could be beneficial to the performance of ECGP, as ECGP would be less dependent on module resources to build larger functions.

The PushGP system [26] has been applied to the even-n-parity problem, where $n = 3, 4, 5, 6$. However, owing to the nature of PushGP system and its instruction set, direct comparisons could not be made with GP (with or without ADFs), as PushGP was seen to have a 'considerable advantage' over GP [26]. Therefore, direct comparisons cannot be made with CGP and ECGP either. Poli and Page have also applied their work using a smooth uniform crossover in GP to the even-parity problem [20]. However, we are also unable to compare our results with this approach as a different function set was used. This affects the difficulty of the even-parity problem and would lead to an unfair comparison between the techniques.

Another GP technique, known as Enzyme GP [13], has also been applied to the even-3 and even-4 parity problems. For the even-3 and even-4 parity problems, Enzyme GP had a computational effort of 79,000 and 1,830,000 respectively. CGP and ECGP performed significantly better than Enzyme GP on both parity problems. In fact, CGP performed 2.37 and 12.06 times faster than Enzyme GP, whilst ECGP performed 2.11 and 9.08 times faster than Enzyme GP for the even-3 and even-4 parity problems, respectively. Although the same function set was used by Enzyme GP, CGP and ECGP, the number of components that Enzyme GP was initially allowed to use in the solution was much lower, which may account for the poorer performance.

Table 3.5 The median number of evaluations (ME), median absolute deviation (MAD) and interquartile range (IQR) of CGP and ECGP for various size even-parity problems. The *U* value is from the Mann–Whitney significance test and is highly significant ($P < 0.001$) when denoted by ‡

	CGP			ECGP			
Parity	ME	MAD	IQR	ME	MAD	IQR	*U*
three-bit	5,993	2,936	6,610	5,931	3,804	10,372	1,299
four-bit	30,589	12,942	25,438	37,961	21,124	49,552	1,521
five-bit	136,693	71,236	199,245	119,625	52,483	98,940	1,128
six-bit	577,237	257,574	594,770	227,891	85,794	190,456	‡449
seven-bit	2,156,139	1,029,010	2,343,039	472,227	312,716	603,643	‡185
eight-bit	7,166,369	3,210,224	6,363,465	745,549	500,924	1,108,934	‡39

In addition to the computational-effort figures shown in Table 3.3, Table 3.5 shows the median number of evaluations required by CGP and ECGP to find a solution to each of the evolved parity problems. Both tables show a similar trend for the performance of CGP compared with ECGP. An interesting observation from Table 3.5 is that the median absolute deviation for ECGP becomes much lower than that for CGP as the parity problem increases in difficulty. This indicates that the use of modules in ECGP improves the accuracy of the technique as well. However, the figures for the Mann–Whitney *U* test show that there is no significant difference between CGP and ECGP on the three-, four- and five even-parity problems but the difference between CGP and ECGP on the six-, seven- and eight even-parity problems is highly significant ($P < 0.001$). This is an encouraging result, as it further supports the hypothesis that the automatic acquisition, evolution and reuse of modules improves the performance of the technique.

3.3 Digital-Adders

The digital-adder is a problem, which has emerged from research in the field of evolvable hardware. It is now a commonly used benchmark for GP techniques with multiple outputs. It is considerably harder than the even-parity problem, as the digital-adder problem has multiple outputs. Also, the number of inputs increases much faster than in the even-parity problem as the difficulty of the test problem is increased. A n-bit digital-adder has two n-bit inputs and a one-bit carry-in, giving a total of $2n+1$ inputs. Likewise, it also has an n-bit sum output and a one-bit carry-out, giving a total of $n+1$ outputs. An example of a two-bit adder is shown in Fig. 3.14.

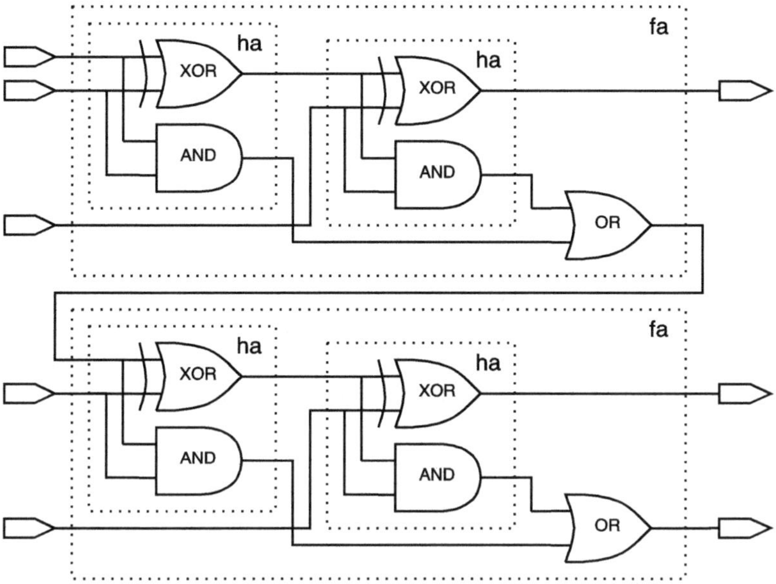

Fig. 3.14 An example of the two-bit adder circuit constructed using the XOR function. Some of the smaller adder circuits, the half adder (ha) and the full one-bit adder (fa), which are used as building blocks in the circuit, are highlighted with dotted boxes.

The function set used is identical to that for the even-parity problem (AND, NAND, OR, NOR). The function set was chosen because the digital-adder is also a modular problem. The digital-adder is easily constructed from halfadders or smaller fulladders (as seen in Fig. 3.14), which, in turn, are more compactly built when an XOR function is allowed. Thus the situation is reminiscent of the even-parity problem. The maximum module size for ECGP was once again set to five nodes.

Computational-effort figures for the results for CGP and ECGP on the one-bit, two-bit and three-bit digital-adder problems were calculated using the formula in Eqn. 3.1 and are shown in Table 3.6. Unfortunately, GP researchers tend to avoid problems with multiple outputs; therefore we have no figures for GP applied to the adder problem to compare with those for CGP and ECGP. Computational effort figures have been published for Enzyme GP applied to the two-bit adder, but a function set was used which made the problem easier to solve, as no intermediate building blocks (such as the XOR function) had to be constructed. Therefore, we cannot use these figures for a fair comparison.

Table 3.6 The computational-effort (CE) figures for CGP and ECGP applied to the one-bit, two-bit and three-bit digital-adder problems. Also included are the lower (CI_{lower}) and upper (CI_{upper}) bounds of the true computational-effort confidence interval

	CGP			ECGP		
Adder	CI_{lower}	CE	CI_{upper}	CI_{lower}	CE	CI_{upper}
one-bit	25,238	36,486	54,115	30,347	43,203	59,880
two-bit	577,067	834,246	1,237,329	419,097	596,643	826,957
three-bit	6,315,127	8,599,682	12,827,903	2,398,586	3,414,723	4,732,865

All of the 50 runs produced 100% successful solutions for both CGP and ECGP. The results are similar in nature to those found for the even-parity problem. The one-bit adder (which is by far the least complex of the adders to evolve) was evolved quicker using CGP than with ECGP. Once again, this could be attributed to the overhead of the exploration of code in the modules, and also ECGP not being able to reuse the discovered modules correctly. On the harder two-bit and three-bit adders, it can be seen that ECGP starts to outperform CGP. Also, as the problem scales in complexity, the speed-up between CGP and ECGP increases as well. This can be attributed to ECGP finding useful partial solutions and then re-using them throughout the genotype, exploiting any modularity found in the adder.

On closer inspection of the evolved modules, it can be seen that the modules represent mainly the XOR and the half-adder functions, which are both partial solutions for the adder problem. In a similar manner to the evolved modules for the even-parity problem (see Sect. 3.2.6.6), we found that the XOR and half-adder functions were constructed in various ways, and varied greatly in complexity. Also, the majority of the nodes in the genotype of an ECGP individual encoding a solution represented modules rather than primitive functions. This implies that modules are more appealing than primitive functions in terms of obtaining a high fitness score. The same phenomenon was also seen in the solutions to the even-parity problem.

Table 3.7 The median number of evaluations (ME), median absolute deviation (MAD) and interquartile range (IQR) of CGP and ECGP for the one-bit, two-bit and three-bit digital-adder problems. The U value is from the Mann–Whitney significance test and is highly significant ($P < 0.001$) when denoted by‡

	CGP			ECGP			
Adder	ME	MAD	IQR	ME	MAD	IQR	U
one-bit	5,923	3,054	7,122	8,825	4,396	8,545	1,449
two-bit	132,565	76,228	178,335	124,251	49,926	99,291	1,060
three-bit	1,943,585	996,482	2,174,500	733,909	307,694	599,132	‡415

From Table 3.7 it can be seen that the figures for the median number of evaluations for CGP and ECGP follow a similar trend to the computational-effort figures in Table 3.6. Also, the degree of variance in the results for ECGP becomes lower than for CGP, as the difficulty of the problem increases, indicating that the ECGP figures could be more accurate than those of CGP. Table 3.7 also shows that only the performance difference between CGP and ECGP on the three-bit adder is highly significant, indicating that ECGP seems to perform better on harder problems than CGP does.

3.4 Symbolic Regression

Symbolic regression was first suggested as a suitable problem for GP by Koza in his first book on GP [11]. Since then it has been widely adopted as a benchmark for testing advancements in the GP field [17, 19].

Given two equal-sized data sets, the aim of the symbolic regression problem is to find a symbolic function which maps each point in the first data set to the corresponding point in the second data set, within a certain degree of error. For example, the solution to the symbolic regression problem $f(x) = x^3 - x^2 + x$ would map each point in the x data set to the corresponding point in the $f(x)$ data set.

The function set used for the symbolic regression problem typically comprises the functions: addition, subtraction, multiplication and protected division (division by zero returns a result of 1). This function set was used for both experiments described in this section. However, occasionally other functions such as sine, cosine, log and exponential are used in the function set when the symbolic regression problem requires it [11].

The fitness function used to distinguish between individuals in the population was the absolute error of the function over all the points in the input set. This is the sum of the absolute differences between the calculated output of each individual and the value for the point in the output set, for all points contained in the output set. The criterion for successfully finding a solution was that the absolute error of each point was within 0.01 of the corresponding point in the output set.

In this section, we describe attempts to evolve solutions to symbolic regression problems with generating functions $x^6 - 2x^4 + x^2$ and $x^5 - 2x^3 + x$, shown in Fig. 3.15. Both problems were evaluated using a data set of 50 randomly chosen points from the range $[-1,1]$. The maximum module sizes allowed for ECGP were three, five and eight. This was also an experiment to see if a maximum module size of five always gave the best performance.

The computational-effort figures for CGP and ECGP applied to the two symbolic regression problems tested are shown in Table 3.8. For comparison, the computational-effort figures for GP with and without ADFs, taken from [12], are shown in Table 3.9.

In all 50 runs, CGP and ECGP found 100% solutions to both problems. The computational-effort figures in Table 3.8 show that both CGP and ECGP outper-

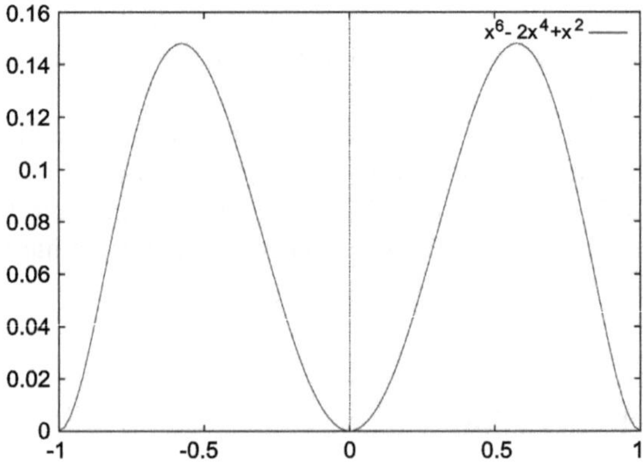

(a) $x^6 - 2x^4 + x^2$

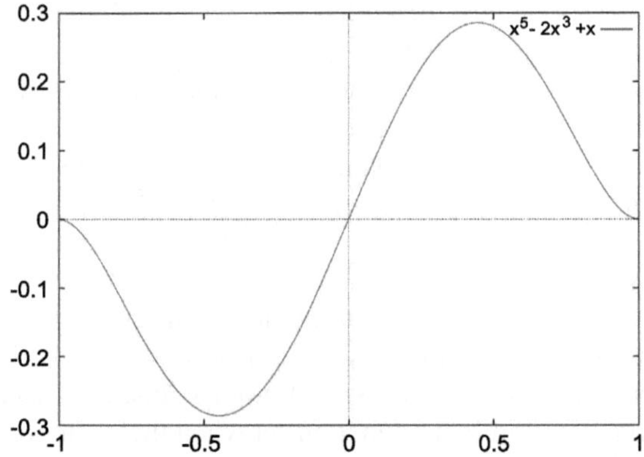

(b) $x^5 - 2x^3 + x$

Fig. 3.15 The symbolic regression functions $x^6 - 2x^4 + x^2$ (a) and $x^5 - 2x^3 + x$ (b) within the range $[-1, 1]$.

formed GP with and without ADFs on both of the symbolic regression problems by a minimum of 15.3 times for the sextic polynomial and 5.5 times for the quintic polynomial.

Comparing the computational-effort figures for CGP and ECGP, it can be seen that CGP is better than or approximately equal to that of ECGP in all of the experiments. The statistics in Table 3.10 once again show a contradictory result between the figures for computational-effort and the median number of evaluations. For the sextic-polynomial problem, CGP performs better than ECGP with any maximum module size, which once again could possibly be attributed to underestimation by the computational-effort formula. The Mann–Whitney test shows that none of the experiments are significant, so it could be possible to conclude that the use of mod-

Table 3.8 The computational-effort (CE) figures for CGP and ECGP (with a maximum module size of 3, 5 and 8) applied to symbolic regression problems defined by $x^6 - 2x^4 + x^2$ and $x^5 - 2x^3 + x$. Also included are the lower (CI_{lower}) and upper (CI_{upper}) bounds of the true computational-effort confidence interval

	CGP			ECGP-3		
Function	CI_{lower}	CE	CI_{upper}	CI_{lower}	CE	CI_{upper}
$x^6 - 2x^4 + x^2$	34,884	55,692	91,860	15,322	54,353	199,557
$x^5 - 2x^3 + x$	6,687	36,708	209,122	29,136	59,551	122,798

	ECGP-5			ECGP-8		
Function	CI_{lower}	CE	CI_{upper}	CI_{lower}	CE	CI_{upper}
$x^6 - 2x^4 + x^2$	32,539	71,077	153,015	35,249	76,997	165,760
$x^5 - 2x^3 + x$	14,142	35,896	92,082	25,011	72,075	212,461

Table 3.9 The computational-effort figures for GP, with and without ADFs, applied to symbolic regression problems defined by $x^6 - 2x^4 + x^2$ and $x^5 - 2x^3 + x$

Function	GP	GP with ADFs
$x^6 - 2x^4 + x^2$	1,440,000	1,176,000
$x^5 - 2x^3 + x$	396,000	1,200,000

Table 3.10 The median number of evaluations (ME), median absolute deviation (MAD) and interquartile range (IQR) of CGP and ECGP (with maximum module sizes 3, 5 and 8) for the symbolic regression problems defined by $x^6 - 2x^4 + x^2$ and $x^5 - 2x^3 + x$. The U value is from the Mann–Whitney significance test, where CGP is compared with ECGP for each of the maximum module sizes. All values are in thousands

	CGP			ECGP-3			
Problem	ME	MAD	IQR	ME	MAD	IQR	$U-3$
$x^6 - 2x^4 + x^2$	12.7	10.9	64.1	29.7	25.1	279.4	1.5
$x^5 - 2x^3 + x$	32.2	31.0	525.6	25.9	24.4	296.8	1.2

	ECGP-5			ECGP-8				
Problem	ME	MAD	IQR	ME	MAD	IQR	$U-5$	$U-8$
$x^6 - 2x^4 + x^2$	43.7	41.0	275.5	38.8	37.0	299.4	1.5	1.5
$x^5 - 2x^3 + x$	38.9	38.4	411.3	167.5	164.4	664.9	1.3	1.4

ules in ECGP produces no benefit to performance on this type of problem. This possibly highlights a lack of modularity in symbolic regression problems, thereby removing any advantage that modular approaches, such as ECGP might have over non-modular approaches such as CGP. However, it does highlight the problem of selecting a suitable maximum limit for the size of the modules.

From the computational-effort results, it can be seen that on the two occasions when ECGP performed on a par with CGP, the maximum limit for the size of the modules was different. In the other ECGP experiments, the cost of choosing an unsuitable maximum limit for the size of the modules could decrease the performance of the program by up to a factor of two compared with CGP. From the experiments described previously in this chapter, if a problem is too simple, CGP tends to find a solution to the problem while ECGP is still exploring the search space, as the overhead associated with the additional evolutionary operators of ECGP increases the exploration time required. However, when a problem reaches a certain level of complexity, the acquisition, evolution and reuse of modules in ECGP causes ECGP to start to outperform CGP for most of the smaller choices for the maximum size limit of the modules.

Parallel distributed GP (see Sect. 1.2.5) has also been applied to the $x^6 - 2x^4 + x^2$ symbolic regression problem. The results in [19] show that parallel distributed GP has a computational-effort of 91,000, which is much better than that for GP with or without ADFs but is still outperformed by CGP and ECGP with any of the three maximum module sizes tested. This result is interesting, as it highlights the performance benefit of implicit reuse in the representations of graph-based GP systems such as parallel distributed GP, CGP and ECGP.

The PushGP system has also been applied to the $x^6 - 2x^4 + x^2$ symbolic regression problem, where the effect of various operators for controlling the program size [6] was investigated. However, in this instance, CGP and ECGP seemed to significantly outperform the PushGP system on the $x^6 - 2x^4 + x^2$ symbolic regression problem (by approximately eight times). This suggests that the instruction set used by PushGP, which gave it an unfair advantage over GP on the even-parity problem, may not be so well suited to this problem.

3.5 Lawnmower Problem

The lawnmower problem was first introduced by Koza in his second book [12] to test the effectiveness of automatically defined functions by exploiting the modularity of the lawnmower problem. Since then, it has been used as a benchmark problem by many other researchers in the testing of new GP techniques and representations [19].

In the lawnmower problem, the objective is to guide a lawnmower around a grass lawn, which consists of $n \times m$ squares (where n and m are user-defined parameters), and mow all the grass. The lawnmower moves around the lawn one square at a time, and cuts all the grass in each square it visits. The lawnmower is allowed to revisit a square of the lawn as many times as it likes, but the grass in a square can only be cut by the lawnmower once; therefore when the lawnmower revisits squares of the lawn, it loses efficiency. However, the lawn is a 'magic' lawn: when the lawnmower moves off a square on any side of the lawn, it reappears on the square on the opposite side. The lawnmower always starts in the centre square of the lawn and starts off by facing in a northward direction. The lawn is cut when every square has been visited by the lawnmower. An example is shown in Fig. 3.16.

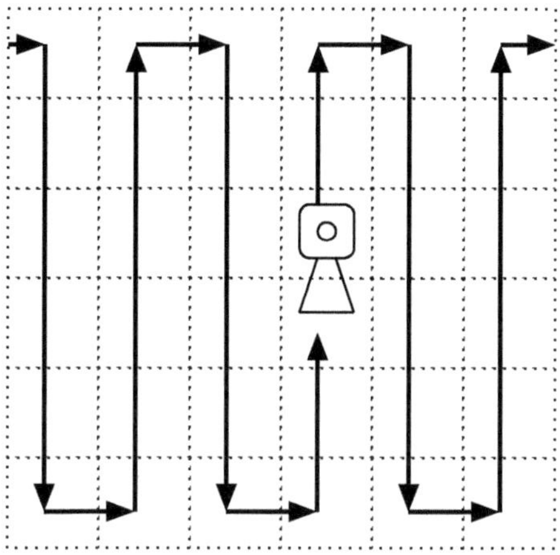

Fig. 3.16 A possible route for the lawnmower to solve the lawnmower problem on a 6×6 lawn.

The movement of the lawnmower is controlled by a CGP or ECGP program. The program has three *inputs* which it can use: *move*, which moves the lawnmower one square forward on the lawn in the direction the lawnmower is facing, and cuts all of the grass in that square; *turn*, which rotates the lawnmower 90° clockwise in the current square on the lawn; and *random constant*, which stores a randomly distributed vector for the entire run of the form $[x,y]$, where $0 \leq x < n$ and $0 \leq y < m$. In conjunction with the operations just described, the move and turn inputs also return the vectors $[0,0]$, so that mathematical operations can take place on any combination of inputs. For further details, see [12].

The function set for the program consists of *v8a*, which takes two vectors and returns the result of the addition of these two vectors; *frog*, which takes a vector $[x,y]$ and jumps the lawnmower to another square on the lawn, a distance of x squares in the horizontal direction and y squares in the vertical direction away, and returns the vector $[x,y]$; and, finally, *progn*, which takes two inputs and executes everything from the first input and then everything from the second input, before returning the resulting vector from the second input.

The fitness function for this problem is defined as the number of squares on the lawn which are left uncut by the lawnmower after the evolved program has been run once. For this problem, we minimize the fitness value, as a lawn on which all the squares are cut has a fitness score of zero, and is a solution to the problem. ECGP was allowed a maximum module size of five.

The computational-effort figures for CGP and ECGP applied to the lawnmower problem are shown in Table 3.11. For comparison, the computational-effort figures for parallel distributed GP (PDGP) and GP (with and without ADFs), taken from [19] and [12], respectively are also shown in Table 3.12.

Table 3.11 The computational-effort (CE) figures for CGP and ECGP applied to the lawnmower problem with various lawn sizes. Also included are the lower (CI_{lower}) and upper (CI_{upper}) bounds of the true computational-effort confidence interval

	CGP			ECGP		
Size	CI_{lower}	CE	CI_{upper}	CI_{lower}	CE	CI_{upper}
32	941	1,282	1,912	854	1,282	1,768
48	950	1,602	2,028	819	1,602	1,839
64	1,688	2,403	3,331	950	1,602	2,028
80	1,411	1,922	2,867	1,140	1,922	2,433
96	1,520	2,562	3,243	1,067	1,602	2,210
112	1,881	2,562	3,822	1,411	1,922	2,867
128	2,351	3,202	4,776	1,646	2,242	3,344
144	1,881	2,562	3,822	1,280	1,922	2,651
160	2,351	3,202	4,776	1,280	1,922	2,651
176	2,116	2,882	4,299	982	1,922	2,207
192	2,586	3,522	5,254	1,520	2,562	3,243
208	2,116	2,882	4,299	1,494	2,242	3,093
224	2,821	3,842	5,731	1,920	2,882	3,975
240	3,056	4,162	6,208	1,881	2,562	3,822
256	2,089	3,522	4,458	1,411	1,922	2,867

For all 50 independent runs, CGP and ECGP produced 100% successful solutions when applied to the lawnmower problem. For all lawn sizes of the lawnmower problem, it can be seen that the performance of CGP and ECGP starts off fairly evenly for the smaller lawn sizes but as the lawn size increases, ECGP starts to perform better than CGP. However, the results for CGP and ECGP on the lawnmower problem show a strange correlation compared with the results for the other techniques on the lawnmower problem and with the previous results for CGP and ECGP in this chapter. Normally, as the problem difficulty increases, the computational-effort increases, but the results for CGP and ECGP on the lawnmower problem do not follow this trend. Instead, the computational-effort figures for CGP and ECGP oscillate, so a harder problem sometimes appears easier to solve. This is shown more clearly in Fig. 3.17. A possible explanation for this could be related to the CGP decoding procedure, as each node in CGP and ECGP produces a set of instructions rather than

Table 3.12 The computational-effort figures for parallel distributed GP (PDGP) and GP with and without ADFs for the lawnmower problem with various lawn sizes. Also included are the speed-up figures when comparisons are made with CGP and with ECGP

Size	PDGP (1)	GP (2)	GP with ADFs (3)	Speed-up CGP&(1)	Speed-up CGP&(2)	Speed-up ECGP&(3)
32	4,000	19,000	5,000	3.1	14.8	3.9
48	5,000	56,000	9,000	3.1	35.0	5.6
64	5,000	100,000	11,000	2.1	41.6	6.9
80	5,000	561,000	17,000	2.6	291.9	8.8
96	6,000	4,692,000	20,000	2.3	1,831.4	12.5
112	6,000	–	–	2.3	–	–
128	7,000	–	–	2.2	–	–

performing a calculation (as in the previous problems in this chapter). This allows the first few nodes of a CGP genotype, in conjunction with the implicit reuse found in the CGP representation, to behave like ADFs, as they each produce a block of instructions, which can be reused. In ECGP, this would be similar to having ADFs inside the modules. Therefore, the oscillations in the computational-effort figures could be directly related to the usefulness of the first few nodes in a CGP or ECGP genotype and the amount that each node is reused.

On examining Fig. 3.17, a general but noisy trend can be seen for CGP and ECGP, in which the computational-effort does increase with problem difficulty. This follows the results for the previous problems discussed in this chapter. It can also be seen that the performance speed-up for ECGP grows with problem difficulty, suggesting that ECGP could perform even better on larger problems. This speed-up can be attributed to the discovery and reuse of subroutines, which allow the lawnmower to cut several grass squares covering an area of the lawn, and then allowing the same pattern to be repeated elsewhere on the lawn. This supports the previous findings for ECGP in this chapter. The statistics in Table 3.13 also support the findings from the computational-effort figures and show that 80% of the experiments can be classed as showing some form of significance from the Mann–Whitney test. It could also be said that the significance of the result increases roughly with problem difficulty, as the results that are classed as highly significant are from the hardest problems.

Comparing the computational-effort figures for CGP with PDGP (up to a lawn size of 128) and GP without ADFs (up to a lawn size of 96) shown Table 3.12, it can clearly be seen that CGP performs better than the two other techniques. CGP performs between 2.2 and 3.1 times faster than PDGP and between 14.8 and 1831.4 times faster than GP without ADFs. In fact, CGP even outperforms GP with ADFs on this problem. This result emphasizes the performance gain associated with using a graph-based representation (as in CGP and PDGP), rather than a tree based representation (as in GP). It can also be seen from comparing the two techniques that include a form of ADF (ECGP and GP with ADFs), that ECGP performs between 3.9 and 12.5 times faster than GP with ADFs. Notice also that the speed-up grows

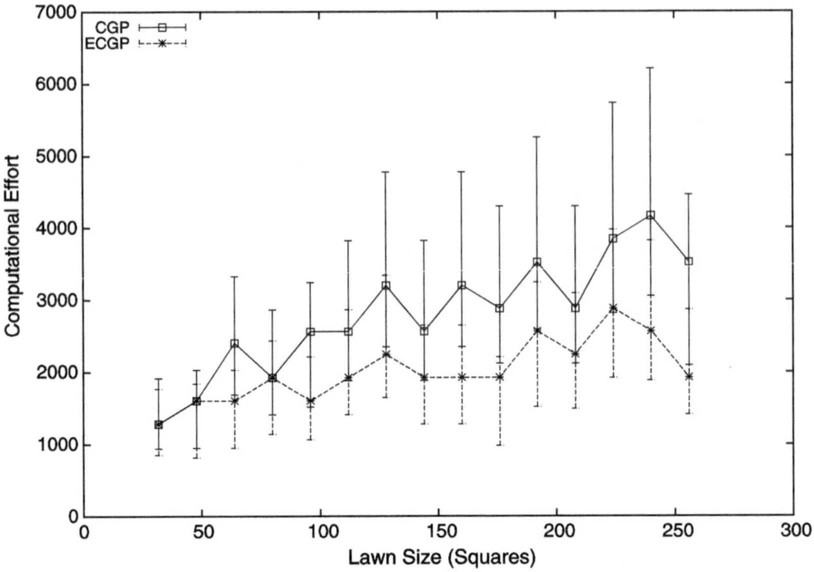

Fig. 3.17 The unusual correlation between the computational-effort figures of CGP and ECGP and increasing lawn size for the lawnmower problem. The general trend for both CGP and ECGP is that the computational-effort required increases with lawn size. The error bars indicate the confidence interval for the true computational-effort.

with the size of the lawn, indicating that ECGP may perform even better than GP with ADFs on larger problems.

Another technique that has been tested on the lawnmower problem is Spector's GP with Automatically Defined Macros (ADMs) [25]. Spector found that GP with ADMs actually performed worse than GP with ADFs on the 64 square lawnmower problem, producing a computational-effort figure of 18,000. Comparing this result for GP with ADMs with those for CGP and ECGP in Table 3.11 shows a performance speed-up of 7.5 and 11.25 times for CGP and ECGP, respectively.

3.6 Alternative ECGP Operators

3.6.1 Cone-Based and Age-Based Module Creation

In [12], Koza described the automatic definition and reuse of functions that are frequently appearing patterns in a chromosome. Turning these patterns into regular building blocks (modules) for evolution automatically decomposes the problem

Table 3.13 The median number of evaluations (ME), median absolute deviation (MAD) and interquartile range (IQR) for CGP and ECGP for the lawnmower problem with various lawn sizes. The U value is from the Mann–Whitney significance test and is marginally significant ($P < 0.05$) when denoted by *, significant ($P < 0.01$) when denoted by † and highly significant ($P < 0.001$) when denoted by ‡

	CGP			ECGP			
Size	ME	MAD	IQR	ME	MAD	IQR	U
32	307	124	206	257	96	186	1,046
48	361	150	289	301	94	207	1,011
64	467	200	414	349	130	233	†831
80	523	182	347	387	142	275	*924
96	623	204	338	427	126	260	‡777
112	581	198	393	493	148	303	1,046
128	711	266	662	561	236	524	*922
144	659	178	393	493	164	295	†833
160	639	260	592	475	108	195	†828
176	625	204	426	529	122	250	*901
192	781	206	436	579	176	407	*916
208	717	254	531	511	196	451	*935
224	751	234	731	643	244	472	*955
240	855	292	810	561	180	462	‡688
256	827	200	446	591	158	330	‡648

and allows hierarchical problem solving. Since Koza relied on trees to represent the chromosome, his implementation of module creation focused on subtrees, which are convex structures corresponding to subfunctions. In contrast, the CGP and ECGP representation models essentially describe a directed acyclic graph (DAG). A subgraph of a DAG is, however, not necessarily convex. The original ECGP model selects a node randomly and then adds nodes with contiguous node numbers to form a module. Such modules do not represent typical hardware building blocks; the nodes within these modules might even be completely unconnected. In this section we discuss two alternative techniques, cone-based and age-based module creation [10]. While cone-based module creation focuses on convex subgraphs that actually represent typical hardware building blocks, age-based module creation extends the original random selection of nodes with the concept of ageing.

3.6.1.1 Cone-based Module Creation

Figure 3.18 presents the standard ECGP module creation technique that aggregates nodes with contiguous node numbers to form modules. Cone-based module creation selects nodes that form cones and thus might better correspond to subfunctions, especially in the area of digital circuits. Cones are a widely used concept in circuit synthesis, especially in look-up table mapping for Field-Programmable Gate Arrays

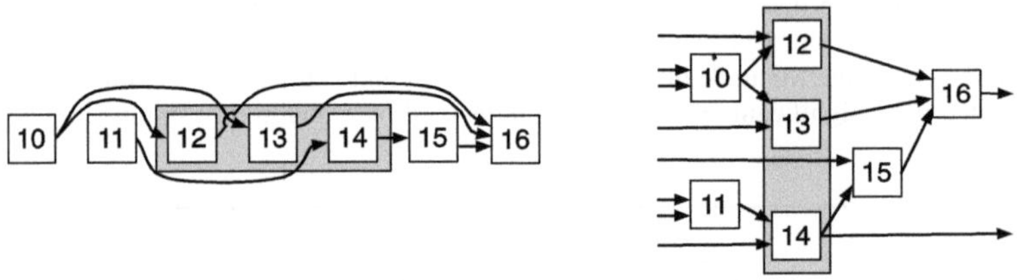

Fig. 3.18 The original ECGP module creation technique(left) can lead to subfunctions which are atypical in digital design (right).

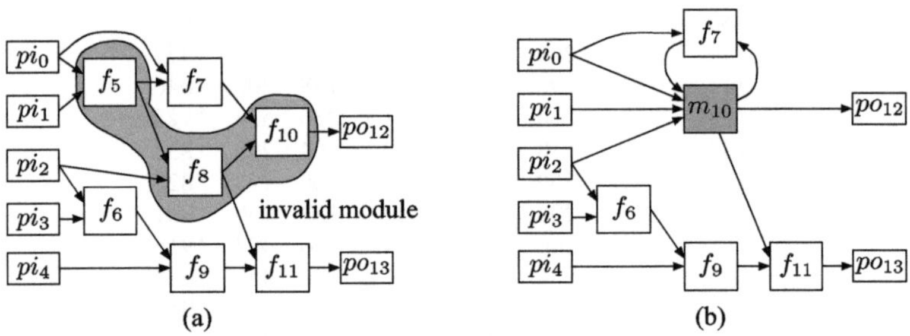

Fig. 3.19 Cones with reconvergent paths are invalid (a), as they can lead to combinational feedback loops (b).

(FPGAs see, for example, [5]). Given a node f_r in the DAG, a cone rooted at f_r consists of f_r itself plus some predecessor nodes such that for any node f_i in the cone there exists a path from f_i to f_r that is entirely in the cone. Thus, a cone is a convex graph. For example, in Fig. 3.19a the node set (f_{11}, f_9, f_8) forms a cone. Note that while a cone has a distinct root node f_r, it can have several outputs. The rationale behind cone-based module creation is that many useful substructures in classically engineered circuits are cones, for example the sum and carry functions of a full-adder.

To generate module candidates, we randomly select a primitive node f_i and create a cone rooted at f_i with a number of nodes randomly chosen between n_{min} and n_{max}. One subtlety in generating cones is that we have to avoid what are called *reconvergent paths* in logic synthesis. To discuss several subtypes of cones consider again the circuit in Fig. 3.19a. The node set (f_{11}, f_9, f_6) is called a fan-out-free cone because the fan-out (output connections) of every node except the root node stays within the cone. Such cones are certainly safe candidates for module creation. The node set (f_{11}, f_9, f_8, f_5) does not form a fan-out-free cone, as the outputs of f_5 and f_8 leave the cone. Nevertheless, this node set forms a valid module. In contrast, the node set (f_{10}, f_8, f_5), which is highlighted in Fig. 3.19a does not form a valid module, as the output of f_5 leaves and re-enters the cone. If this cone was turned into a module, the resulting circuit, shown in Fig. 3.19b, would contain a combinational feedback loop. The path formed by nodes (f_5, f_7, f_{10}) is called reconvergent with respect to the cone (f_{10}, f_8, f_5).

Reconvergent paths are specific to the cone-based module creation technique. Neither the age-based technique (see Sect. 3.6.1.2) nor the original ECGP method can create such paths. In contrast to the original ECGP module creation technique, which relies on contiguous node numbers, we have used breadth-first search, starting with the cone's root node, to avoid reconvergent paths. Resuming the example of Fig. 3.19a, a cone of size 3 rooted at node f_{10} would be formed by the nodes (f_{10}, f_7, f_8). Module creation stops when a randomly chosen number of nodes has been aggregated or when a module is hit. There are no modules within modules. As in the original ECGP technique, our cone-based module creation technique ensures that inputs to a new module come only from nodes with smaller node numbers, and outputs of the new module connect only to nodes with higher node numbers. After the nodes for a new module have been selected the nodes are compacted and renumbered to form a coherent block, which then can be compressed to a module by the standard ECGP compress operator.

3.6.1.2 Age-based Module Creation

Age-based module creation aggregates primitives nodes that have persisted in the chromosome for a higher number of generations. The rationale behind age-based module creation is that aged nodes are likely to contribute directly or indirectly to an individual's success and should therefore be preferred over randomly selected nodes.

We assign to each primitive node f_i an attribute $age(f_i)$. The age is incremented by one in each generation and set to zero when the node is selected for mutation or compression. The age of primitive nodes within modules remains unchanged; modules themselves do not have an age. We form module candidates by aggregating primitive nodes, restricting the number of nodes by lower (n_{min}) and upper (n_{max}) bounds. The average age of a module candidate m_j is then given by

$$\text{age}(m_j) = \frac{\sum_{f_i \in m_j} age(f_i)}{|m_j|}.$$

Our age-based module creation technique relies on a two-stage binary tournament to select a module. That is, we generate a module candidate by the following procedure. First, we select a random primitive node f_i and a number of primitive nodes $n, n_{min} \leq n \leq n_{max}$, randomly. Then, we extend the module from f_i to nodes with smaller node numbers until we hit a module or aggregate exactly n primitive nodes. We create another module using a different random primitive node f_i and draw the one with higher average age. If both modules have the same average age, we draw one module randomly. This step is repeated once to derive the final module.

We have also experimented with selecting the module candidate with the maximum average age. This requires the formation and evaluation of a larger number of module candidates. However, picking the module with the maximum average age has proved inferior to the two-stage binary tournament scheme for all test problems.

An explanation for this lies in the fact that maximizing average module age tends to generate modules with a very small number of high-aged nodes. It seems that while using node age as a guide to steer module creation is highly effective (see Sect. 3.6.1.3), the technique is rather sensitive to the size of modules.

3.6.1.3 Benchmarks, Metrics and Results

Table 3.14 ECGP parameters for the parity, multiplier and classifier benchmarks

Parameter	Parity	Multiplier	EMG classifier
Chromosome length (nodes)	50	200	200
Number of inputs n_i	3/4/5	4/6	200
Number of outputs n_o	1	4/6	1
Function set	2-input LUT: AND NAND, OR, NOR	4-input LUT: AND AND_{inv}, OR, XOR	4-input LUT any function
Mutation rate	0.03	0.03	0.03
One-point mutation probability	0.6	0.6	0.6
Compress/expand probability	0.1/0.2	0.1/0.2	0.1/0.2
Module mutation probability	0.1	0.1	0.1
Module size (n_{min}, n_{max} nodes)	2,...,8	2,...,10	2,...,10

We compared the original, age-based and cone-based ECGP module creation techniques on even-parity and multiplier benchmarks. Additionally, we included classifiers for electromyographic (EMG) signals as test problems. In this classification application, skin-attached sensors collected electric signals from contracting muscles to control a prosthetic hand [7]. The test data was recorded from four muscles of a volunteer's forearm. A sequence of eight contractions (movements) with 20 repetitions of each was measured. The typical signal for a movement was composed of a 9s relaxation phase and a 5s contraction phase. For 2s of the contraction phase, we removed the DC offset and applied RMS smoothing to obtain the feature vectors. The resulting data set consisted of 144 strings of 200 bits each. Based on that data, we tried to evolve a classifier circuit for the movement 'open hand'.

Table 3.14 shows the ECGP model parameters, including the chromosome length, the number of inputs and outputs, and the function set for the nodes. For the parity function, we used two-input look-up tables (LUTs) to model the nodes, but restricted the function set to a few Boolean functions. For the multipliers, we used four-input LUTs (4-LUTs) as node models but again restricted the function set to the functions AND, OR, XOR, and AND_{inv}, which is an AND with one input inverted. Finally, for the EMG classifiers we use 4-input LUTs without any restriction on the node function. As an optimization algorithm, we employed a $1+4$ evolutionary strategy (ES). The corresponding parameters and the parameters for module creation are also shown in Table 3.14. We conducted all experiments with our modular framework for evolutionary hardware design presented in [9].

We chose the computational-effort (CE) as presented by Koza [12]. As the evaluation metric metric states the expected value for the number of fitness evaluations required to reach the optimization goal with a probability of z. To determine the CE metric, we repeated all experiments 50 times and set z to 99%. The CE metric cannot be applied directly to the classifier benchmark. Classifiers differ from arithmetic circuits in that there is no simple correctness measure. Typically, classifiers are evolved with training data and then run on test data to determine metrics such as the classification rate. As we wanted to investigate and compare the computational-effort for evolving a classifier and not the generalization capabilities of the ECGP model, we measured the classifiers' fitness on the training data set and defined the model to be correct when the classification rate on the training data exceeded 85% or 95%.

Table 3.15 Computational-effort figures for an $1+4$ ES with random, age-based and cone-based module creation

	Random	Age-based		Cone-based	
Problem	Absolute	Absolute	Relative	Absolute	Relative
2×2 multiplier	66,623	51,961	**−22.0%**	49,052	**−26.4%**
3×3 multiplier	8,840,574	6,001,917	**−32.1%**	3,638,120	**−58.9%**
Even-3-parity	81,122	49,160	**−39.4%**	87,915	+8.4%
Even-4-parity	477,880	494,295	+3.4%	265,796	**−44.4%**
Even-5-parity	1,825,645	1,385,244	**−24.1%**	1,112,691	**−39.1%**
85% EMG classifier	18,260	14,743	**−19,3%**	23,855	+30.7%
95% EMG classifier	510,147	314,311	**−38,4%**	873,319	+71.2%

Table 3.15 presents a comparison of the various module creation techniques. The table reports the computational-effort in absolute numbers and also relative to the original random module creation technique. A negative relative effort denotes an improvement. From the experimental results, we can make the following observations: (i) Age-based module creation is highly effective. For six out of the seven test problems, age-based module creation lowered the computational-effort in comparison with the previous method, with improvements ranging between some 20% and 40%. The one exception was the even-4-parity function, where the computational-effort increases slightly by 3.4%. (ii) The overall results for the cone-based module creation technique are somewhat inconclusive. However, looking at the various test problems, we note that for the evolution of multipliers and for larger parity functions, cone-based module creation proved highly beneficial. In contrast, for evolving EMG classifiers, the cone-based approach did not work at all. Intuitively, the identification of cones as useful subcircuits is hampered if the function is rather small or is a single-output function. In the first case there is not sufficient potential for creating cones, whereas the second case lacks reusability of a cone for different

outputs. Multipliers are very regularly structured functions that are neither particularly small nor single-output functions. From the experimental data it is clear that cone-based module creation is effective for multipliers, and is especially more effective than age-based module creation. In contrast, EMG classifier circuits are random logic functions, which might explain the unsatisfactory performance of cone-based module creation for this class of problems.

3.6.2 Cone-Based Crossover

Based on the cone-based module creation technique, we investigated the efficiency of a cone-based crossover operator for the ECGP hardware representation model. This crossover operator selects a convex sub-DAG in one parent and replaces it with a convex sub-DAG extracted from another parent. Crossover implies the use of a genetic algorithm (GA) instead of the evolutionary strategy typically applied in ECGP. GAs are of interest for two reasons. First, a GA exchanges partial solutions within the population which can help escape local optima. Specifically, a GA scheme with an increased population might reveal a more stable convergence behaviour compared to an ES. Second, one might be interested in evolving circuits that are not only functionally correct but also fast and small. Multi-objective evolutionary algorithms (MOEAs) are an intriguing approach to such optimization problems. Modern Pareto-based MOEAs evolve a set of diverse individuals in a single run and rely on an operator that exchanges partial solutions of individuals.

The cone-based crossover operator works as follows. In the first step, we form a cone in a donor chromosome by randomly selecting a root node and a size between n_{min} and n_{max}. This procedure is similar to the cone-based module creation of Sect. 3.6.1.1 except that we treat both primitive nodes and modules as atomic nodes. Note that at this point, a cone can contain modules. In the second step, we randomly select a root node in a recipient chromosome and try to form a cone of exactly the same size as the donor's cone. Depending on the actual DAGs, the resulting cone in the recipient can be smaller than the donor's cone. The third step comprises the formation of two sets, a set p that contains nodes of the donor's cone which have output connections to nodes outside the cone, and another set q that contains the nodes of the recipient which connect to nodes within the recipient's cone. Figure 3.20 displays an example. The donor's cone consists of three nodes. As all these nodes provide cone outputs, we derive $p = \{f_{10}, f_{11}, f_{22}\}$. The recipient's cone receives inputs from three nodes, and we derive $q = \{f_5, f_7, f_8\}$. The fourth step actually transplants the donor's cone into a clone of the recipient, forming a new recombined chromosome. This process preserves all node types. Specifically, nodes in the donor's cone which are modules of type I or II (see Sect. 3.2) remain modules of type I or II, respectively. The module descriptions of the recombined chromosome are updated accordingly. In the final step, dangling inputs of the transplanted module and the recipient chromosome are randomly connected to the nodes in lists p

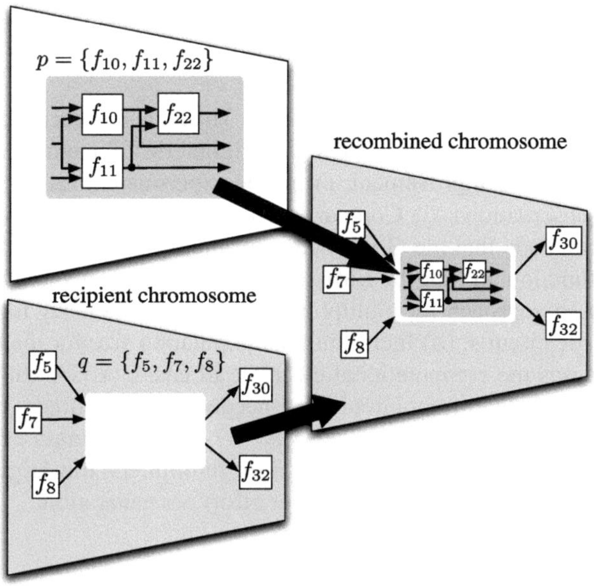

Fig. 3.20 Cone-based crossover: a cone of a donor chromosome is transplanted into a clone of a recipient chromosome.

and q, respectively. If the resulting chromosome still contains unconnected inputs, they are connected randomly to preceding nodes.

Table 3.16 Computational-effort figures for $1+4$ ES versus GA with different population sizes

	1+4 ES	GA, |population|=5		GA, |population|=50	
Problem	Absolute	Absolute	Relative	Absolute	Relative
2×2 multiplier	66,623	64,111	**−3.8%**	102,593	+54.0%
3×3 multiplier	8,840,574	2,518,964	**−71.5%**	39,064,742	+341.9%
Even-3-parity	81,122	382,036	+470.9%	186,898	+130.4%
Even-4-parity	477,880	6,294,678	+1217.2%	6,482,504	+1256.5%
85% EMG classifier	18,260	19,859	+8,8%	28,825	+57,9%
95% EMG classifier	510,147	576,988	+13.1%	695,794	+36.4%

To evaluate the cone-based crossover scheme experimentally we used the same set-up as in Sect. 3.6.1.3 and compared the 1+4 ES scheme with a standard, elitism-based GA configured with a small (GA-5) and a large population (GA-50). The elitism rate for the GA was 5.0%, with at minimum one selected individual. In GA-5, the best individual proceeded directly to the next generation. In GA-50, the three best individuals proceeded directly to the next generation, which left 47 remaining fitness evaluations. The cone-based crossover operator considered cones with a size of up to 20 nodes (primitive nodes and modules) and was applied with a probability of 1.0%.

Table 3.16 presents the experimental results. The table reports the computational-effort in absolute numbers and relative to the original ECGP method. A negative relative effort denotes an improvement. From the experimental results, we can make the following observations: (i) Comparing the $1+4$ ES to a GA with population size of 5, we conclude that the GA is better for multipliers and dramatically worse for the parity function and for the EMG classifiers. This points to the effectiveness of the cone-based approach for multipliers and to its inefficiency for single-output and random logic circuits. (ii) Increasing the population size for the GA to 50 substantially increases the computational-effort in all cases. A GA with a population size of 50 also evolved correct circuits but needed far more fitness evaluations. In each generation, GA-50 performed $11.75\times$ $(47/4)$ more fitness evaluations than did ES and GA-5. As the results show, even for multipliers, this larger potential for recombination does not outweigh the higher effort per generation.

3.7 Modular Cartesian Genetic Programming (MCGP)

Section 3.2 showed that the use of modules in ECGP significantly improves the performance of the technique compared with non-modular CGP on a number of problems. However, ECGP has so far used modules containing primitive functions. Therefore a module did not contain another module. This restriction did not occur in Angeline and Pollack's original module acquisition method [1].

In Sect. 3.2, the performance of ECGP was compared with two forms of GP that included ADFs. However, this comparison was somewhat unfair. Both GP and EP used an ADF which was allowed to call another ADF in its function set. This could give GP and EP an advantage. Indeed, we found that EP performed better than ECGP, and was more scalable on the even-parity problem. This suggests that allowing modules to call other modules might also confer an advantage on ECGP, and, in particular, improve its scalability.

In this section, one possible approach to the construction of a module hierarchy in ECGP is discussed in Sect. 3.7.1, followed by the application of the proposed technique to some of the problems already seen in Sect. 3.2. Comparisons are made between the performance of the modified ECGP and the original form.

3.7.1 Multi-level Module Hierarchy Representation

To distinguish between ECGP and the version of ECGP with a *multi-level* module hierarchy, the latter is called *modular CGP* (MCGP).

To overcome the run time issues associated with modules existing inside other modules that was mentioned in the previous section, it is necessary to prevent modules from being promoted to a higher level in the hierarchy (thereby removing the possibility of loops), whilst still allowing them to evolve. Therefore, a module's position in the hierarchy must remain *fixed*, and the module should be allowed only to contain modules lower than itself in the hierarchy. This has the effect of transforming the topology of the module hierarchy into a feed-forward directed graph, which is the same type of topology as used in the CGP representation. An example of the module hierarchy used is shown in Fig. 3.21.

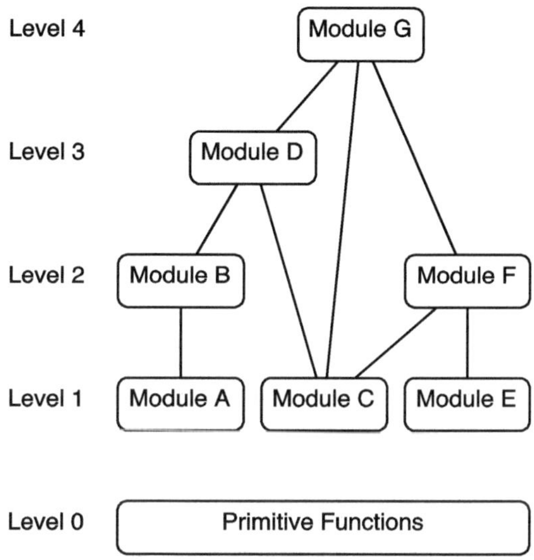

Fig. 3.21 The directed-graph module hierarchy topology used in MCGP. A line between two modules indicates that the higher-level module contains the lower-level module. For example, module G contains modules D, C and F.

In order to achieve this, the concept of *level types* was introduced to MCGP. A level type simply indicates which level of the module hierarchy a module belongs to. It is assigned to a module when it is created by the compress operator. During the compress operation, the potential contents of a module are examined, and the level type of the module is assigned based on the level types of the potential module's contents. For example, if the potential contents of a module contain only primitive functions, a level-1 module and a level-2 module, the new module is assigned a level type of 3 when it is created. The level type assigned to a new module is always one higher than the highest-level module it contains. In the event that a module contains only primitive functions when it is created, it is assigned a level type of 1, the lowest level type available for a module. For completeness, level 0 of the hierarchy is the primitive functions. A module's level type remains unchanged for the duration of the module's life span. The maximum number of levels available in the module hierarchy of MCGP can therefore be assigned by the user.

In MCGP, the module genotype has to be slightly modified to allow the inclusion of level types. The module header section of the module genotype now has an extra integer to store the module's level type. Therefore, the total size of the module header requires five integers instead of four. An example of the new module header is shown in Fig. 3.22.

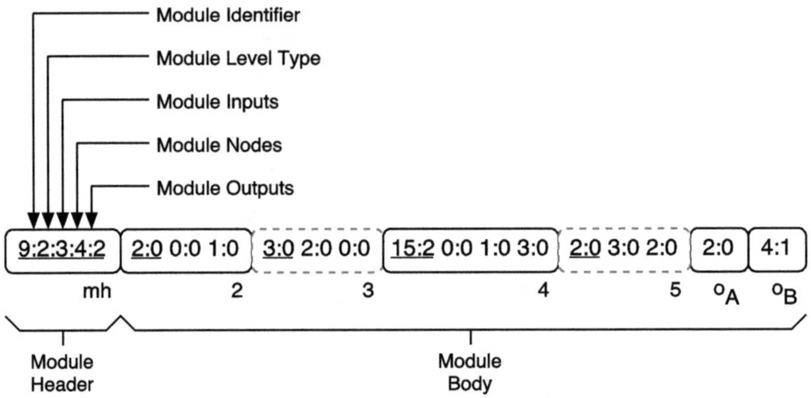

Fig. 3.22 An example of a module genotype in MCGP with the modified module header.

When the function of a node contained in a module is to be mutated through the action of the module point mutation operator, the level type of the module is used to determine the function set allowed. In order to prevent a module being promoted in the module hierarchy, it should not contain any modules of higher level than itself. Therefore, the function of a node contained in a module can be mutated only to a primitive function or a lower-level module in the hierarchy. A general list of the function sets available for each level type is shown in Table 3.17. The relationship between the level type and function set of a module in MCGP is similar to how the function sets are specified for ADFs in GP. However, the main difference ADFs are used is that each function (including other ADFs) in the function set is explicitly specified, whereas in MCGP, the function set states only the level type of the modules allowed and not the actual function. This allows the automatic acquisition and evolution of modules during the evolutionary process and does not constrain the module function set to a specific size.

The combination of the compress and the module point-mutation operators working together ensures that only higher-level modules can be created by the compress

Table 3.17 The function sets available to the modules in each level of the module hierarchy in MCGP

Level type	Function set
1	Primitive functions
2	Primitive functions and level 1 modules
3	Primitive functions and level 1, level 2 modules
⋮	⋮
n	Primitive functions and level 1, level 2, level 3, ... , level $n-1$ modules

operator and that a module can contain only modules from a lower level of the module hierarchy.

Another modification which had to be made to the modules relates to the constraints which govern the numbers of module inputs and module outputs. It was seen in Sect. 3.2.2.1 that a module was allowed i module inputs, where $2 \leq i \leq 2n$ and n is the number of nodes in the module genotype. Likewise, a module was allowed o outputs, where $1 \leq o \leq n$. In both constraints, it is assumed that the nodes in the module represent only two input primitive functions. However, when a module hierarchy is used and modules are allowed within other modules, these two constraints become invalid. This is due to the increased number of inputs and outputs that a node possesses when it represents a module. Also, the number of inputs and outputs increases with the number of levels in the module hierarchy. Therefore dynamic constraints on the numbers of module inputs and module outputs are required.

In MCGP, the dynamic constraint on the number of module inputs i, is defined by $2 \leq i \leq k$, where $k = \sum_{j=0}^{n} x_{inputs,j}$, n is the number of nodes in a module and $x_{inputs,j}$ is the number of inputs of the jth node in the module. Likewise, the dynamic constraint for the number of module outputs o is defined by $1 \leq o \leq l$, where $l = \sum_{j=0}^{n} x_{outputs,j}$ and $x_{outputs,j}$ is the number of outputs of the jth node in the module. These dynamic constraints allow the numbers of module inputs and module outputs to increase and decrease within a range that is dependent upon the contents of the module. This happens through the module mutation operators (described in Sect. 3.2.4). If a node represents a module, the information about the number of inputs and outputs is read from the module header, whereas the inputs and outputs of a primitive function are previously known.

The MCGP genotype is allowed to use modules in the same way as does an ECGP genotype, except that the function of a node in an MCGP genotype can be mutated to represent a primitive function or a module of any level type contained in the module list. This allows the reuse of higher-order functions which have been constructed in the module hierarchy. As in ECGP, when a module is reused, the rules for node types are obeyed (see Sect. 3.2).

In ECGP, the modules found in the fittest individual's genotype are promoted to the next generation so they can be reused. This is also the case in MCGP. However, in MCGP it is also possible that modules may occur only inside other modules, and not occur in the genotype of the fittest individual. When this happens, the modules found inside other modules are maintained in the module list but are not available for reuse. The reason for this is that the evolutionary strategy has removed these modules from the genotype owing to the individual having poor fitness or when the module did not affect the fitness of the individual. Therefore, it would seem unwise to reuse a module which does not contribute towards high fitness on it own.

3.7.2 Benchmark Experiments

In order to benchmark the performance of MCGP against ECGP and CGP, the proposed approach was tested on the even-parity and digital-adder problems. Both experiments used the parameters described in Sect. 3.2.6.1 except for two modifications.

The previous section discussed the shortcomings of the computational-effort statistic, in particular the requirement for a large number of runs to increase accuracy. The previous experiments used only 50 independent runs, which is at the lower end of the number of runs recommended for evaluating the computational-effort. Therefore, in the experiments described in this section, the number of independent runs was increased to 200, and this allows us to examine the accuracy of the computational-effort statistic used earlier. The second parameter changed was the mutation rate. In the experiments described here, a 4% mutation rate was used to see how it affected the results.

The new parameter introduced to MCGP for the number of levels allowed in the module hierarchy was investigated for the values 1, 2 and 3. To differentiate between the various level hierarchies used in MCGP they will be denoted here, as $MCGP_1$, $MCGP_2$ and $MCGP_3$, respectively. The results for the even-parity and digital-adder problems are also presented using the statistics described in Sect. 3.2.

3.7.2.1 Even-parity

The figures for the computational-effort statistic and the bounds for its confidence interval were calculated for CGP and $MCGP_{1-3}$, and are shown in Table 3.18.

In all of the experiments, CGP and $MCGP_{1-3}$ produced 100% successful solutions. From the results in Table 3.18, it is possible to see that $MCGP_1$ (equivalent to ECGP) performs better than CGP on all of the even-parity problems, $MCGP_2$ performs better than CGP for all even-parity problems larger than 4, and $MCGP_3$ performs better than CGP for all even-parity problems larger than 5. This indicates that the overhead of module acquisition associated with MCGP increases in a way that dependd on the number of levels allowed in the module hierarchy. The reason for the increase is attributed to the increased size and complexity of the building

Table 3.18 The computational-effort (CE) figures for CGP and $MCGP_{1-3}$, applied to various-size even-parity problems. Also included are the lower (CI_{lower}) and upper (CI_{upper}) bounds of the true computational-effort confidence interval

	CGP			$MCGP_1$		
Parity	CI_{lower}	CE	CI_{upper}	CI_{lower}	CE	CI_{upper}
three-bit	78,180	92,163	110,213	77,366	91,203	109,065
four-bit	279,119	357,441	494,024	264,090	312,322	378,284
five-bit	1,398,643	1,654,082	2,003,419	556,739	712,961	985,394
six-bit	3,488,352	4,125,442	4,996,723	946,430	1,212,001	1,675,124
seven-bit	7,178,623	9,192,961	12,705,723	1,629,484	2,086,721	2,884,087
eight-bit	20,949,947	24,696,963	29,533,936	2,100,013	2,689,281	3,716,894

	$MCGP_2$			$MCGP_3$		
Parity	CI_{lower}	CE	CI_{upper}	CI_{lower}	CE	CI_{upper}
three-bit	86,859	102,722	124,417	87,077	110,891	139,784
four-bit	272,248	348,641	481,862	325,847	417,281	576,730
five-bit	529,502	678,081	937,185	745,525	954,721	1,319,534
six-bit	897,578	1,149,441	1,588,659	999,281	1,279,681	1,768,665
seven-bit	1,683,584	2,156,001	2,979,840	2,212,585	2,833,441	3,916,139
eight-bit	2,594,780	3,322,881	4,592,601	2,665,247	3,413,121	4,717,323

blocks created, which in turn requires a greater amount of time for MCGP to discover useful blocks that it can use effectively.

Previous experiments (see Sect. 3.2.6.6) showed that the performance of ECGP scaled better with problem difficulty than did CGP. The same was found with MCGP. $MCGP_1$ performs between 1.01 and 9.18 times better than CGP for the even-n-parity problem when $n = 3,4,5,6,7,8$; $MCGP_2$ performs between 1.03 and 7.43 times better than CGP for the even-n-parity problem when $n = 4,5,6,7,8$; and $MCGP_3$ performs between 1.73 and 7.24 times better than CGP for the even-n-parity problem when $n = 5,6,7,8$.

The variation in computational-effort for MCGP when different numbers of levels are allowed in the module hierarchy is interesting. $MCGP_1$ and $MCGP_2$ are very similar over all of the even-parity problems (see Table 3.18). However, on the even-5 and even-6 parity problems, $MCGP_2$ can be seen to have an advantage over $MCGP_1$. This suggests that the two-level hierarchy is able to construct larger level-2 modules using a combination of level-1 modules and primitive functions in order to find a solution. The results are promising and indicate that a multi-level module hierarchy gives MCGP the potential to improve the performance of the technique. However, it does appear that allowing MCGP too large a number of levels hinders the performance of the technique, compared with using a smaller module hierarchy. This can be seen from the results for $MCGP_3$, although on harder problems it may be the case that a higher-level hierarchy may be beneficial to the technique's performance. $MCGP_3$ can in principle construct a solution to the even-8-parity problem in a single level-3 module. Therefore, using a three-level hierarchy allows solutions to be built which are unnecessary for the test problems under study.

The computational-effort figures in Table 3.18 are also supported by the statistics in Table 3.19 and the results of the Mann–Whitney U test shown in Table 3.20. Once $MCGP_{1-3}$ performs better than CGP, the U values are classed as significant ($P < 0.01$), or highly significant ($P < 0.001$) on all subsequent even-parity problems. There was only one instance when CGP performed better than $MCGP_3$, on the even-4-parity problem, which was classed as highly significant. This is an encouraging result, as it further supports the hypothesis that the automatic acquisition, evolution and reuse of modules improves the performance of the technique when a module hierarchy is used and the problem is above a certain level of difficulty.

Table 3.19 The median number of evaluations (ME), median absolute deviation (MAD) and interquartile range (IQR) of CGP, $MCGP_{1-3}$ for the even n parity problem, where $n = 3,4,5,6,7,8$

	CGP			$MCGP_1$		
Parity	ME	MAD	IQR	ME	MAD	IQR
three-bit	15,117	9,036	21,317	14,485	8,370	19,709
four-bit	79,995	42,456	91,203	57,473	27,946	65,712
five-bit	315,713	183,784	401,113	155,369	81,406	189,699
six-bit	793,923	391,046	836,156	297,585	162,856	331,257
seven-bit	1,920,743	1,106,102	2,397,459	345,639	227,436	547,461
eight-bit	4,440,799	2,249,250	5,386,688	491,835	305,648	694,937

	$MCGP_2$			$MCGP_3$		
Parity	ME	MAD	IQR	ME	MAD	IQR
three-bit	16,287	10,056	23,421	18,101	12,658	28,618
four-bit	89,839	38,620	82,091	112,359	49,942	99,826
five-bit	204,395	85,350	169,001	266,053	98,424	195,856
six-bit	327,693	154,642	335,815	382,843	159,208	320,067
seven-bit	519,507	276,264	551,086	635,483	309,152	594,332
eight-bit	662,173	382,278	787,857	686,121	342,940	726,482

In order to see how the higher mutation rate and increased number of runs affected the performance of CGP and $MCGP_1$ (equivalent to ECGP), we now compare the computational-effort figures in Table 3.18 with those in Sect. 3.2.

From Tables 3.3 and 3.18, it is possible to observe an unusual trend. For the even-n-parity problem, where $n = 3,4,5,6$ for CGP and $n = 3,4,5,6,7$ for ECGP, the computational-effort figures presented in Sect. 3.2.6.6 are better. These used a lower mutation rate and a smaller number of runs. However, for the even-n-parity problem, where $n = 7,8$ for CGP and $n = 8$ for ECGP, the computational-effort figures presented in this section are better. These used a higher mutation rate and a greater number of runs.

Table 3.20 The U values from the Mann–Whitney significance test for comparing CGP and $MCGP_1$ (U_1), $MCGP_2$ (U_2) and $MCGP_3$ (U_3) on the even-n-parity problem, where $n = 3,4,5,6,7,8$. The U values are significant ($P < 0.01$) when denoted by † and highly significant ($P < 0.001$) when denoted by ‡

Parity	U_1	U_2	U_3
three-bit	19,623	20,366	21,719
four-bit	†16,671	21,608	‡24,428
five-bit	‡12,135	‡13,847	17,967
six-bit	‡7,484	‡8,383	‡9,482
seven-bit	‡3,781	‡4,955	‡6,471
eight-bit	‡1,423	‡1,841	‡1,998

Comparing the median numbers of evaluations in Tables 3.19 and 3.5 supports these findings. This implies that it is more likely that the difference between the two experiments was caused by the change in the mutation rate rather than by the increased number of runs. This affected the computational-effort calculation (although the number of runs could still affect the accuracy of the median). From the comparison of the present results with Sect. 3.2, it appears as though a higher mutation rate is suitable for harder problems, whereas a lower mutation rate is better for easier problems. This once again possibly highlights the difficulty of choosing a good mutation rate in CGP. Further investigations are required to decide if the difference between the results is definitely caused by the mutation rate and what effect the increased number of runs actually has on the performance of CGP.

3.7.2.2 Digital-Adder

CGP and $MCGP_{1-3}$ were applied to a number of digital-adder problems. The computational-effort was calculated for 200 runs and the results are shown in Table 3.21. In addition, true computational-effort confidence intervals were calculated according to (3.4) from Sect. 3.2.

As in the results described in the previous section, CGP and $MCGP_{1-3}$ produced 100% successful solutions to all of the digital-adder problems. Examining the computational-effort figures in Table 3.21 shows that $MCGP_1$ performs better than CGP on all three adder problems, $MCGP_2$ performs better than CGP on the two- and three-bit digital-adder problems, and $MCGP_3$ only performs better than CGP on the hardest of the test problems. The speed-up in performance also increases with problem difficulty, which was also found in the digital-adder experiments described in Sect. 3.3.

From the results of the previous experiment (Sect. 3.7.2.1), we can infer that the use of a multiple-level hierarchy is detrimental to the performance of MCGP on simple problems. However, once the test problem becomes sufficiently difficult, the use of a module hierarchy with a higher number of levels becomes beneficial to the

Table 3.21 The computational-effort (*CE*) figures for CGP and $MCGP_{1-3}$ applied to various size digital-adder problems. Also included are the lower (CI_{lower}) and upper (CI_{upper}) bounds of the true computational-effort confidence interval

	CGP			$MCGP_1$		
Adder	CI_{lower}	CE	CI_{upper}	CI_{lower}	CE	CI_{upper}
one-bit	86,630	105,606	127,360	84,373	101,605	121,912
two-bit	925,394	1,094,402	1,325,537	809,829	980,005	1,168,798
three-bit	4,355,875	5,175,684	6,197,861	2,853,075	3,363,363	4,022,088

	$MCGP_2$			$MCGP_3$		
Adder	CI_{lower}	CE	CI_{upper}	CI_{lower}	CE	CI_{upper}
one-bit	102,242	125,447	153,114	143,875	177,928	219,430
two-bit	818,514	968,002	1,172,441	1,344,954	1,598,084	1,913,699
three-bit	2,528,311	3,237,761	4,474,956	3,383,366	4,001,282	4,846,341

performance of MCGP, when it is compared with CGP. This is shown by the results for $MCGP_{2-3}$ on the two- and three-bit adder problems.

When the figures in Table 3.21 are examined it is also possible to observe how the performance of $MCGP_{1-3}$ becomes similar as the problem difficulty increases. This could indicate either that the use of a higher-level module hierarchy is beneficial and will eventually overtake the performance of MCGP with a lower-level module hierarchy, or that the performance of MCGP with any module hierarchy will eventually converge on a similar performance curve. Both of these suggestions require further investigations on much harder problems to confirm or reject these observations.

A noticeable difference between the computational-effort figures in Table 3.21 and the results of the even-parity experiment in the previous section is that $MCGP_2$ performs better than $MCGP_1$ on the two- and three-bit adder problems. This is the first sign that the use of modules within modules improves the performance of MCGP. This could be attributed to the fact that the adder problem can be constructed in a more hierarchical fashion than is the case with the even-parity problem. Large full-adders can be constructed from smaller full-adders, which in turn can be constructed from half-adders, which can be constructed from XOR functions. This allows the possibility for the module hierarchy to evolve and construct modules representing all of these different functions at different levels of the hierarchy. However, further investigations are required to verify this point, as the statistical figures in Table 3.22 contradict the computational-effort figures in Table 3.21, and claim that $MCGP_1$ still performs the best. However, the figures in Table 3.22 do support the fact that $MCGP_{1-3}$ perform better than CGP on the harder test problems.

As a further indication of the accuracy of the comparison, a Mann–Whitney significance test was performed on the run data between CGP and $MCGP_{1-3}$, the results of which are shown in Table 3.23. From the results, it is possible to deduce two interesting facts. Firstly, the results for the three-bit adder are highly significant

Table 3.22 The median number of evaluations (ME), median absolute deviation (MAD) and interquartile range (IQR) for CGP and $MCGP_{1-3}$ for the n-bit adder problem, where $n = 1,2,3$

	CGP			$MCGP_1$		
Adder	ME	MAD	IQR	ME	MAD	IQR
one-bit	16,083	9,540	21,403	17,161	10,726	24,787
two-bit	200,451	112,230	231,545	158,289	84,666	224,401
three-bit	920,259	415,240	1,070,797	669,677	308,048	672,577

	$MCGP_2$			$MCGP_3$		
Adder	ME	MAD	IQR	ME	MAD	IQR
one-bit	18,323	11,860	29,924	28,079	17,120	41,793
two-bit	199,517	82,506	187,804	265,613	116,502	319,488
three-bit	699,207	310,124	685,242	835,197	367,444	810,213

for $MCGP_{1-2}$, and marginally significant for $MCGP_3$. This adds further support to the findings in Table 3.21. However, the results for $MCGP_{2-3}$ for the two-bit adder problem are also classed as highly significant, but show that CGP performs better. This contradicts the figures in both Table 3.21 and Table 3.22.

Table 3.23 The U values from the Mann–Whitney significance test when CGP is compared with $MCGP_1$ (U_1), $MCGP_2$ (U_2) and $MCGP_3$ (U_3) on the n-bit adder problem, where $n = 1,2,3$. The U values are marginally significant ($P < 0.05$) when denoted by * and highly significant ($P < 0.001$) when denoted by ‡

Adder	U_1	U_2	U_3
one-bit	19,965	21,859	25,236
two-bit	21,438	‡24,432	‡28,836
three-bit	‡14,902	‡15,192	*17,732

Comparing the results of this experiment with those in Sect. 3.3, it is once again possible to observe that the results of this experiment are much worse for the simpler one- and two-bit adder problems but are better for the three-bit adder problem. This was also seen when the results in the previous section (Sect. 3.7.2.1) were compared with those in Sect. 3.2.6.6. The difference in the results can be attributed to the mutation rate parameter being changed from 3% in Sect. 3.2 to 4% in this section. The higher mutation rate seems to be beneficial to the performance of CGP and $MCGP_{1-3}$ on harder problems but detrimental to the performance of both techniques on the simpler problems. The same trend was also seen for the even-parity problem in the previous section. This once again highlights the sensitivity of the mutation rate parameter.

3.8 Multi-chromosome Cartesian Genetic Programming (MC-CGP)

The previous sections in this chapter have been concerned with incorporating a form of ADF to CGP in order to improve the performance of the underlying technique. In this section, an alternative approach to improving the performance of CGP or ECGP on multiple-output problems is discussed. The proposed approach is based on the idea of using multiple chromosomes within a single genotype to split a complex multiple-output problem into many simple single-output subproblems, which are co-evolved. This approach is known as *multi-chromosome CGP* or *multi-chromosome ECGP*, depending on which underlying technique is used. Also, a multi-chromosome evolutionary strategy is introduced, which behaves similarly to an 'intelligent' crossover operator during the selection process of the evolutionary cycle.

3.8.1 Multi-chromosome Representation

The difference between a CGP or an ECGP genotype (described earlier in Sects. 2.2 and 3.2) and a Multi-chromosome CGP or multi-chromosome ECGP genotype is that the multi-chromosome CGP or multi-chromosome ECGP genotype is divided into a number of equal-length sections called chromosomes. The number of chromosomes present in the genotype of an individual is dictated by the number of program outputs required by the problem, as each chromosome is connected to a *single* program output. This allows large problems with multiple outputs (normally encoded in a single genotype) to be broken down into many smaller problems (each encoded by a chromosome) with a single output. The idea is that this approach should make the problem easier to solve. Evolving the outputs to a problem incrementally has already been shown to improve evolvability [27], but in the work described in this chapter all of the outputs were evolved simultaneously. By allowing each of the smaller problems to be encoded in a chromosome, the whole problem is still encoded in a single genotype but the interconnectivity between the smaller problems (which can cause obstacles in the fitness landscape) has been removed.

Each chromosome contains an equal number of nodes, and is treated as a genotype of an individual with a single program output. The inputs of each node encoded in a chromosome are allowed to connect only to the outputs of earlier nodes encoded in the same chromosome or any program input (terminals). This creates a form of compartmentalization in the genotype which supports the idea of removing the interconnectivity between the smaller problems encoded in each chromosome. An example of a multi-chromosome CGP genotype for the two-bit multiplier problem is shown in Fig. 3.23. The two-bit multiplier problem has four outputs, so it is broken down into four smaller problems. Each of the smaller problems has one output and is encoded in a single chromosome.

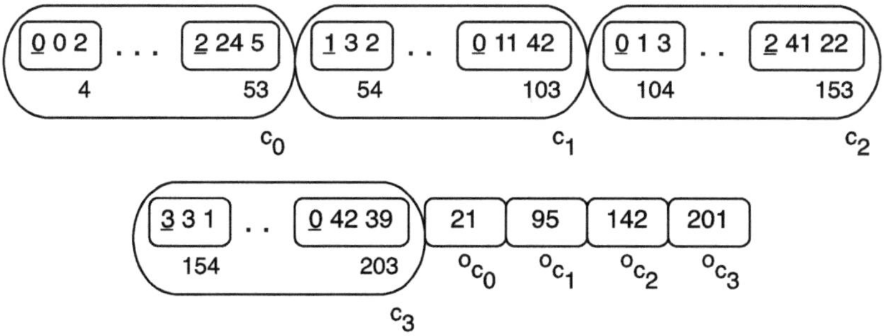

Fig. 3.23 A multi-chromosome CGP genotype encoding a two-bit multiplier (four outputs, o_{c_0}–o_{c_3}) containing four chromosomes (c_0–c_3), each consisting of 50 nodes.

The multi-chromosome approach to CGP and ECGP shares some similarities with another GP technique known as Parisian GP [4], which is inspired by the Michigan approach to classifier systems, in that both techniques form a solution to a problem from subsolutions. However, in Parisian GP, an individual represents only part of a solution, and the whole solution is made up of a set of individuals from the population. This differs from the multi-chromosome approach to CGP and ECGP, as each chromosome encodes a solution to a distinct subproblem and the solution to the entire problem is contained in a single individual, which consists of a number of chromosomes, each encoding a different subproblem. Another difference between the two techniques is that Parisian GP uses two separate fitness functions, a local fitness function to assess each individual's contribution, and a global fitness function to evaluate how well the set of individuals solves the problem. In contrast, the multi-chromosome approach to CGP and ECGP uses a single fitness function to evaluate how well each chromosome solves the subproblem it has been assigned. When all of the subproblems are solved, the entire problem is also solved.

All the chromosomes for an individual are contained in a single genotype; therefore a single module list is used in multi-chromosome ECGP, which stores modules created in any chromosome contained in the genotype (rather than having an individual module list for each chromosome, which remains to be investigated). This allows the sharing and reuse of genetic information associated with high fitness between the chromosomes in the genotype of an individual. Therefore a partial solution which is present in all of the chromosomes needs only to be constructed in one chromosome, and then reused in the other chromosomes. In theory, this could reintroduce interconnectivity between the smaller problems encoded in each chromosome, by exploiting the similarities between the chromosomes. Reusing relevant information from one chromosome in a different chromosome is similar to connecting a section of a genotype pertaining to one subproblem to another section of a genotype pertaining to a different subproblem.

3.8.2 Multi-chromosome Evolutionary Strategy

Rather than assigning a single fitness value to a number of program outputs, as in single-chromosome CGP, a fitness value is assigned to the output of each chromosome in multi-chromosome CGP, as each chromosome's output is also a program output. Therefore, if an individual with a multi-chromosome genotype has n program outputs, the individual's genotype contains n chromosomes, and the individual has n fitness values. This allows each chromosome of an individual to be compared with the corresponding chromosome of other individuals in the population, by using a variation of the (1 + 4) evolutionary strategy (described earlier in Sect. 3.2.5) called the *(1 + 4) multi-chromosome evolutionary strategy*.

The (1 + 4) multi-chromosome evolutionary strategy selects the best chromosome at each position from all of the individuals in the population and generates a new best-of-generation individual containing the fittest chromosome at each position. The new best-of-generation individual may not have existed in the population, as it is a combination of the best chromosomes from all the individuals, so it could be thought of as a 'super' individual. The multi-chromosome version of the (1 + 4) evolutionary strategy therefore behaves as an intelligent multi-chromosome crossover operator, as it selects the best parts from all the individuals. The overall fitness of the new individual will also be better than or equal to the fitness of any individual in the population from which it was generated. An example of the multi-chromosome evolutionary strategy is shown in Fig. 3.24, and an outline of the $(1+4)$ multi-chromosome evolutionary strategy is shown in Procedure 3.2.

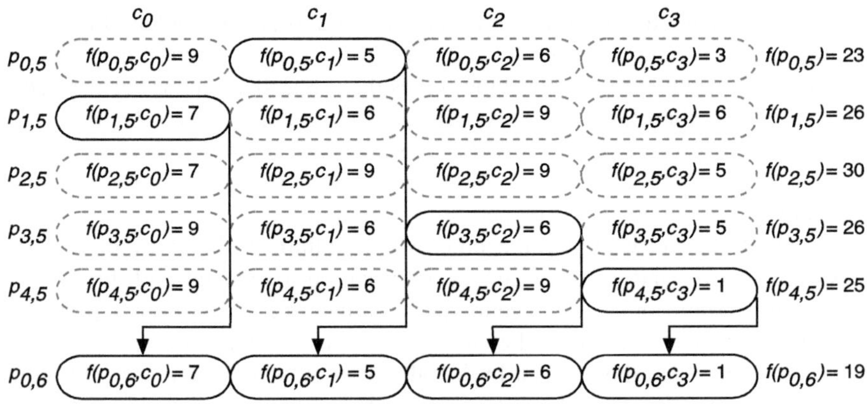

Fig. 3.24 The $(1+4)$ multi-chromosome evolutionary strategy used in multi-chromosome CGP and multi-chromosome ECGP. $p_{x,g}$, parent x at generation g; c_y, chromosome y; $f(p_{x,g}, c_y)$, fitness of chromosome y in parent x at generation g; $f(p_{x,g})$, fitness of parent x at generation g.

Procedure 3.2 The $(1+4)$ multi-chromosome evolutionary strategy

```
for all i such that 0 ≤ i < 5 do
    Randomly generate individual i
end for
Select the fittest individual, which is promoted as the parent
while a solution is not found or the generation limit is not reached do
    for all i such that 0 ≤ i < 4 do
        Mutate the parent to generate offspring i
    end for
    Generate the fittest individual using the following rules:
    for all j such that 0 ≤ j <number of chromosomes do
        if an offspring chromosome j has a better or equal fitness than the parent chromosome j
        then
            Offspring chromosome j is promoted as the new parent chromosome j
        else if many offspring chromosome j's have an equal fitness which is a better or equal
        fitness than the parent chromosome j then
            A randomly selected offspring chromosome j is promoted as the new parent chromo-
            some j
        else
            The parent chromosome j is promoted
        end if
    end for
end while
```

3.8.3 Benchmark Experiments

The performance of the multi-chromosome and the single-chromosome versions of CGP and ECGP was tested on a number of multiple-output digital-circuit problems, shown in Table 3.24 with their corresponding number of inputs and outputs. The adder and multiplier problems were the same as those used for testing the performance of ECGP in Sect. 3.2. The 3:8-bit de-multiplexer converts a signal consisting of three components (three inputs), which has already been compressed by a multiplexer, back into the original uncompressed signal consisting of eight components (eight outputs). The 4×1-bit comparator takes four one-bit integers (four inputs) and compares every possible pair combination of them to find out if the first number of the pair is less than, equal to or greater than the second number of the pair (six pair combinations each with three outputs, totalling 18 outputs overall). An example is shown in Fig. 3.25a to illustrate the point. The final problem tested was a possible implementation of a three-bit arithmetic logic unit (ALU), which takes two three-bit integers and a low and a high carry-in (totalling eight inputs) and performs the functions of addition, subtraction, multiplication and protected division, all in parallel, on the two three-bit integers to produce two four-bit numbers from addition and subtraction, a six-bit number from multiplication and a three-bit number from protected division (17 outputs overall). This is illustrated in Fig. 3.25b.

Table 3.24 The digital circuit problems used to test the performance of the single and multi-chromosome versions of CGP and ECGP. The abbreviation for each problem is shown in parentheses

Digital circuit	Number of inputs	Number of outputs
2-bit adder (Add)	5	3
3-bit adder (Add)	7	4
2-bit multiplier (Mul)	4	4
3-bit multiplier (Mul)	6	6
3 : 8-bit de-multiplexer (DeMUX)	3	8
4×1-bit comparator (Comp)	4	18
3-bit arithmetic logic unit (ALU)	8	17

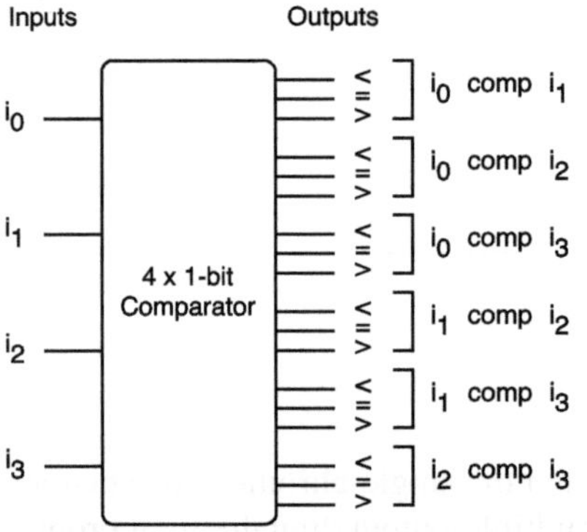

(a) 4×1-bit comparator

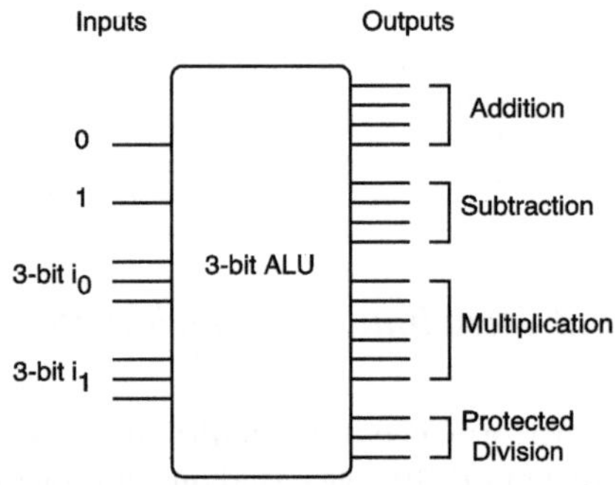

(b) three-bit ALU

Fig. 3.25 Examples of a 4×1-bit comparator (a) and the three-bit ALU (b), showing the inputs and outputs of each circuit. In part (a), each of the six output blocks constitutes a comparison, and produces a result of less than (<), equal to (=) or greater than (>) at one of the three outputs for that block.

The parameters used for the multi-chromosome and single-chromosome versions of CGP and ECGP were the same as those used for testing the performance of ECGP in Sect. 3.2 except for two changes. Firstly, a new parameter was introduced for the *initial chromosome size*, which was set at 100 nodes (300 genes). Secondly, the setting for the *initial genotype size* had to be altered, to allow the genotype to contain different numbers of chromosomes. The new setting for the *initial genotype size* parameter was equal to *initial chromosome size* $\times$ *number of chromosomes*.

The digital-multiplier and ALU problems used a function set consisting of AND, AND (one input inverted), OR and XOR, whilst the other problems used the function set AND, NAND, OR and NOR. Both of these function sets were used previously for the experiments described in Sect. 3.2.

Once again, the computational-effort statistic was used for comparing the techniques as described in Sect. 3.2. The computational-effort figures for both the multi-chromosome and single-chromosome versions of CGP and ECGP were calculated over 50 independent runs and are shown in Table 3.25. The computational-effort figures for CGP and ECGP for the adder and multiplier problems were taken from the previous experiments described in [31]

Table 3.25 The computational-effort (CE) figures for CGP, ECGP, multi-chromosome CGP (MC-CGP) and multi-chromosome ECGP (MC-ECGP). Also included are the lower (CI_{lower}) and upper (CI_{upper}) bounds of the confidence interval for the true computational-effort

	CGP			ECGP		
Problem	CI_{lower}	CE	CI_{upper}	CI_{lower}	CE	CI_{upper}
two-bit Add	577,067	834,246	1,237,329	419,097	596,643	826,957
three-bit Add	6,315,127	8,599,682	12,827,903	2,398,586	3,414,723	4,732,865
two-bit Mul	24,675	33,602	50,123	35,720	48,642	72,558
three-bit Mul	16,448,737	24,152,005	33,867,501	3,664,438	4,990,082	7,443,564
3 : 8-bit DeMUX	44,180	75,000	89,742	28,435	48,400	57,761
4 × 1-bit Comp	2,304,080	3,922,000	4,680,272	870,222	1,548,600	1,717,113
three-bit ALU	–	–	–	–	–	–

	MC-CGP			MC-ECGP		
Problem	CI_{lower}	CE	CI_{upper}	CI_{lower}	CE	CI_{upper}
two-bit Add	82,718	140,800	168,025	142,171	242,000	288,791
three-bit Add	700,668	1,286,000	1,442,657	722,831	1,230,400	1,468,285
two-bit Mul	5,971	11,200	12,362	12,466	22,400	24,963
three-bit Mul	513,220	873,600	1,042,503	509,695	867,600	1,035,343
3 : 8-bit DeMUX	2,346	4,400	4,858	3,761	6,400	7,640
4 × 1-bit Comp	5,876	10,000	11,936	11,281	19,200	22,915
three-bit ALU	1,016,914	1,908,000	2,105,517	909,883	1,548,800	1,848,244

For all of the digital-circuit problems tested, CGP, ECGP, multi-chromosome CGP and multi-chromosome ECGP produced 100% successful solutions, except for the three-bit ALU, where CGP and ECGP failed to find a solution after 20 million generations. This gives a good indication of how difficult it is to evolve a solution to the three-bit ALU problem.

Comparing the results for CGP and multi-chromosome CGP, it is clear to see the use of multiple chromosomes to break down the test problems into smaller, simpler problems provides a distinct advantage. Multi-chromosome CGP significantly outperforms CGP on all of the problems tested. Multi-chromosome CGP improves the performance by amounts between approximately 3 and 392 times compared with CGP (see Table 3.25). It is also worth noting that the speed-up increases

Table 3.26 The median number of evaluations (ME), median absolute deviation (MAD), and interquartile range (IQR) for CGP, ECGP, multi-chromosome CGP (MC-CGP) and multi-chromosome ECGP (MC-ECGP)

	CGP			ECGP		
Problem	ME	MAD	IQR	ME	MAD	IQR
two-bit Add	132,565	76,228	178,335	124,251	49,926	99,291
three-bit Add	1,943,585	996,482	2,174,500	733,909	307,694	599,132
two-bit Mul	6,197	4,130	8,489	7,485	3,304	9,875
three-bit Mul	4,030,201	2,181,656	6,110,863	854,579	257,354	773,789
3 : 8-bit DeMUX	13,797	4,842	9,635	7,809	3,210	7,618
4 × 1-bit Comp	770,475	335,916	672,160	271,553	93,906	178,257
three-bit ALU	–	–	–	–	–	–

	MC-CGP			MC-ECGP		
Problem	ME	MAD	IQR	ME	MAD	IQR
two-bit Add	29,239	13,350	25,663	43,117	21,330	50,734
three-bit Add	174,625	82,050	222,178	222,889	90,642	199,633
two-bit Mul	1,721	662	1,872	2,809	1,410	3,586
three-bit Mul	140,643	52,722	102,960	129,767	54,178	97,426
3:8-bit DeMUX	845	274	524	985	402	1,103
4 × 1-bit Comp	1,693	704	1,442	3,097	1,620	3,156
three-bit ALU	304,345	127,312	246,988	320,975	86,894	211,884

Table 3.27 The comparative U values from the Mann–Whitney significance test for CGP, ECGP, multi-chromosome CGP (MC-CGP) and multi-chromosome ECGP (MC-ECGP). The U values are classed as significant ($P < 0.01$) when denoted by a †, and highly significant ($P < 0.001$) when denoted by a ‡

Problem	CGP Vs. MC-CGP	ECGP Vs. MC-ECGP	MC-CGP Vs. MC-ECGP
two-bit Add	‡133	‡395	‡1766
three-bit Add	‡96	‡282	1345
two-bit Mul	‡315	‡460	†1702
three-bit Mul	‡0	‡59	1195
3 : 8-bit DeMUX	‡0	‡6	1521
4 × 1-bit Comp	‡0	‡0	‡1783
three-bit ALU	–	–	1327

with problem complexity on the adder and multiplier problems, implying that multi-chromosome CGP may perform even better on larger, more complex problems of this nature. A similar trend is also noticed when the computational-effort figures for ECGP and multi-chromosome ECGP are compared with a performance increase of between approximately 2.2 and 81 times. These results are supported by the statistics in Table 3.26. The results of the Mann–Whitney significance test in Table 3.27 also support the findings from Table 3.25, as every comparison between the single- and multi-chromosome approaches to CGP and ECGP are classed as highly significant, indicating that there is only a very small probability ($P < 0.001$) that the results of the single- and multi-chromosome techniques are from the same distribution.

The variance between the speed-up times for multi-chromosome CGP and multi-chromosome ECGP (compared with CGP and ECGP) for different problems appears to be related to the number of problem outputs. Notice how the speed-up increases between the two-bit and three-bit adder and multiplier problems as the number of outputs increases in both problems. The biggest speed-up recorded was found on the 4×1-bit comparator problem, which was also the problem with the most outputs. This suggests that problem complexity and the number of program outputs are directly linked, implying that the multi-chromosome approach is less affected by an increase in problem complexity than is the single-chromosome approach.

To rule out the possibility that the increased overall genotype length in the multi-chromosome approach was responsible for the performance difference between the single- and multi-chromosome versions of CGP and ECGP, further runs of CGP and ECGP were done on the adder and multiplier problems, where the initial genotype length was equivalent to that in the multi-chromosome approach. The computational-effort figures from these runs are shown in Table 3.28. By comparing the figures in Table 3.28 with the previous results for CGP and ECGP in Table 3.25, it is possible to see that giving CGP and ECGP extra resources by increasing the genotype length does improve the performance of the techniques on the adder problem, but is actually detrimental to the performance on the multiplier problem. However, the performance on the adder problem is still much worse than that of the multi-chromosome approach for both CGP and ECGP. This provides further evidence that breaking down a complex problem into many smaller problems that are co-evolved is beneficial to performance.

The noticeable speed-up caused by the use of multiple chromosomes clearly indicates that breaking down these complex, difficult problems into smaller, simpler problems, where all interconnections between the smaller problems have been severed, makes the whole problem much easier to solve. The multi-chromosome approach could therefore be used to evolve much harder multiple-output problems (such as digital circuits), which CGP and ECGP currently fail to solve. The only drawback of the multi-chromosome approach is that the solutions are much larger (in terms of the number of gates used) than the optimal solution. However, our objective in this section is not to find efficient solutions; our main concern is with improving performance. The larger solutions appear to be a result of severing the

Table 3.28 The computational-effort (*CE*) figures for CGP and ECGP with the same initial genotype length (*len*), in terms of nodes, as in the multi-chromosome approach. Also included are the lower (CI_{lower}) and upper (CI_{upper}) bounds of the confidence interval for the true computational-effort

		CGP			ECGP		
Problem	*len*	CI_{lower}	*CE*	CI_{upper}	CI_{lower}	*CE*	CI_{upper}
two-bit Add	300	275,645	469,200	559,917	182,824	311,200	371,370
three-bit Add	400	4,811,660	8,190,400	9,773,914	1,272,473	2,166,000	2,584,772
two-bit Mul	400	30,550	52,000	62,057	24,675	42,000	50,123
three-bit Mul	600	35,231,021	65,416,400	78,074,575	4,173,192	7,103,600	8,476,994

interconnections between the smaller problems, as early sections of the evolved solution which would normally be reused later in the solution are replicated.

Comparing the results of multi-chromosome CGP and multi-chromosome ECGP shows that multi-chromosome ECGP performs worse than multi-chromosome CGP on some of the problems where ECGP performs better than CGP. This suggests that breaking problems down into smaller problems reduces the complexity of the problem to such a degree that the overhead of module acquisition in multi-chromosome ECGP increases the time taken to find a solution. A similar trend was previously observed in Sect. 3.2 when ECGP was applied to simple problems. The results of the Mann–Whitney significance test in Table 3.27 support this theory, as it is only the U values from the simpler problems (two-bit adder, two-bit multiplier and 4×1-bit comparator) that are classed as significant or highly significant. However, the computational-effort figures show multi-chromosome ECGP does perform better than multi-chromosome CGP when a problem reaches a higher-level of difficulty (for example the three-bit multiplier). This is also supported by the statistics in Table 3.26. This suggests that the sharing of information between chromosomes is beneficial to the performance of multi-chromosome ECGP. Therefore, even if each subproblem was encoded in a separate CGP program and all of the programs were evolved in parallel, multi-chromosome ECGP would still perform better owing to the reuse of partial solutions between chromosomes.

3.8.3.1 Investigating the Multi-chromosome Evolutionary Strategy

To see how much of an impact the multi-chromosome evolutionary strategy had on the results, further experiments were carried out on the two-bit and three-bit adder problems using multi-chromosome CGP and multi-chromosome ECGP with the $(1+4)$ evolutionary strategy used in CGP instead of the $(1+4)$ multi-chromosome evolutionary strategy. This has the effect of grouping all of the chromosomes within a genotype together, and treating the individual like a CGP genotype that has been compartmentalized. Therefore, good chromosomes are not allowed to be exchanged between individuals in the selection process to form and promote a 'super' individual.

Table 3.29 The computational-effort (CE) figures for multi-chromosome CGP (MC-CGP$^\diamond$) and multi-chromosome ECGP (MC-ECGP$^\diamond$) with a $(1+4)$ evolutionary strategy. Also included are the lower (CI_{lower}) and upper (CI_{upper}) bounds of the confidence interval for the true computational-effort

	MC-CGP$^\diamond$			MC-ECGP$^\diamond$		
Adder	CI_{lower}	CE	CI_{upper}	CI_{lower}	CE	CI_{upper}
two-bit	140,262	249,600	276,764	160,041	289,200	343,155
three-bit	3,938,076	7,008,000	7,770,571	1,972,744	3,358,000	4,007,230

Comparing the results in Tables 3.25 and 3.29 clearly shows that the use of a $(1+4)$ evolutionary strategy with multi-chromosome CGP and multi-chromosome ECGP, does not perform as well as multi-chromosome CGP and multi-chromosome ECGP with the $(1+4)$ multi-chromosome evolutionary strategy. This implies that the use of the multi-chromosome evolutionary strategy to select the fittest individual in the population (by selecting the fittest chromosomes from each position) is beneficial to the performance of multi-chromosome CGP and multi-chromosome ECGP, in contrast to the selection of the fittest individual based on the individual's overall fitness (the sum of all of its chromosome fitness values). However, multi-chromosome CGP and multi-chromosome ECGP with a $(1+4)$ evolutionary strategy does perform marginally better than the single-chromosome versions of CGP and ECGP with a $(1+4)$ evolutionary strategy, respectively. This implies that the multi-chromosome evolutionary strategy is not solely responsible for the improvement in performance, but that the use of a multi-chromosome representation, as opposed to a single chromosome representation (as in CGP and ECGP), also improves the performance of both CGP and ECGP.

References

1. Angeline, P.J., Pollack, J.: Evolutionary Module Acquisition. In: Proc. Conference on Evolutionary Programming, pp. 154–163. MIT Press (1993)
2. Chellapilla, K.: A Preliminary Investigation into Evolving Modular Programs without Subtree Crossover. In: Proc. Conference on Genetic Programming, pp. 23–31. Morgan Kaufmann (1998)
3. Christensen, S., Oppacher, F.: An Analysis of Koza's Computational Effort Statistic for Genetic Programming. In: Proc. European Conference on Genetic Programming, *LNCS*, vol. 2278, pp. 182–191. Springer (2002)
4. Collet, P., Lutton, E., Raynal, F., Schoenauer, M.: Polar IFS, + Parisian Genetic Programming = Efficient IFS Inverse Problem Solving. Genetic Programming and Evolvable Machines **1**(4), 339–361 (2000)

5. Cong, J., Ding, Y.: Combinational Logic Synthesis for LUT Based Field Programmable Gate Arrays. ACM Transactions in Design Automation of Electronic Systems **1**(2), 145–204 (1996)
6. Crawford-Marks, R., Spector, L.: Size Control Via Size Fair Genetic Operators in the PushGP Genetic Programming System. In: Proc. Genetic and Evolutionary Computation Conference (GECCO). Morgan Kaufmann (2002)
7. Glette, K., Gruber, T., Kaufmann, P., Torresen, J., Sick, B., Platzner, M.: Comparing Evolvable Hardware to Conventional Classifiers for Electromyographic Prosthetic Hand Control. In: Proc. NASA/ESA Conference on Adaptive Hardware and Systems (AHS'08), pp. 32–39. IEEE Computer Society (2008)
8. Karatsuba, A., Ofman, Y.: Multiplication of Multiplace Numbers on Automata. Dokl. Akad. Nauk SSSR **145**(2), 293–294 (1962)
9. Kaufmann, P., Platzner, M.: MOVES: A Modular Framework for Hardware Evolution. In: Proc. Adaptive Hardware and Systems, pp. 447–454. IEEE Press (2007)
10. Kaufmann, P., Platzner, M.: Advanced Techniques for the Creation and Propagation of Modules in Cartesian Genetic Programming. In: Proc. Conference on Genetic and Evolutionary Computation, pp. 1219–1226. ACM Press (2008)
11. Koza, J.R.: Genetic Programming: On the Programming of Computers by Natural Selection. MIT Press (1992)
12. Koza, J.R.: Genetic Programming II: Automatic Discovery of Reusable Programs. MIT Press (1994)
13. Lones, M.A., Tyrrell, A.M.: Biomimetic Representation with Genetic Programming Enzyme. Genetic Programming and Evolvable Machines **3**(2), 193–217 (2002)
14. Mann, H.B., Whitney, D.R.: On a Test of Whether One of 2 Random Variables is Stochastically Larger than the Other. Annals of Mathematical Statistics **18**, 50–60 (1947)
15. Miller, J.F.: An Empirical Study of the Efficiency of Learning Boolean Function using a Cartesian Genetic Programming Approach. In: Proc. Genetic and Evolutionary Computation Conference, pp. 1135–1142. Morgan Kaufmann (1999)
16. Miller, J.F., Smith, S.L.: Redundancy and Computational Efficiency in Cartesian Genetic Programming. IEEE Transactions on Evolutionary Computation **10**(2), 167–174 (2006)
17. Miller, J.F., Thomson, P.: Cartesian Genetic Programming. In: Proc. European Conference on Genetic Programming, *LNCS*, vol. 1802, pp. 121–132. Springer (2000)
18. Niehaus, J., Banzhaf, W.: More on Computational Effort Statistics for Genetic Programming. In: Proc. European Conference on Genetic Programming, *LNCS*, vol. 2610, pp. 164–172. Springer (2003)
19. Poli, R.: Parallel Distributed Genetic Programming. In: D. Corne, M. Dorigo, F. Glover (eds.) New Ideas in Optimization, pp. 403–432. McGraw-Hill (1999)
20. Poli, R., Page, J.: Solving High-order Boolean Parity Problems with Smooth Uniform Crossover, Sub-machine Code GP and Demes. Genetic Programming and Evolvable Machines **1**(1), 37–56 (2000)
21. Rosca, J.P.: Genetic Programming Exploratory Power and the Discovery of Functions. In: Proc. Conference of Evolutionary Programming, pp. 719–736. MIT Press (1995)
22. Rosca, J.P.: Towards Automatic Discovery of Building Blocks in Genetic Programming. In: Working Notes for the AAAI Symposium on Genetic Programming, pp. 78–85. AAAI (1995)
23. Rosca, J.P., Ballard, D.H.: Learning by Adapting Representations in Genetic Programming. In: Proc. International Conference on Evolutionary Computation, pp. 407–412 (1994)
24. Schwefel, H.P.: Kybernetische Evolution als Strategie der experimentellen Forschung in der Strömungstechnik. Master's thesis, Technical University Berlin (1965)
25. Spector, L.: Simultaneous Evolution of Programs and their Control Structures. In: Advances in Genetic Programming II. MIT Press (1996)
26. Spector, L., Robinson, A.: Genetic Programming and Autoconstructive Evolution with the Push Programming Language. Genetic Programming and Evolvable Machines **3**(1), 7–40 (2002)
27. Torresen, J.: Evolving Multiplier Circuits by Training Set and Training Vector Partitioning. In: Proc. International Conference on Evolvable Systems (ICES), *LNCS*, vol. 2606, pp. 228–237. Springer (2003)

28. Walker, J.A., Miller, J.F.: Evolution and Acquisition of Modules in Cartesian Genetic Programming. In: Proc. European Conference on Genetic Programming, *LNCS*, vol. 3003, pp. 187–197. Springer (2004)
29. Walker, J.A., Miller, J.F.: Improving the Evolvability of Digital Multipliers using Embedded Cartesian Genetic Programming and Product Reduction. In: Proc. International Conference on Evolvable Systems, *LNCS*, vol. 3637, pp. 131–142. Springer (2005)
30. Walker, J.A., Miller, J.F.: Investigating the Performance of Module Acquisition in Cartesian Genetic Programming. In: Proc. Genetic and Evolutionary Computation Conference, vol. 2, pp. 1649–1656. ACM Press (2005)
31. Walker, J.A., Miller, J.F.: Automatic Acquisition, Evolution and Re-use of Modules in Cartesian Genetic Programming. IEEE Transactions on Evolutionary Computation **12**, 397–417 (2008)
32. Walker, M., Edwards, H., Messom, C.: Confidence Intervals for Computational Effort Comparisons. In: Proc. European Conference on Genetic Programming, *LNCS*, vol. 4445, pp. 23–32. Springer (2007)
33. Walker, M., Edwards, H., Messom, C.: The Reliability of Confidence Intervals for Computational Effort Comparisons. In: Proc. Genetic and Evolutionary Computation Conference, pp. 1716–1723. ACM (2007)
34. Wilcoxon, F.: Individual Comparisons by Ranking Methods. Biometrics Bulletin **1**, 80–83 (1945)

Chapter 4
Self-Modifying Cartesian Genetic Programming

Simon L. Harding, Julian F. Miller and Wolfgang Banzhaf

4.1 Introduction

In some forms of genetic programming there is little distinction between genotype and phenotype. For instance, in tree-based GP [11], the genotype is a LISP expression representing a compilable LISP program, which is the phenotype. In biology, the genotype is the genetic instructions encoded in the DNA of each cell, while the phenotype is generally considered to be the physical form of the organism. This should not be confused with the behaviour of the organism. As we have seen in Chap. 2, in CGP the distinction between genotype and phenotype is much more explicit, since the genotype is decoded to a more compact form, which is the phenotype. In the decoding process, genetic instructions may be ignored, since they are not referred to in the mapping from program inputs to outputs. We have also seen that the presence of these 'junk' genes is highly beneficial to the evolutionary process (see Sect. 2.7).

Turning our attention to biology we see that in many cases (if not all), time is an essential part of the genotype–phenotype mapping. This is perhaps clearest when we consider the development of multicellular organisms. In these systems, the 'distance' between the genotype and the phenotype is enormous, as there may be a large number of cells in a phenotype. Thus, the phenotype is constructed over time, through the interaction of the genotype and the environment. Including the notion of time in the genotype–phenotype mapping may be beneficial [1].

Self-modifying Cartesian genetic programming (SMCGP) has included time by allowing new kinds of functions into the genotype. These are self-modification (SM) functions, i.e. instructions that cause alteration of the code itself. When these SM functions are obeyed, the genotype changes into a new form (i.e. develops into a new phenotype). The new phenotype can also contain SM functions, so that when these are obeyed, a further phenotype can be created. As we will see, we refer to each of these phases of obeying SM instructions as an 'iteration'.

J.F. Miller (ed.), *Cartesian Genetic Programming*, Natural Computing Series,
DOI 10.1007/978-3-642-17310-3_4, © Springer-Verlag Berlin Heidelberg 2011

Another aspect of biological cells is that they are responsive to an arbitrary number of inputs. So, in addition to time, we have also introduced into CGP *functions* that acquire program inputs and outputs. When used in concert with SM functions, these allow SMCGP programs to acquire more inputs and deliver more outputs. These two new features allow the possibility of CGP being able to find general solutions to problems. For instance, in CGP one can attempt to solve a particular parity problem, that is, one with a given number of inputs, whereas SMCGP allows the possibility of evolving a program which – when iterated – produces an infinite sequence of parity functions, each one having one more input than in the previous iteration. Further to this, it has already been shown that SMCGP genotypes can *provably* produce general solutions to a number of problems (e.g. parity, binary addition, and computing pi and e) [6, 7].

4.1.1 Discovering Mathematical Results Using Genetic Programming

Evolving provable mathematical results is a rarity in evolutionary computation. Streeter and Becker used tree-based GP to discover mathematical approximations to well-known mathematical series such as the harmonic series and also new Padé approximations to mathematical functions [20]. Although these were interesting formulae, they did not find exact analytical results that could be shown to converge in the limit, but they noted that finding such results would be a ‘striking and exciting application of genetic programming’.

Schmidt and Lipson used GP to discover the known Hamiltonians and Lagrangians of mechanical systems purely by using GP symbolic regression techniques on data acquired through motion tracking [16]. Schmidt and Lipson also investigated using GP to solve iterated function problems (i.e. $f(f(x)) = g(x)$) and showed that one evolved function provably made $f(f(x))$ converge to $x^2 - 2$ in the limit [17].

Spector et al. showed that GP could be used to evolve hitherto unknown algebraic expressions that are important in the mathematics of finite algebras and are orders of magnitude shorter than those that could be produced by prior mathematical methods [18].

SMCGP has been able to find exact analytical results. In [6, 7], it has been shown that mathematical functions have been evolved that converge rapidly to pi and e in the limit of large iterations.

4.2 Overview of Self-Modification

Procedure 4.1 gives a high-level overview of the process of mapping a genotype to a phenotype in SMCGP. The first stage of the mapping is the modification of the genotype. This happens through the use of evolutionary operators acting on the genotype. The developmental steps in the mapping are outlined in lines 3–8 of the algorithm. The first step is to make an exact copy of the genotype and call it the phenotype at iteration 0. After this, the self-modification operators are applied to produce the phenotype at the next iteration. Development stops when either a predefined iteration limit is achieved or it turns out that the phenotype has no self-modification operations that are active.

At each increment, the phenotype is evaluated and its fitness calculated. The underlying assumption here is that one is trying to solve a series of computational problems, rather than a single instance as is usual in GP. For instance, this might be a series of parity functions, ever-closer approximations to pi, or the natural numbers of the Fibonacci sequence. If the problem, however, has only a single instance (i.e. a classification problem), we can take a fixed number of iterations (either a user-defined parameter or evolved) and evaluate the single phenotype. Another possibility would be to iterate until no self-modification rules are active.[1]

It is important to note that there are various ways in which there may be no active self-modification operations. Firstly, no self-modification operations may exist in the phenotype. Secondly, self-modification operations may be present but be non-coding. Thirdly, the self-modification operations may not be 'activated' when the instructions encoded in the phenotype are executed. These various conditions will be discussed in the detailed description in the following sections.

Procedure 4.1 Overview of genotype, phenotype and development

1: Generate genotype
2: Copy genotype to phenotype. Iteration, $i = 0$
3: **repeat**
4:     Apply self-modification operations to phenotype i
5:     increment i
6:     Calculate fitness increment, f_i
7: **until** ((i equals number of iterations required) **OR** (No self-modification functions to do))
8: Evaluate phenotype fitness F from fitness increments, f_i

4.3 SMCGP and Its Relation to CGP

The genetic representation in SMCGP has much in common with the representation used in CGP. The genotype encodes a graph (usually acyclic) that includes a list of node functions used and their connections. The arity of all functions is chosen to be equal to that of the largest-arity function in the function set. So, as in CGP, functions of lower arity ignore extraneous inputs. However, there are important differences.

[1] This has not been investigated, but is likely to be problematic as the number of iterations could be very large.

Firstly, SMCGP genotypes represent a linear string of nodes. That is to say, only one row of nodes is used (in contrast to CGP, which can have a rectangular grid of nodes). Other important differences are discussed in the following sections. The evolutionary algorithm that we have used with SMCGP was the $1+4$ evolutionary strategy which is detailed in Sect. 2.6.3.

4.3.1 Self-Modification Operators

The most significant difference is the addition of self-modification (SM) operators to the function set. These operators can be used in the genotype in the same manner as the more conventional computational operators, but at run time they provide different functionality. When the SMCGP phenotype is run, the nodes in the graph are parsed in a similar way to CGP. The graph is executed recursively by following nodes from the output nodes to the terminals (inputs). When computational functions are called, then – as usual – they operate on the data coming into the node.

When an SM node is called, the process changes slightly. If an SM node is 'activated'; then its self-modification instructions are added to a list of pending manipulations which is called the *To-Do* list. The modifications in this list are then performed between iterations. Note that SM active instructions are added to the *To-Do* list in a left-to-right traversal of the encoded graph. It was decided that SM nodes should be activated in a way that is dependent on the data presented to them. If the non-SM functions were all numerical functions in nature, SM nodes are activated when the first input is greater than the second.

When solely Boolean functions are used, SM nodes are always considered to be activated.[2] The size of the *To-Do* list is chosen by the user. This importance of this is discussed further in Sect. 4.3.7.

Many SM operators are imaginable, and Table 4.1 lists the currently available operators. In the table, we can see that the operators also require arguments. These come from the genotype and are described in Sect. 4.3.3. It is also worth noting that the indices for SM operations are defined relative to the current node. This relative addressing is a fundamental part of SMCGP and is discussed in Sect. 4.3.4.

Clearly, Table 4.1 lists many functions! One of the important unresolved issues in SMCGP is deciding on a minimal SM function set.

4.3.2 Computational Functions

The computational functions used in SMCGP are typical of such functions in GP in general; however, some functions are particular to SMCGP. A nominal list of func-

[2] This is an implementation convenience that stems from the parallel implementation, where multiple bits are packed into a single word. This makes it less obvious what an analogous activation operation to that used with numerical functions should be.

Table 4.1 Definition of the self-modification functions. P_i are the evolved arguments of the self-modification functions; x represents the absolute position of the node in the graph, where the left-most node has position 0; and c_{ij} is the jth connection gene on the node at position i

Basic	
Delete (DEL)	Delete the nodes between (P_0+x) and (P_0+x+P_1).
Add (ADD)	Add P_1 new random nodes after (P_0+x).
Move (MOV)	Move the nodes between (P_0+x) and (P_0+x+P_1) and insert after (P_0+x+P_2).
Duplication	
Overwrite (OVR)	Copy the nodes between (P_0+x) and (P_0+x+P_1) to position (P_0+x+P_2), replacing existing nodes in the target position.
Duplication (DUP)	Copy the nodes between (P_0+x) and (P_0+x+P_1) and insert after (P_0+x+P_2).
Duplicate preserving connections (DU2)	Copy the nodes between (P_0) and (P_0+P_1) and insert after (P_0+P_2).
Duplicate preserving connections (DU3)	Copy the nodes between (P_0+x) and (P_0+x+P_1) and insert after (P_0+x+P_2). When copying, this function modifies the c_{ij} of the copied nodes so that they continue to point to the original nodes.
Duplicate and scale addresses (DU4)	Starting from position (P_0+x) copy (P_1) nodes and insert after the node at position (P_0+x+P_1). During the copy, the c_{ij} of copied nodes are multiplied by P_2.
Copy to stop (COPYTOSTOP)	Copy from x to the next COPYTOSTOP or STOP function node, or the end of the graph. Nodes are inserted at the position the operator stops at.
Stop marker (STOP)	Marks the end of a COPYTOSTOP section.
Connection modification	
Shift connections (SHIFTCONNECTION)	Starting at node index (P_0+x), add P_2 to the values of the c_{ij} of next P_1 nodes.
Shift connections 2 (MULTCONNECTION)	Starting at node index (P_0+x), multiply the c_{ij} of the next P_1 nodes by P_2.
Change connection (CHC)	Change the $(P_1 \bmod 3)$th connection of node P_0 to P_2.
Function modification	
Change function (CHF)	Change the function of node P_0 to the function associated with P_1.
Change argument (CHP)	Change the $(P_1 \bmod 3)$th argument of node P_0 to P_2.
Miscellaneous	
Flush (FLR)	Clears the contents of the *To-Do* list

tions is provided in Table 4.2. A node CONST returns the first argument supplied to it. Using CONST in the function set obviates the need to have terminals providing constants (as is typical in GP). INDX outputs the position of the node in the SMCGP graph where nodes are numbered sequentially from the left, with the leftmost node being 0. INCOUNT returns the number of inputs are available to the program.

Table 4.2 Nominal computational functions. This assumes that all nodes are supplied with two inputs, a and b. Functions are protected from invalid input values where necessary

Function	Operation
NOP	No operation
DADD	$a+b$
DSUB	$a-b$
DMULT	$a \times b$
DDIV	a/b
CONST	constant (defined by P_0)
AVG	$(a+b)/2$
DSQRT	$\sqrt{a}$
DRCP	$1/\sqrt{a}$
DABS	$\|a\|$
TANH	$\tanh(a)$
TANH2	$\tanh(a+b)$
FACT	$!a$
POW	a^b
COS	$\cos(a)$
SIN	$\sin(a)$
MIN	$\min(a,b)$
MAX	$\max(a,b)$
IFLTE	if $(a < 0)$ return b, else 0
INDX	current node index
INCOUNT	number of inputs

4.3.3 Arguments

Each node in the SMCGP genotype contains three floating-point numbers. These numbers are evolved and are used in several ways by the SMCGP phenotype. The SM functions require several arguments to specify how graph modifications are to be carried out (see Table 4.1). These arguments are simply numbers, and their values are taken from the node's arguments. As the SM function arguments have to be integers, the values are truncated. However, as we saw, the arguments are also used by the CONST function. This allows SMCGP to evolve programs that contain numerical constants thus giving the program access to an arbitrary number of numerical constants. The arguments can take any value; however, when they are interpreted as parameters for self-modifying operations, values that are too large will be truncated to within the range of allowed offsets.

During iteration of the genotype, the arguments can be altered by the SM function CHP ('change parameter'). This, in principle, allows storing of the state (i.e. a memory), since a phenotype could pass information to the phenotype at the next iteration through a collection of constant values.

4.3.4 Relative Addressing

The SM operators' ability to move, delete and duplicate sections of the graph means that the classical CGP approach of labelling nodes becomes cumbersome. Classical CGP uses *absolute addressing*, so that each node has an address and nodes reference each other using these addresses (this is what connection genes are – see Sect. 2.2). In SMCGP, the structure of the phenotype program is dynamic, and to use this approach would require significant overhead in relabelling the graph at each iteration.

To simplify the representation, we therefore replace the fixed addresses with relative addresses. Now, instead of a node containing an absolute address of another node, it specifies how many nodes back from its position are required to make a connection. The connections in the genotype are now defined as positive integers that are greater than 0 (which prevents cycles). It is worth noting that classical CGP could easily have chosen this method of addressing and indeed it might have advantages in embedded CGP, since it could avoid the need to readdress the CGP fragments inside modules (see Sect. 3.2.2.1).

When the graph is run, the interpreter can calculate where a node gets its input values from by just subtracting the connection value from the current address of the node. If the node addresses a value that is not in the graph (i.e. connects too far back), then a default value is returned (in the case of numeric applications this is 0).

The arguments of SM operators are also defined relative to the current node (see Table 4.1). The relative addressing allows subgraphs to be placed or duplicated in the graph whilst retaining their semantic validity. This means that subgraphs could represent the same subfunction, but act on different inputs. This can be done without recalculating any node addresses, thus maintaining validity of the whole graph. So subgraphs can be used as functions in the sense of the ADFs of standard GP.

4.3.5 Input and Output Nodes

Most, if not all, genetic programming implementations have a fixed number of inputs. This certainly makes sense when there is a constant or bounded number of inputs over the lifetime of a program. However, it does prevent the program from scaling to larger problem sizes by increasing the number of inputs it uses – and this in turn may prevent general solutions from being found.

Similarly, most GP has a fixed number of outputs. In tree-based GP, there is typically a single output. In classical CGP, a number of input nodes are placed at one end of the graph, and these are used as the starting point for the recursive interpretation of the program.

To allow an arbitrary number of inputs and outputs, SMCGP introduces several new functions into the basic function set. These are shown in Table 4.3.

The interpreter now keeps track of an input pointer, which points to a particular input in the array of input values. Calling the function INP returns the value that

Table 4.3 Input and output functions. P_0 is the first argument gene

Function	Operation
INP	Return input pointed to by current_input, increment current_input
INPP	Return input pointed to by current_input, decrement current_input
SKIPINP	Return input pointed to by current_input, current_input = current_input + P_0
OUTPUT	Return data provided

the pointer is currently on, and then moves the pointer to the next value. When the pointer runs out of inputs, it resets to the first input. Similarly, the INPP function returns an input but then moves the pointer to the previous value. Sometimes it may not be convenient or useful to move by only one input at a time. The SKIPINP function therefore moves the pointer a number of places (specified by truncating the first floating-point argument in the node), and then returns that input.

By duplicating INP, INPP and SKIP nodes, the SMCGP phenotypes can acquire more inputs when they are iterated.

Variable numbers of outputs are handled using a similar strategy. A function OUTPUT allows the interpreter to find which nodes to use as output nodes at run time. Again, the location and number of OUTPUT nodes can change over the run time of a program. When the graph is run, the interpreter starts at the beginning of the graph and iterates over the nodes until it finds the appropriate number of OUTPUT nodes. It then evaluates (recursively) from these nodes.

For the outputs, the interpreter has features that allow it to cope when the number of OUTPUT nodes is different from the required number of outputs. If there are more OUTPUT nodes found than are needed, the excess nodes are simply ignored. If there are too few (or none) the interpreter starts using nodes from the end of the graph as outputs. This ensures that programs (of sufficient size) are always ‘viable’.

4.3.6 A Simple Example

An example genotype is given in Fig. 4.1. The figure also shows, purely schematically, some phenotypes arising at different iterations.

4.3.7 Discussion

The length of the *To-Do* list is an important parameter in SMCGP. If it is chosen to be a large value, it can allow phenotypes to grow rapidly in subsequent iterations. In most experiments conducted so far (see [6]), the length of the list has been kept

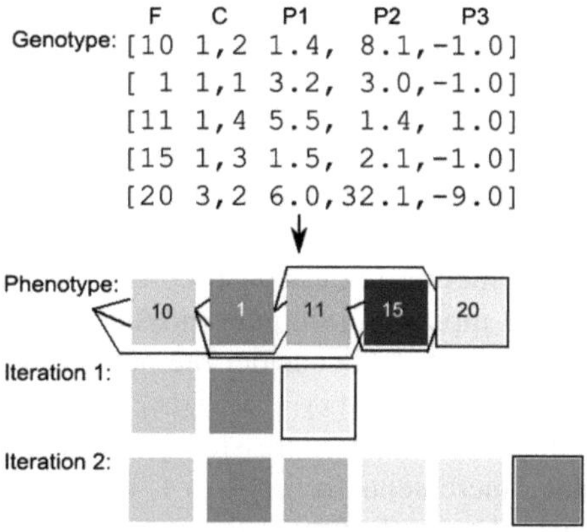

Fig. 4.1 The genotype maps directly to the initial graph of the phenotype. The genes control the number of nodes and the type and connectivity of each of the nodes. The phenotype graph is then iterated to perform computation and produce subsequent graphs. Function genes are denoted by F, connection genes by C and argument genes by P_i. The nodes in the phenotype that act as outputs are outlined.

no larger than two. At present, it is not known how the choice of this parameter affects the efficiency of the evolutionary search. When the length of the *To-Do* list is greater than one, the programs encoded in the phenotype become very difficult to understand, and so far it has only been possible to supply formal proofs of the generality of solutions when the length is one.

Another issue worthy of discussion is deciding when to apply SM operations. At present, they are added to a *To-Do* list and the SM instructions are carried out one by one until the end of the list is reached, thus producing a new phenotype. Indeed, this is what we mean by an iteration of the phenotype. However, another approach could be to carry out SM operations singly, producing a new phenotype each time, which could be evaluated in some way. This would be a more continuous form of operation and it could, at least in principle, allow more opportunities to assess the fitness of the genotype. This is effectively what happens when the length of the *To-Do* list is one.

At present, SM functions are mixed in the genotype with computational functions; thus they must act on numerical inputs. At present, we have given them passive computational roles (i.e. pass the second input). In principle, one could envisage a separation of the two aspects of computation and self-modification and introduce two separate chromosomes: a computational chromosome (CC) and a self-modification chromosome (SC). The SC would act on the CC to produce a new computational phenotype. This would be one iteration. Then, at the next iteration, the SC would be applied to the phenotype, and so on. By allowing the SM nodes in the SC to be assumed to have a single nominal input, one could use CGP to determine a graph (the SM phenotype) that would determine which SM operations would be applied to the CC (i.e. the SC would have non-coding nodes). This could potentially make the genotype–phenotype mapping easier to reason about.

The issue of SM activation, as it stands, is problematic. In non-Boolean problems we allow the data to decide whether an SM node is active or not. Actually, in most problems thus far studied, this has not caused an issue, as only a single input has been applied at each iteration. However, we can envisage a problem which is a generalization of a symbolic regression problem, in which each instance of the problem has a table of inputs and outputs (possibly representing an unknown symbolic function). To illustrate this, consider the problem of determining the following sequence of polynomials: $f_1(x) = x$, $f_2(x) = x+1$, $f_3(x) = x^2+x+1$, $f_4(x) = x^3+x^2+x+1, \ldots$. For each expression f_i, we could generate a table of values by assuming that x increments from -1 to 1, in intervals of 0.01 (say). Thus the problem to solve would be to determine the sequence of polynomials that give zero error in each instance. If SM functions were mixed with computational functions (as at present in SMCGP) and a data-dependent SM activation method was used, then different phenotypes could emerge with each data instance! Of course, evolution can find a way around this scenario when needed, perhaps by making both inputs to a SM node point to the same location. In some circumstances, for example as described in [3], the only way that the SMCGP could possibly find a learning-algorithm is to have self-modification that depends on the data. In SMCGP, if the phenotype needs to have different developmental outcomes based on the values of the computation, this is possible. It is also possible for SMCGP to ignore this feature. Further, for some problems, it can just be disabled (such as with the Boolean functions). Therefore the possibility of having data dependence seems desirable, as it provides extra functionality when it is needed, but otherwise can be ignored.

4.3.8 And Back to CGP

It is interesting to note that some of the changes to the representation are also applicable to classical CGP, and some to GP in general. The input/output strategy, relative addressing and evolved constants are all compatible with classical CGP. It remains to be seen how useful they are. However, since CGP has a fixed-length genotype even if INPUT and OUTPUT functions are included, the numbers of inputs and outputs will be bounded.

4.4 Solving Computational Problems with SMCGP: Parity

In this section we illustrate SMCGP using an example problem. The problem is how to evolve a program (or circuit) that computes even-n-parity. The parity circuit is a Boolean circuit that takes a bit string and returns TRUE if there are an even number of 1s in that string, and FALSE otherwise (note that the case of no bits is treated as an even number).

The even-n-parity problem has been a benchmark in GP ever since John Koza discussed the problem in his first book [11]. The problem is to evolve an even-n-parity function using the Boolean two-input functions AND, NAND, OR, and NOR. Koza tackled up to 11-parity [12] using a GP system with ADFs, and found them difficult to evolve. Without ADFs, his approach failed to evolve parity circuits beyond five inputs [11]. The largest evolved parity circuit we found in the literature was 22 bits [15]. It should be noted that Poli and Page used all 16 two-input Boolean functions in their function set, whereas in the work described in this section, we used just AND, NAND, OR and NOR.

Spector also examined the even-parity problem, however he used a different function set [19]. He found better scaling behaviour than Koza did, on even-parity functions up to six inputs. Other approaches have looked at finding general solutions. Huelsbergen evolved machine-language programs that could iterate over the bits in a string and from this, parity could be easily determined [10]. The solutions would be suitable for any length of bit string. Recursion has also been successfully used to solve the parity problem [24, 23, 25]. These approaches produced programs rather than circuits to solve the problem. They also used high-level programming constructs rather than purely Boolean logical primitives.

However, we will look at a slightly different problem, namely that of evolving all even-n parity functions up to even-N parity (where N may be infinite, in which case we have a general solution). We call this problem the ‘all-even-n-parity problem”. In [2, 5], we investigated using SMCGP to evolve large even-parity circuits. We found that some of the evolved programs were able to generate arbitrary-size parity circuits. In [6], we gave a formal proof of the generality of one of the evolved solutions. This is reproduced in Sect. 4.4.3.

4.4.1 Definition of Fitness

With SMCGP, instead of evolving a circuit for a particular input size, we can use the technique to find a program that will generate parity functions of any size. To do this, we evolve a program that produces a two-input parity circuit on the first iteration, produces a three-input parity circuit on the second, and so on up to N bits. The fitness is the number of correctly predicted output bits. We stop testing an individual if it fails to produce a correct solution for a given input size. Procedure 4.2 details the fitness function.

We accumulated fitness over 19 iterations ($LIMIT = 19$), starting with even-2-parity and ending on even-20-parity. Subsequently we examined whether the best solutions were correct up to 24 inputs to check for signs of generalization.

Procedure 4.2 Fitness function

```
Fitness, F = 0
Copy genotype to phenotype. Iteration, i = 0
repeat
    BREAK = FALSE
    Apply self-modification operations to phenotype i
    increment i
    Calculate fitness on test case, f_i, by counting number of incorrect bits
    if f_i ≠ 0 then
        BREAK = TRUE
    end if
    F = F + f_i
until i = LIMIT OR BREAK
```

4.4.2 Results

Table 4.4 shows the average number of evaluations required to evolve a program for a given number of inputs. The success rate was 100%. The results are based on 50 trials per function set.

In Table 4.5, the results are compared with results obtained from previous CGP representations and Koza's figures for GP with ADFs [12]. The SMCGP results are clearly highly competitive. It should be remembered that SMCGP is solving the *all-even-n-parity problem* rather than a series of single instances. We have included Koza's figures for reference. Koza calculated the computational effort for a 99% success rate and so represented the number of evaluations assuming the most favourable number of runs and number of generations. More detailed comparisons between CGP and other methods have been published previously in [22]. There, it was seen that embedded Cartesian genetic programming was highly competitive with other GP methods. Poli and Page evolved solutions to even-parity for up to 22 inputs [15]; however, they only gave numbers of evaluations for a single evolutionary run when the number of inputs was 13, 15, 17, 20 and 22. Thus we could not make a comparsion.

After evolution, solutions were tested for inputs of up to 24 bits. It was found that all solutions generalized to problems of this size. Most solutions had generalized to all inputs when even-8 parity was reached. This is reflected in the results in Table 4.4, where the number of evaluations required to solve a problem stops increasing because once a solution is found to generalize, no further evolution is required.

Examining the comparative results in Table 4.5, we can see that SMCGP scales much better than CGP, ECGP and GP.

Table 4.4 Average number of evaluations required to find a program that will solve parity up to a given number of bits (50 runs)

Input bits	Average evaluations
3	247,753
4	275,663
5	278,635
6	298,104
7	318,376
8	322,843
9	322,843
10	322,843
11	322,851
12	322,851
13	322,866
14	322,866
15	322,866
16	322,866
17	322,870
18	322,870
19	322,874
20	322,874

Table 4.5 Comparison with previous work on evolving parity. 'GP' means Koza's tree-GP (with ADFs) [12], 'ECGP' means embedded CGP [21], and 'CGP' is conventional CGP. With the exception of the figures from Koza, the figures show the average number of evaluations required to find a given-sized parity circuit. Results for higher numbers of inputs are not available for CGP or ECGP. The figures from Koza represent computational effort so they represent the minimum number of evaluations required to achieve 99% success. The minimum was selected from the 'ideal' number of runs and number of generations. A dash indicates that figures are unavailable

Input bits	SMCGP	CGP	ECGP	GP
4	275,663	81,728	65,296	176,000
5	278,635	293,572	181,920	464,000
6	298,104	972,420	287,764	1,344,000
7	318,376	3,499,532	311,940	1,440,000
8	322,843	10,949,256	540,224	–

4.4.3 A General Solution to Computing Even-Parity

It is instructive to examine an evolved solution to even-n-parity and how it can be shown that the evolved genotype represents a general solution.

The genotype in Fig. 4.2 was evolved with a *To-Do* list length of 1. The 20-node genotype had only seven active nodes. The inactive nodes are shown as unconnected smaller squares. The nodes INPP at positions 0 and 2 obtain inputs x_1 and x_0, respectively. Three Boolean functions BNOR, BAND and BOR appear at positions 4, 5 and 6, respectively. The OUTPUT function obtains the single output from the BOR node. The only active SM node is DUP at position 1. It carries arguments which cause it to copy eight nodes, beginning at the node on its left (INPP), and insert them immediately after itself. The action of the DUP node is shown using a curved line with an arrow emanating from the box. Since the genotype has no connections that are to the left of the first node, when DUP copies it disconnects the first two nodes in the generated phenotype. These appear at the beginning (left) of the new phenotype (iteration 1) at positions 0 and 1. It is important to note that in this phenotype, a previously inactive node (BNAND) at position 9 becomes active.

We can see that the phenotype at iteration 0 computes even-2 parity as follows. Denote the outputs of node i by z_i. The symbol $\oplus$ denotes the exclusive OR operation. When two or more Boolean arguments are side by side (as if being multiplied), it is assumed that the Boolean AND (BAND) operation is applied to the arguments (e.g. xy is equivalent to BAND (x,y)). An overbar represents inversion.

$$\begin{aligned} z_0 &= x_1, \\ z_1 &= x_1, \\ z_2 &= x_0, \\ z_4 &= \mathrm{BNOR}(z_2, z_1) = \overline{x_0 + x_1} = \overline{x_0}\overline{x_1} = (1 \oplus x_0)(1 \oplus x_1), \\ z_5 &= \mathrm{BAND}(z_1, z_2) = x_1 x_0, \\ z_6 &= \mathrm{BOR}(z_5, z_4) = z_5 + z_4 = z_5 \oplus z_4 \oplus z_4 z_5. \end{aligned} \tag{4.1}$$

Substituting for z_5 and z_4, expanding and then cancelling terms we obtain

$$\begin{aligned} z_6 &= x_1 x_0 \oplus (1 \oplus x_0)(1 \oplus x_1) \oplus x_1 x_0 (1 \oplus x_0)(1 \oplus x_1), \\ z_6 &= x_1 x_0 \oplus 1 \oplus x_1 \oplus x_0 \oplus x_1 x_0 \oplus x_1 x_0 (1 \oplus x_1 \oplus x_0 \oplus x_1 x_0), \\ z_6 &= x_1 \oplus x_0 \oplus 1. \end{aligned} \tag{4.2}$$

Thus $z_{16} = z_6$ is the even-2 parity function. When the eight duplicated nodes are inserted into the genotype just before position 2, they cause the activation of the BNAND node at position 9 in the new phenotype. This inverts the function computed by the eight duplicated nodes in the genotype. So the output of this block of nodes (denoted by A) is $x_2 \oplus x_1$, since the INPP functions return the inputs in descending order.

Now we turn our attention to the second iteration. When DUP inserts nodes 2–9 after itself, nodes 4–9 are shifted right (becoming nodes 12–17) in the second-iteration phenotype (enclosed in a box labeled B in Fig. 4.2). We now prove that this set of nodes carries out the exclusive OR of its input (emanating from the NAND node, which we call y) with the input variable (in this case x_1):

$$\begin{aligned} z_{12} &= x_1, \\ z_{14} &= \mathrm{BNOR}(x_{12}, y) = \mathrm{BNOR}(x_1, y) = (1 \oplus x_1)(1 \oplus y), \\ z_{15} &= \mathrm{BAND}(y, z_{12}) = \mathrm{BAND}(y, x_1) = x_1 y, \\ z_{16} &= \mathrm{BOR}(z_{15}, z_{14}) = z_{15} + z_{14} = z_{15} \oplus z_{14} \oplus z_{14} z_{15}, \\ z_{17} &= \mathrm{BNAND}(z_{16}, z16) = z_{16} \oplus 1. \end{aligned} \tag{4.3}$$

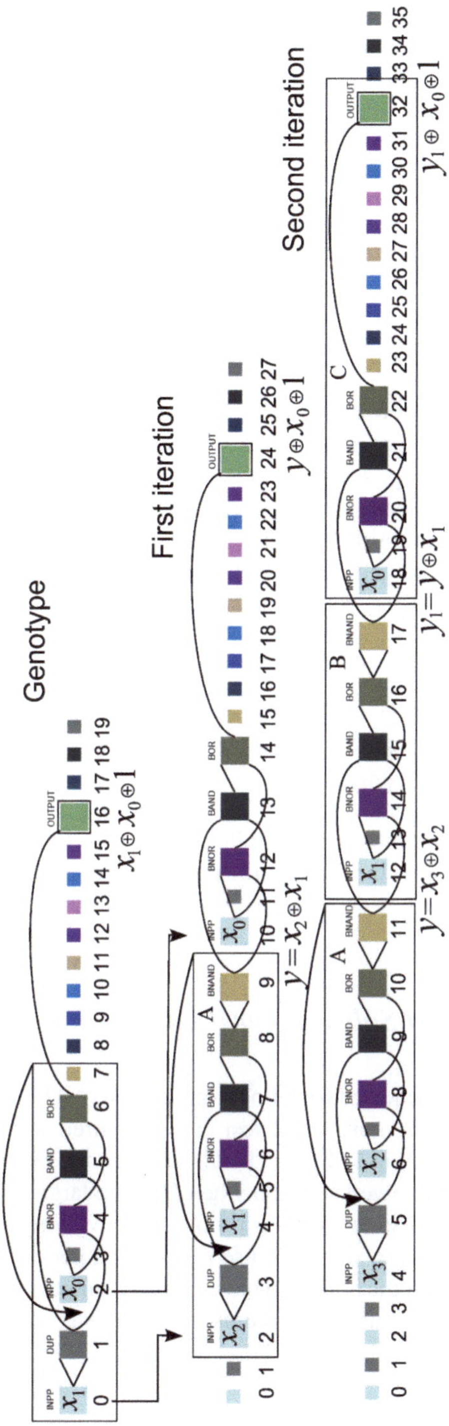

Fig. 4.2 An evolved genotype, iterated twice, that computes even-parity.

Substituting for z_{14} and z_{15} in z_{16} and then noting that in the last term, when $x_1 y$ multiplies $(1 \oplus x_1)$, we obtain $(x_1 y \oplus x_1 y)$, which is zero, we can simplify these equations as follows:

$$\begin{aligned} z_{16} &= x_1 y \oplus (1 \oplus x_1)(1 \oplus y) \oplus x_1 y(1 \oplus x_1)(1 \oplus y), \\ z_{16} &= x_1 y \oplus (1 \oplus x_1)(1 \oplus y) = x_1 y \oplus 1 \oplus x_1 \oplus y \oplus x_1 y, \\ z_{16} &= 1 \oplus x_1 \oplus y, \\ z_{17} &= x_1 \oplus y. \end{aligned} \tag{4.4}$$

Since we have seen that the nodes in section C compute the odd-parity of the supplied input y and the acquired input (by INPP), we find that at iteration 2 the phenotype computes $y_1 \oplus x_0 \oplus 1 = y \oplus x_1 \oplus x_0 \oplus 1 = x_3 \oplus x_2 \oplus x_1 \oplus x_0 \oplus 1$. This is the even-4-parity function. To construct a proof by induction, we will assume that for n inputs the phenotype computes even-n-parity.

The upper diagram in Fig. 4.3 shows the even-n-parity function E_n. This is the inductive hypothesis. We have already seen that the function enclosed in box A produces at the next iteration the two disconnected nodes and the function in A, $y = x_n \oplus x_{n-1}$, followed immediately by the function in box B, $y_{n-2} = y \oplus x_{n-2}$. Thus the new phenotype, E_{n+1}, generates the function

$$\begin{aligned} E_{n+1} &= y_{n-2} \oplus E_n \oplus x_{n-1} \oplus x_{n-2}, \\ E_{n+1} &= y \oplus x_{n-2} \oplus E_n \oplus x_{n-1} \oplus x_{n-2}, \\ E_{n+1} &= x_n \oplus x_{n-1} \oplus x_{n-2} \oplus E_n \oplus x_{n-1} \oplus x_{n-2}, \\ E_{n+1} &= x_n \oplus E_n. \end{aligned} \tag{4.5}$$

Thus the inductive hypothesis also applies to the $n+1$th iteration. We have already seen that at iteration 2, the form of the phenotype obeys the inductive hypothesis. Hence the general case is proved.

4.4.4 Why GP Cannot Solve General Parity Without Iteration

The number of test cases required to define even-n-parity is 2^n; from this, it follows in a simple manner that the number of test cases required to define all the even-parity functions from 2 to N is $4(2^{N-1} - 1)$. Clearly, it soon becomes impossible to evaluate the fitness of an evolved solution when N is too large.

SMCGP is able to address the scaling problem by evolving a program that can generate instances of arbitrary size. The use of function nodes (i.e. INP, INPP and SKIP) to acquire inputs and the ability to have multiple outputs (using the OUTPUT function) mean that as programs alter themselves they can use as many inputs as needed.

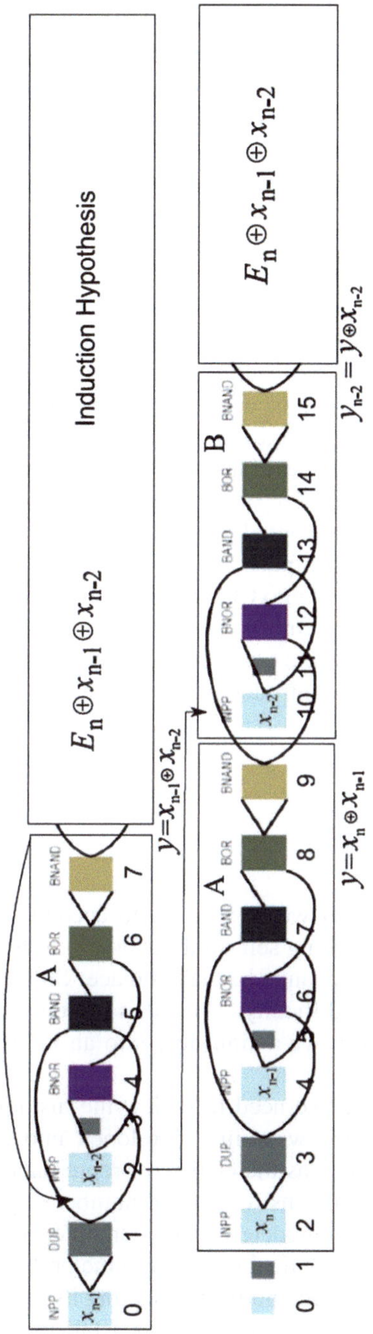

Fig. 4.3 The inductive hypothesis in diagrammatic form.

A statically sized CGP program can use only a fixed number of inputs – or, with the addition of functions such as INP, as many inputs as there are nodes in the graph. If the GP method does not have recursion or iteration (which is common), then it cannot acquire an arbitrary number of inputs. That is to say, it is always defined to have a specific number of inputs (i.e. even-6 or even-9). Since SMCGP has iteration built into the genotype–phenotype mapping, it can handle sequences of problems such as parity and many other problems [6, 7, 4].

It is worth contrasting evolved programs that have iteration in the form of feedback loops with SMCGP solutions. A program to do this would need to supply a bit-string and the number of bits in it. A general solution to even-n-parity can be obtained by reading in $n-1$ bits and EXORing them, and then, EXNORing with the last bit. In circuit terms, such a program would be a *sequential* circuit (it would have to be clocked). SMCGP solutions are different from this. The genotype is iterated n times to produce a *non-sequential circuit* that computes the parity. In other words, SMCGP produces a sequence of parallel parity circuits none of which have iteration in them. This has advantages and disadvantages. An advantage is that a parallel non-sequential circuit is faster than a sequential circuit with the same function. A disadvantage is that the size of the circuit is dependent on the number of inputs, whereas the size of a sequential program (or circuit) solving the problem is not dependent on the number of inputs.

Feedback could be introduced into CGP quite easily but it has only recently been given attention (see Sect. 2.9). One way would be to simply clock the feedback signals back to wherever they were connected (this could be done with a flip-flop). Clearly, feedback could also be introduced into SMCGP. It would be interesting to investigate the utility of this.

4.5 SM vs GP vs GA

Besides the many changes that we have made to accommodate a developmental approach to CGP by implementing self-modification in SMCGP, there is one other key issue where a substantial change has taken place: SMCGP trains on problems of different size and variable difficulty. Thus, it is not only the representation that has changed, but also the way we assign fitness to an individual in the course of evolution.

To understand this change, we need to review the historical changes that took place when genetic algorithms were first developed into genetic programming. Again, a key point at the time was the change of representation. From a bit-string representation of genetic algorithms (or a fixed number of parameters), the representation was changed to accommodate variable-length expression trees. This was a key innovation, allowing an individual to adjust to the problem size by growing (or sometimes shrinking) or more succinctly, by adapting its representation size.

As it was later pointed out and examined in detail by Langdon and Poli [13, 14], below a certain threshold of program size, an exact solution to a programming problem is not possible at all. Above that critical threshold, however, there exist exact solutions to the problem, the more the larger the size of the program. This insight has relevance not only to genetic programming but also to all methods of automatic programming imaginable.

The other innovation of genetic programming was that evolution would be performed not on one particular problem configuration but on a multitude. In GP parlance, a set of fitness cases was used to determine the fitness of the program. This is in line with the functionality of a program which requires it to react to many situations, not just one. A program could perform relatively well by fulfilling certain fitness cases perfectly and others only weakly, or by performing equally well on all fitness cases (but without fulfilling any perfectly). The trade-off between these two types of solutions gave rise to generalization measurements, with a program doing the former being classified as a specialist, and a program doing the latter being classified as a generalist.

The main point here is that the measurement of fitness based on many fitness cases made the evolution of a more general solution possible in the first place. Without this input being offered to the system, generalization capabilities would have remained elusive to genetic programming, much as they are to a genetic algorithm, which usually performs an optimization for a particular solution only.

At the same time, a set of fitness cases constituted a problem for GP that required its representation to be flexible enough to adapt its complexity. Thus, the problem and its solution correspond to each other in the appropriate way.

In a similar vein, the progress envisioned with SMCGP depends on two interrelated innovations. On the one hand, the self-modification operations, together with the adjustment of the numbers of inputs and outputs, allow an individual solution not only to adapt to problem difficulty once and for all, but also to adapt continuously in the course of its existence. This is the representation aspect of the innovation.

The other aspect, however, is that during training and evolution, fitness cases from different problem dimensions and difficulties are used to determine the fitness of an individual. Thus, for example, in the case of the even-n-parity problem, evolution is fed with fitness cases for a number of problem dimensions $n = 2,3,4,5,\ldots$. As we could see, if this is done in the proper way, SMCGP is able to extract structure from the fitness feedback that allows it to solve not only a single dimension of parity problems, e.g. $n = 5$, but to solve the problem for the case of general n.

In a way, SMCGP can generalize on a higher level: it can generalize from simple cases of a problem to more difficult ones, provided it has been shown in the course of evolution what this will entail in terms of developing a structure.

Many people have, for many years, maintained that GP is not really what we need since it would not be able to generalize appropriately, in the way a human can generalize from a set of sample problems to the general structure. This, however, is precisely what SMCGP can do, thanks to its corresponding features of SM operations, ability to choose input and output dimensions, and training with problem instances of varying difficulty.

Will SMCGP remain the only GP system that can do this? We doubt it. Sooner or later, self-modifying linear or tree GP systems will be designed, and the training method of using problem instances of varying degree for fitness evaluation will be applied to those representations. In fact, it might be interesting to apply this training method to existing classical representations to study what effect the additional information offered in those examples has on a solution.

4.6 Implementing Incremental Fitness Functions

The fitness function described in Sect. 4.4.1 is typical of those used so far in SM-CGP. But how does the incremental fitness function affect evolution?

In the experiment described in this section, we examined the even-parity problem but introduced a variation of the fitness function where the fitness function was no longer incremental.

We evolved circuits up to a certain number of test sets, and then checked to see if the solutions generalized to more test sets. For the incremental fitness function, the evaluation was aborted if the solution failed to correctly predict all bits in one test set. For the alternative fitness function, the number of test sets that the fitness function saw was fixed. This fitness function continued to evaluate iterations even if the previous iteration was unsuccessful.

From Table 4.6 we see that for smaller problems, the incremental fitness function requires more evaluations than does a fixed-sized fitness function. However, the incremental function soon becomes very efficient at finding solutions. The dip in the average number of evaluations shows where the solutions begin to exhibit generalization (see also Fig. 4.4).

Table 4.6 The average number of evaluations required to find a solution to the parity problem. Programs need to grow from two inputs through to the maximum number of inputs

Inputs	Fixed	Incremental
2	5,068	6,547
3	14,131	20,630
4	26,006	30,692
5	56,600	39,152
6	108,685	57,430
7	122,506	58,313
8	167,623	44,298
9	180,048	39,049
10	204,034	51,483

However, from the results in Table 4.7 and Fig. 4.5, we see that the fixed-sized fitness function is more likely to produce general solutions when they are evolved for larger-input problems. For a smaller maximum number of inputs, the incremental fitness function produces more general solutions, but still with low probability.

In addition to the increased number of evaluations required with the fixed-size fitness function, there is another penalty to pay in terms of the number of test sets

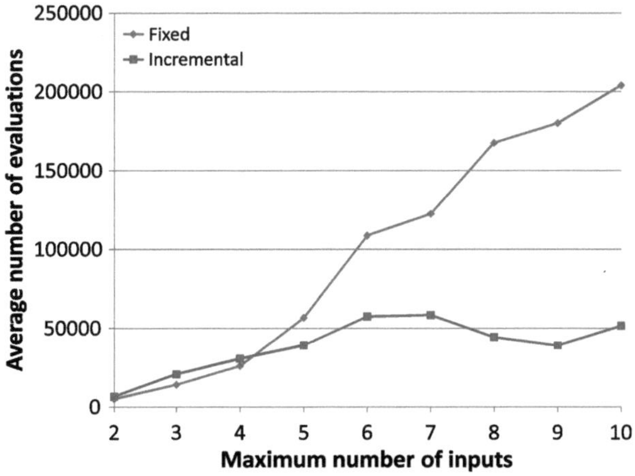

Fig. 4.4 The average number of evaluations required to find a solution to the parity problem. Programs need to grow from 2 inputs through to the maximum number of inputs.

Table 4.7 Percentage of solutions that, when evolved up to a given input size, continue to generalize to 20 inputs

Inputs	Fixed	Incremental
2	0	0
3	2	4
4	27	38
5	53	69
6	80	67
7	90	83
8	95	89
9	94	84
10	94	88

that need to be evaluated. Table 4.8 shows that the incremental fitness function requires far fewer test sets to be processed and in turn, we can deduce that far fewer test cases are required (as the earlier truth tables are small).

To conclude, we see that the incremental fitness function does not actively help generalization. However, the performance benefits in terms of the amount of the tests that need to be performed are likely to outweigh this. In future work, we will investigate if this behaviour holds for other problems.

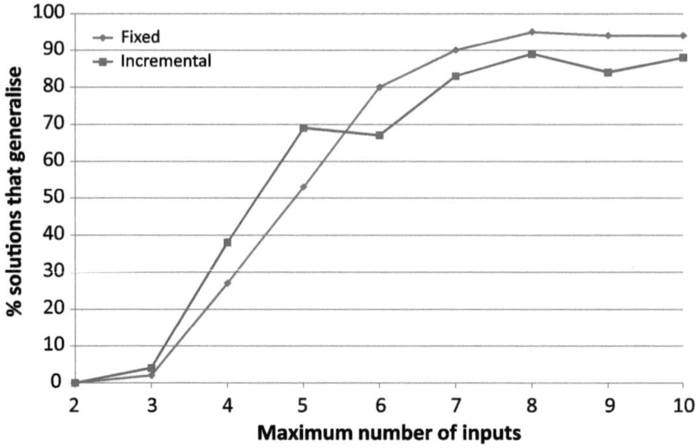

Fig. 4.5 Percentage of solutions that, when evolved up to a given input size, continue to generalize to 20 inputs.

Table 4.8 Average number of test sets evaluated when the solution is evolved up to a given input size

Inputs	Fixed	Incremental
2	5086	6565
3	28281	26969
4	78037	41912
5	226417	54829
6	543441	82278
7	735053	86203
8	1173376	65050
9	1440402	56945
10	1836328	78538

4.7 Conclusions

SMCGP has shown itself to be a capable and versatile extension to CGP. In the examples shown here and discussed elsewhere, we see that the addition of self-modification adds useful functionality and can be used to solve problems that a fixed-representation GP system cannot. Furthermore, we see that the implementation and representation are relatively straightforward. SMCGP can output human-readable programs that can be proved to behave in certain ways. Additionally, work has shown that the addition of self-modification does not harm the evolvability of CGP [4]. This gives us confidence that the SMCGP approach may be a suitable replacement for the classical CGP model.

Currently, a new version of SMCGP is under development that aims to simplify SMCGP (for example by having a more streamlined representation and function set) whilst at the same time increasing the developmental richness (by moving to a 2D representation). Early results on digital circuits [9] and mathematical results [8] are promising.

4.8 Acknowledgements

WB and SH gratefully acknowledge funding from Atlantic Canada's HPC network ACENET; from the Canadian Foundation of Innovation, New Opportunities Grant number 204503; and from NSERC under the Discovery Grant Program RGPIN 283304-07.

References

1. Banzhaf, W., Miller, J.F.: The Challenge of Complexity, chap. 11, pp. 243–260. Kluwer Academic (2004)
2. Harding, S., Miller, J.F., Banzhaf, W.: Self-modifying Cartesian Genetic Programming. In: Proc. Genetic and Evolutionary Computation Conference, pp. 1021–1028 (2007)
3. Harding, S., Miller, J.F., Banzhaf, W.: Development and Learning Using Self-modifying Cartesian Genetic Programming. In: Proc. Genetic and Evolutionary Computation Conference, pp. 699–706. ACM Press (2009)
4. Harding, S., Miller, J.F., Banzhaf, W.: Self-modifying Cartesian Genetic Programming: Fibonacci, Squares, Regression and Summing. In: Proc. European Conference on Genetic Programming, *LNCS*, vol. 5481, pp. 133–144 (2009)
5. Harding, S., Miller, J.F., Banzhaf, W.: Self-modifying Cartesian Genetic Programming: Parity. In: Proc. IEEE Congress on Evolutionary Computation, pp. 285–292 (2009)
6. Harding, S., Miller, J.F., Banzhaf, W.: Developments in Cartesian Genetic Programming: Self-modifying CGP. Genetic Programming and Evolvable Machines **11**, 397–439 (2010)
7. Harding, S., Miller, J.F., Banzhaf, W.: Self-modifying Cartesian Genetic Programming: Finding algorithms that calculate pi and e to arbitrary precision. In: Proc. Genetic and Evolutionary Computation Conference, pp. 579–586. ACM (2010)
8. Harding, S., Miller, J.F., Banzhaf, W.: SMCGP2: Finding Algorithms That Approximate Numerical Constants Using Quaternions and Complex Numbers. In: Proc. Genetic and Evolutionary Computation Conference Companion, pp. 197–198. ACM (2011)
9. Harding, S., Miller, J.F., Banzhaf, W.: SMCGP2: Self-modifying Cartesian Genetic Programming in Two Dimensions. In: Proc. Genetic and Evolutionary Computation Conference, pp. 1491–1498. ACM (2011)
10. Huelsbergen, L.: Finding General Solutions to the Parity Problem by Evolving Machine-Language Representations. In: J.R. Koza, W. Banzhaf, K. Chellapilla, K. Deb, M. Dorigo, D.B. Fogel, M.H. Garzon, D.E. Goldberg, H. Iba, R. Riolo (eds.) Proc. Conference on Genetic Programming, pp. 158–166. Morgan Kaufmann (1998)
11. Koza, J.R.: Genetic Programming: On the Programming of Computers by Natural Selection. MIT Press (1992)
12. Koza, J.R.: Genetic Programming II: Automatic Discovery of Reusable Programs. MIT Press (1994)

13. Langdon, W.B.: Scaling of program fitness spaces. Evolutionary Computation **7**(4), 399–428 (1999)
14. Poli, R., Langdon, W.B.: Foundations of Genetic Programming. Springer (2002)
15. Poli, R., Page, J.: Solving High-Order Boolean Parity Problems with Smooth Uniform Crossover, Sub-Machine Code GP and Demes. Genetic Programming and Evolvable Machines **1**(1–2), 37–56 (2000)
16. Schmidt, M., Lipson, H.: Distilling free-form natural laws from experimental data. Science **324**, 81–85 (2009)
17. Schmidt, M., Lipson, H.: Solving iterated functions using genetic programming. In: Proc. Conference Companion on Genetic and Evolutionary Computation, pp. 2149–2154. ACM (2009)
18. Spector, L., Clark, D.M., Lindsay, I., Barr, B., Klein, J.: Genetic programming for finite algebras. In: Proc. Genetic and Evolutionary Computation Conference, pp. 1291–1298. ACM (2008)
19. Spector, L., Robinson, A.: Genetic programming and autoconstructive evolution with the Push programming language. Genetic Programming and Evolvable Machines **3**, 7–40 (2002)
20. Streeter, M., Becker, L.A.: Automated Discovery of Numerical Approximation Formulae via Genetic Programming. Genetic Programming and Evolvable Machines **4**, 255–286 (2003)
21. Walker, J.A., Miller, J.F.: Investigating the Performance of Module Acquisition in Cartesian Genetic Programming. In: Proc. Genetic and Evolutionary Computation Conference, pp. 1649–1656. ACM (2005)
22. Walker, J.A., Miller, J.F.: Automatic Acquisition, Evolution and Re-use of Modules in Cartesian Genetic Programming. IEEE Transactions on Evolutionary Computation **12**, 397–417 (2008)
23. Wong, L.M.: Evolving Recursive Programs by Using Adaptive Grammar Based Genetic Programming. Genetic Programming and Evolvable Machines **6**, 421–455 (2005)
24. Wong, M.L., Leung, K.S.: Evolving Recursive Functions for the Even-Parity Problem Using Genetic Programming. In: P.J. Angeline, K.E. Kinnear, Jr. (eds.) Advances in Genetic Programming 2, chap. 11, pp. 221–240. MIT Press (1996)
25. Yu, T.: Hierachical Processing for Evolving Recursive and Modular Programs Using Higher Order Functions and Lambda Abstractions. Genetic Programming and Evolvable Machines **2**, 345–380 (2001)

Chapter 5
Evolution of Electronic Circuits

Lukas Sekanina, James Alfred Walker, Paul Kaufmann and Marco Platzner

5.1 Introduction

CGP has been applied many times in the nascent field known as evolvable hardware (EHW) [26, 23, 79, 47]. By evolvable hardware, we mean the usage of evolutionary and other bio-inspired algorithms either for automated hardware design or for dynamic hardware adaptation. While in the first case the goal is to automatically generate innovative solutions, the second case deals with the online modification of reconfigurable hardware. The objective is to improve the performance of a system working in a changing environment or repair a system when faults are present.

In the context of evolutionary circuit design, the term 'innovative' means that the solution exhibits better quality in some way with respect to existing designs of the same category. For example, the solution might occupy a smaller area on a chip, compute faster, provide better precision, have reduced energy consumption, have increased reliability.

However, evolutionary circuit design has not been competitive in a wide area of design problems so far, because of *scalability problems*. From the viewpoint of *scalability of representation*, the problem is that the long chromosomes which are usually required to represent complex solutions imply large search spaces that are typically difficult to search. In many cases, even a well-tuned evolutionary algorithm fails to find an innovative solution in a reasonable time.

Another problem is related to the fitness calculation time. In the case of the evolution of combinational circuits, the evaluation time for a candidate circuit grows exponentially with the number of inputs (assuming that all possible input combinations are tested in the fitness function). Hence, the evaluation time becomes the main bottleneck of the evolutionary approach when complex circuits with many inputs are evolved. This problem is known as the problem of *scalability of evaluation*. In order to avoid the scalability problem, various techniques have been adopted, including functional-level evolution, development, and incremental and modular evolution.

J.F. Miller (ed.), *Cartesian Genetic Programming*, Natural Computing Series,
DOI 10.1007/978-3-642-17310-3_5,

The following sections will demonstrate typical applications of CGP in the area of digital-circuit synthesis. We will also include two examples of adaptive hardware – adaptive signal classifiers and adaptive caches.

5.2 Direct Evolution of Small Combinational Circuits

The evolution of small combinational circuits was the first application of CGP, where it was shown that CGP can produce unconventional but useful designs. We will briefly introduce the circuit synthesis problem and show how to apply CGP to evolve combinational circuits.

5.2.1 Evolutionary vs Conventional Synthesis of Combinational Circuits

Common logic synthesis algorithms operate on a circuit representation. Various models have been devised to represent digital-circuits in a form that is suitable for synthesis algorithms [73]. Among other representations, truth tables, algebraic formulae, and-inverter graphs and binary decision diagrams have been utilized. The synthesis algorithms are capable of transforming the initial circuit representation (which is derived from a behavioural specification) into a circuit representation which is suitable for subsequent circuit fabrication (typically, the goal is to minimize the number of gates, i.e. the area). A key feature of the common synthesis algorithms is that any intermediate transformation produces a circuit which is functionally equivalent to the original circuit. The circuit representation together with the set of transformations determines the space of possible implementations that one can obtain as a result of the synthesis process. In other words, some possibly efficient circuit implementations are not allowed in this process in order to obtain the resulting circuit in a reasonable time.

Evolutionary circuit design allows any transformation (here viewed as mutation or crossover) to be performed on a circuit representation even if the transformation does not lead to a solution functionally equivalent to the specification. By applying genetic operators, the evolution can implicitly discover compact circuit structures unreachable using conventional synthesis. Selection pressure guides the evolution towards better areas of the search space. Hence an evolutionary algorithm combined with assemble-and-test can be used to explore over a much larger area of design space than is possible using a top-down rule-based design algorithm. Miller explains this as follows [39]:

> The [conventional] methods though powerful in that they can handle large numbers of input variables are not adaptable to new logical building blocks and require a great deal of analytical work to produce small optimizations in the representation. Assembling a function from a number of component parts begins in the space of all representations and maps it

into the space of all the truth tables with m input variables. The evolutionary algorithm then gradually pulls the specification of the circuit towards the target truth table. Thus the algorithm works in a much larger space of functions many of which do not represent the desired function. ... this is the only way one can discover radically new designs.

5.2.2 CGP for Logic Synthesis

The basic formulation of the logic synthesis problem for CGP is as follows [39, 70, 32]. Using a given set of gates find a combinational circuit whose function is equivalent to the specification given in a truth table and minimize the number of gates. Additional optimization criteria such as delay or the cost of gates can be introduced in a more advanced problem formulation. CGP can be seeded using existing designs in order to reduce the design time. In some studies, the goal was to evolve a functional solution as quickly as possible from a randomly initialized population without considering the size of the evolved circuits (see e.g. [62]).

CGP can be utilized in its standard version to solve this task. As the numbers of inputs and outputs, the set of gates and the target function are given in the specification, it remains to define the size of the CGP array, the l-back parameter, the population size, the mutation rate and the fitness function. A recommended setting when no domain knowledge is available is $n_c = K$, $n_r = 1$, $l = n_c$, $\lambda = 4$ and the number of mutational genetic changes per chromosome $\mu_g = 1$. The value of K has to be estimated according to the expected circuit size. Numerous experiments have confirmed that having some redundant gates in the genotype is useful for an efficient search. Finally, having only a linear array of gates ($n_r = 1$) and the maximum l allow the creation of arbitrary feed-forward connections among gates.

When CGP is applied to reduce the number of gates in a digital-circuit, it starts with the a fitness function which evaluates the circuit behaviour only. Once one of the candidate circuits conforms to the behavioural specification the number of gates becomes important and is reflected in the fitness value. The fitness value of a candidate circuit is defined as [32]

$$fit1 = \begin{cases} b & \text{when} \quad b < n_o 2^{n_i}, \\ b + (n_c n_r - z) & \text{otherwise}, \end{cases} \tag{5.1}$$

where b is the number of correct output bits obtained as the response for all possible assignments to the inputs, z denotes the number of gates utilized in a particular candidate circuit and $n_c.n_r$ is the total number of available gates. The last term, $n_c n_r - z$ is considered only if the circuit behaviour is perfect, i.e. $b = b_{max} = n_o 2^{n_i}$. We can observe that the evolution has to discover a perfectly working solution first, and at this stage the size of the circuit is not important. Then, the number of gates is optimized. Another goal may be to minimize the number of transistors instead of the number of gates [15]. A multi-objective approach to circuit synthesis will be formulated in Sect. 5.3.

Unfortunately, the evaluation time grows exponentially with the number of inputs. One approach that allows a reduction of the evaluation time – bit-level simulation – has been proposed in [44]. The idea of bit-level simulation (or parallel simulation) is to utilize bitwise operators operating on multiple bits in a high-level language (such as C) to perform more than one evaluation of a gate in a single step. For example, when the combinational circuit under simulation has three inputs and it is possible to concurrently perform bitwise operations over $2^3 = 8$ bits in the simulator, then the circuit can be simulated completely by applying a single eight-bit test vector at each input (see the encoding in Fig. 5.1). In contrast, when this is impossible, then eight three-bit test vectors must be applied sequentially. Current processors allow us to operate with 64-bit operands, i.e. it is possible to evaluate the truth table of a six-input circuit by applying a single 64-bit test vector at each input. Therefore, the speed-up obtained is 64 compared with the sequential simulation. If the circuit has more than six inputs then the speed-up is constant, i.e. 64.

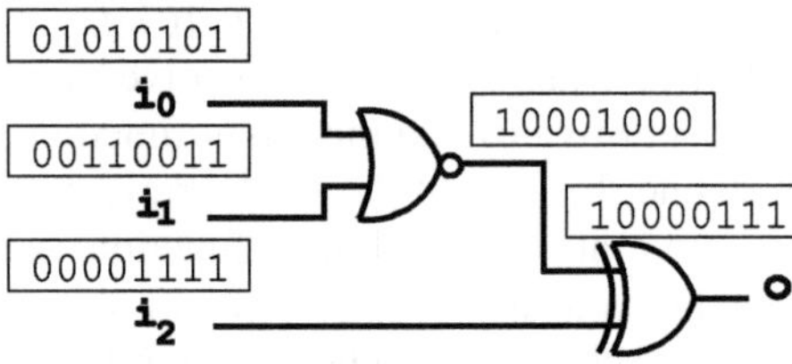

Fig. 5.1 Bit-level simulation of a combinational circuit.

5.2.3 Benchmark Problems

At the gate level, CGP has typically used to evolve small combinational (mostly arithmetic) circuits. Other work that has been done includes evolution of digital filters at the gate level [38, 52, 37], finite state machines [3] and polymorphic circuits (see Sect. 5.4). We will briefly discuss the evolution of small multipliers, which is the most popular benchmark problem for gate-level circuit evolution. Because the direct CGP approach is not scalable, only small multipliers have been evolved so far (four-bit multipliers, i.e. eight-input/eight-output circuits and smaller).

Table 5.1 summarizes the best known results for various multipliers according to [39, 70]. CGP was used with two-input gates; $l = n_c$, $\lambda = 4$, $\mu_g = 3$, and the remaining parameters are given in Table 5.1. The CGP was seeded using conventional designs.

CGP is capable of creating innovative designs for this class of circuits. However, it is important to carefully initialize the CGP parameters. For example, in order to reduce the search space, the function set should contain just the logic functions that are important for multipliers (the solutions denoted as ‘Best CGP’ in Table 5.1

were obtained using $\Gamma = \{x\ AND\ y,\ x\ XOR\ y,\ (not\ x)\ AND\ y\}$). However, the gate $(not\ x)\ AND\ y$ is not usually considered as a single gate in digital design. Its implementation is constructed using two gates: AND and NOT. Hence we also included 'Recalc. CGP' in Table 5.1 which is the recalculated result obtained when $(not\ x)\ AND\ y$ are considered as two gates in the multipliers shown in [70].

Table 5.1 Numbers of two-input gates in multipliers according to [39, 70]

Multiplier	Best conventional	Best CGP	Recalc. CGP	$n_r \times n_c$	Max. generation
2b×2b	8	7	9	1×7	10k
3b×2b	17	13	14	1×17	200k
3b×3b	30	23	25	1×35	20M
4b×3b	47	37	44	1×56	200M
4b×4b	64	57	67	1×67	700M

Results for another setting of the parameters were given in [17]. In all experiments, $n_r = 1$, $l = n_c$, $\lambda = 14$ and μ_g was between 1 and 14 (the mean value was 7), and $\Gamma = \{, or, not, nand, nor, xor, identity, const_1, const_0\}$ where *not* and *identity* are unary functions (taking the first input of the gate) and $const_k$ is constant generator with the value k. The initial population was seeded randomly. Each experiment was repeated 10 times with a 100 million generation limit. Two selection strategies are compared in Table 5.2: the fit1 selection strategy is the standard selection strategy for CGP (see Eqn. 5.1), and the fit2 selection considered a fully functional but not necessarily smallest discovered individual as the parent for the new population [17].

Figure 5.2 shows the best evolved four-bit multiplier obtained using the fit2 selection strategy (CGP was seeded using the best multiplier from [70]). The evolved circuit consists of 56 gates and reduces the number of transistors by 8.7% with respect to the seed [17].

Table 5.2 The best obtained and mean numbers of gates for the multiplier benchmarks when CGP starts from a randomly generated initial population

Circuit	Algorithm	n_c	Gates (best)	Gates (mean)	Mean # gener.	Successful runs
2b × 2b	fit1	7	7	7	2,738	100%
	fit2		7	7	2,777	100%
3b × 2b	fit1	16	13	13	651,297	100%
	fit2		13	13	741,758	100%
3b × 3b	fit1	57	25	27.7	476,812	100%
	fit2		23	23.4	625,682	100%
4b × 3b	fit1	125	46	52.7	2,714,891	100%
	fit2		37	43.1	4,271,179	100%
4b × 4b	fit1	269	110	128.3	29,673,418	90%
	fit2		60	109.4	37,573,311	70%

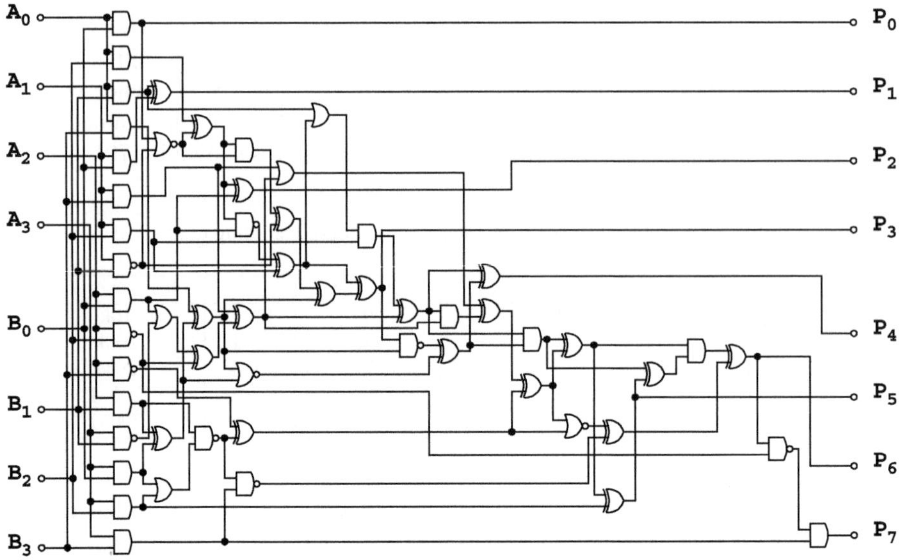

Fig. 5.2 A 4b × 4b multiplier (56 gates, delay 18δ) evolved using CGP seeded by a functional design, where δ is the generic date delay for each gate.

5.2.4 Summary

In the case of digital logic synthesis, CGP has led to innovative designs only for small circuits, mainly because of the very time-consuming and so non-scalable fitness evaluation [62, 54]. In summary, evolutionary circuit design can produce quite compact designs for the cost of a run time, whereas conventional synthesis quickly gets stuck in a local optimum. Unfortunately, the evolutionary approaches currently used are not scalable in this application domain. It is worth noting that using the technique of self-modifying CGP discussed in Chap. 4, certain classes of *arbitrary-sized* circuits can be evolved (i.e. n-bit parity, and n-bit adder). It remains to be seen whether such methods can produce innovative designs for other classes of circuits (e.g. n-bit parallel multipliers). Another promising approach for addressing the scalability issue is described in [68]. An initial population is seeded with a conventionally synthesized circuit design and CGP is used to optimize the design. The technique uses a satisfiability problem solver to decide whether a candidate solution is functionally correct or not. This allows digital circuits with over a hundred inputs to be optimized.

5.3 Multi-objective Evolution of Combinational Circuits

The previous section has shown that CGP has been successfully used to create numerous novel topologies for building-block logic circuits such as full-adders and multipliers. Whilst the circuits are functionally correct in terms of binary output and often minimal in terms of the number of gates, they are not always optimal compared with conventional human designs, as they tend to have longer paths between inputs and outputs and are potentially larger in terms of the number of transistors. In standard logic design, one of the primary goals is to minimize both the circuit area and the delay; fewer large circuits can be fabricated on a single wafer, which results in increased cost, and longer delays result in a decrease in the maximum operating frequency of the device.

In the work described in this section, the conventional CGP algorithm was augmented with a stage which further optimizes circuits, once a functionally-correct design has been found. To achieve this goal, a two-tiered fitness function was used. The first tier was the conventional Boolean-error score based on the binary Hamming distance between the observed output and the target truth table. For each circuit found that was functionally correct, it was then rated for performance on a number of different criteria and sorted into Pareto fronts using a muti-objective fitness function.

5.3.1 *Multi-objective Fitness Function*

The multi-objective strategy chosen was based on the popular NSGA-II (Non-dominated Sorting Genetic Algorithm II). This selection algorithm arranges all the individuals within the population into a series of non-dominated Pareto fronts and allows an unbiased trade-off between the objectives [12]. One characteristic of this selection algorithm is that it promotes a wide spread of results on the primary fronts and avoids filling the population with closely clustered individuals, thus maintaining a diverse set of solutions. Each circuit was simulated in software by calculating its Boolean outputs for every possible input. For each circuit evaluated, the primary fitness measure was the circuit functionality; circuits which were not functionally correct were not evaluated further and not included in the fronts used by NSGA-II. For circuits which passed the functionality test, four fitness scores were calculated which were equally weighted in the fitness calculation.

The first of these objectives was the number of logic gates used; minimizing the number of gates will generally result in the most compact circuit schematic. In other research to optimize designs based on CGP and other evolutionary algorithms, the number of gates has been the fitness criterion against which circuits are optimized. Whilst having a minimum number of gates is a clear goal in terms of creating compact, space-efficient schematics, and is an interesting optimization target, there are a number of reasons why it is not necessarily the best target for an optimized circuit. It must be remembered that not all gates are equal; exclusive-OR and exclusive-

NOR gates contain more transistors than AND/OR gates, which themselves contain more transistors than NAND/NOR gates. The number of transistors impacts on the die area, power consumption and maximum operating speed in a fabricated design. Logic effort theory reveals how NAND gates themselves are preferable to NOR gates in terms of delay, owing to the lower input capacitance [63]. For these reasons, we looked at not just the number of logic gates, but also the number of transistors and the lengths of the paths between input and output.

The secondary objective was the number of transistors used, based on the logic-gate arrangements found within a standard cell library. This meant that each inverter is considered to be two transistors, each NAND and NOR function is four transistors, each AND and OR gate is six transistors and the exclusive-AND and exclusive-OR gates are nine transistors each. Whilst there will be a degree of correlation between gate count and transistor count, there are important differences which may affect which design is considered more useful; a minimized number of gates will generally result in a more compact gate-level schematic and thus might be preferable if the design is to appear in print; however, a minimized number of transistors will result in a more compact fabricated design and thus may be more efficient in actual fabrication.

The third objective was the longest gate-level path length between input and output. For this objective, all gates were equally weighted, with the resultant scores being the total number of gates in the longest path between an output and any of its inputs. The final objective was the longest approximated delay between input and output based on the switching delays within the cell library; in the work described in this section, this was a score based on the count of transistor gates that are traversed in the longest path. A more accurate score could be calculated from logic-effort formulae; however, this would add significantly to the processing time for each evaluation. As with the previous objectives, there will be a degree of correlation between these two objectives; however, minimizing the former will produce the most compact-width circuits when drawn as a gate-level schematic, whilst the latter will minimize the worst-case switching delay in an actual fabricated circuit.

5.3.2 Benchmarks

The CGP algorithm was set to evolve three test-circuits: a three-bit adder, a three-bit multiplier and a seven-segment hexadecimal display driver. The solutions could be built from a complete set of two-input logic gates (NAND, NOR, AND, OR, XNOR and XOR) alongside a single-input inverter function. For each of the test circuits, a number of runs were executed for a predetermined number of generations as described in Table 5.3.

The results for the first functionally correct circuit in each run are shown in Table 5.4. These results demonstrate the output that could be expected for a CGP algorithm without a multi-objective optimization stage, where once a functionally correct circuit is found the program terminates. The table shows the best extracted

Table 5.3 Parameters used by CGP in the experiments in Sect. 5.3.2

Parameter	Value
Number of runs	20
Population size	100
Number of nodes	3
Genotype length (genes)	300
Mutation rate (%)	2
End generation	50,000,000

results alongside the averaged results for all the runs. It is clear from the figures that whilst the CGP algorithm is effective at finding functionally correct results, there is little optimization and some of the results are large in terms of transistor count and circuit area; every run found a functionally correct circuit well in advance of the terminating generation. Table 5.5 contains the best and average scores for the test circuits at the termination of the algorithm. Naturally, a significant improvement in all scores was noticed by the end of each run, with certain populations reaching an optimum and stagnating well before the terminating generation. With the smaller circuits, the same set of results appeared in most of the runs; the larger circuits had more varied runs, with more diverse populations towards the end of the run.

Table 5.4 Results before multi-objective optimization

	Three-bit adder		Three-bit multiplier		Seven-segment display	
	Best	Avg.	Best	Avg.	Best	Avg.
Generation	3,576	20,019	4,974	292,195	10,257	39,070
Gate count	28	38	41	57.9	41	67.5
Transistor count	192	223	225	326	223	384
Longest gate path	7	8	7	9.7	8	11.1
Longest transistor path	20	25.2	20	28.2	18	32.7

Table 5.5 Results after multi-objective optimization

	Three-Bit Adder		Three-Bit Multiplier		Seven-segment Display	
	Best	Avg.	Best	Avg.	Best	Avg.
Gate count	15	15.6	28	31.64	22	26.95
Transistor count	90	94.6	148	177.3	125	146.5
Longest gate path	5	5.6	5	5.86	4	5.05
Longest transistor path	12	13.6	13	15.82	11	13.85

5.3.2.1 Three-Bit Adder

Most of the runs for the three-bit adder circuit discovered circuits which utilized 15 gates and a gate-path depth of 5. One run also discovered the circuit illustrated in Fig. 5.3, which utilized 18 gates and has a deepest gate-path depth of 6, but offers a lower longest transistor-path length score of 12. The observed improvement provided by the multi-objective optimization stage is steady and regular until the circuits with fewer than 18 gates are discovered, at which point progress typically stagnates.

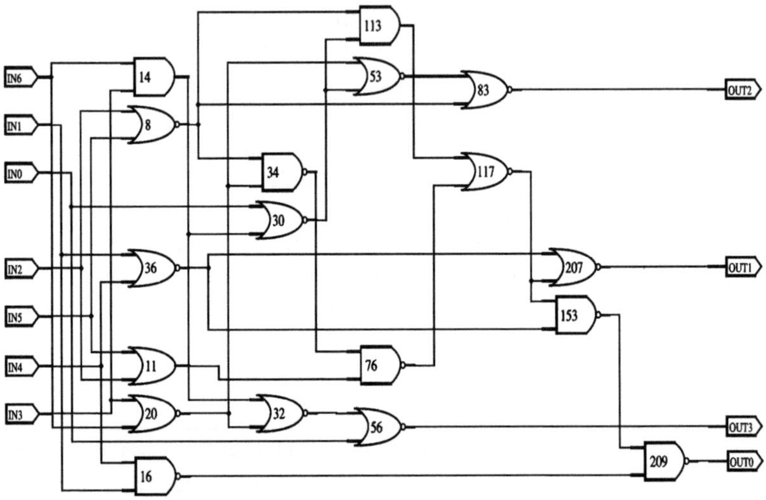

Fig. 5.3 Evolved three-bit adder design with minimum transistor-path length.

5.3.2.2 Three-bit Multiplier

The most compact three-bit multiplier extracted after multi-objective optimization utilized 28 gates, as illustrated in Fig. 5.4. Another optimized design had a longest gate-path length of five gates, but used a total of 31 gates. The results show promise when compared with results from other researchers. A better result in terms of gate count for this circuit has been extracted in work by Wang and Lee [77]; however their evolved design featured a longer propagation gate delay, with a path length of six gates. The Wang and Lee method utilized an adapted version of CGP incorporating a self-adaptive mutation rate control to enhance algorithm performance. Wang and Lee observed a reduction in average gate count from 52.97 to 31.70 with the addition of a multi-objective optimization phase; similar results were observed here with a reduction from 57.90 to 31.64 in average number of gates before and after multi-objective optimization. Some results from Vassilev and Miller [71], which produce a three-bit multiplier from only 23 gates, make use of non-standard gates such as a one-inverted-input AND and use a much smaller library of gates for optimization; however the produced circuit produced has a longest gate path length of eight gates plus an additional two inverted inputs. The most efficient human-designed topologies for this circuit use 30 gates [39].

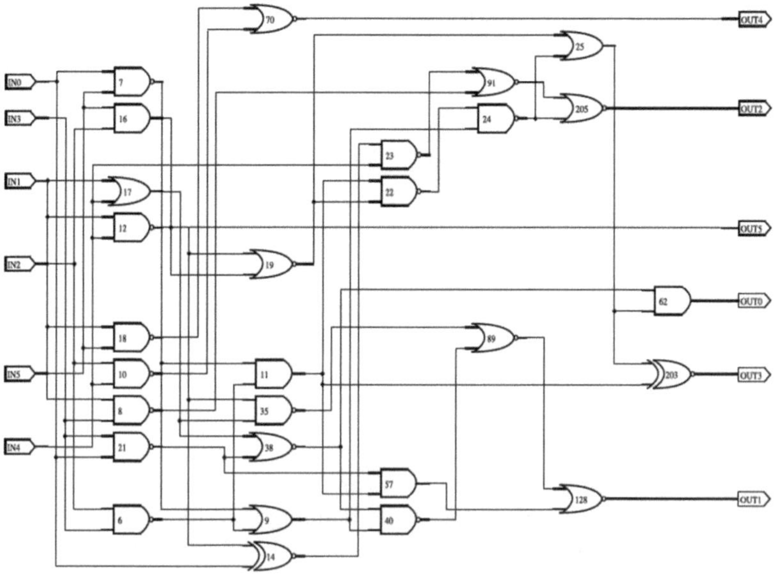

Fig. 5.4 Evolved three-bit multiplier design with minimum gate count.

5.3.2.3 Seven-Segment Display Driver

The most consistent improvement obtained through multi-objective optimization was seen with the seven-segment hexadecimal display driver. Across the 20 runs, the initial circuit that matched the target functionality contained on average 67.5 gates and an average longest gate-path length of 11.1; the best observed circuits had a gate count of 41 and a gate-path length of 8. By the end of the optimization phase, each run had individuals within the population with gate counts under 30 and gate-path lengths of 6 or less, the most gate-efficient circuit had only 22 gates and is shown in Fig. 5.5; another circuit utilized 26 gates but managed a gate-path length of just 4. The best solutions exploit gates shared between most of the output paths that would be hard to identify using traditional logic-design methods such as Karnaugh mapping, and this suggests that multi-objective CGP (MOCGP) could be very beneficial for combinational logic problems with several distinct outputs.

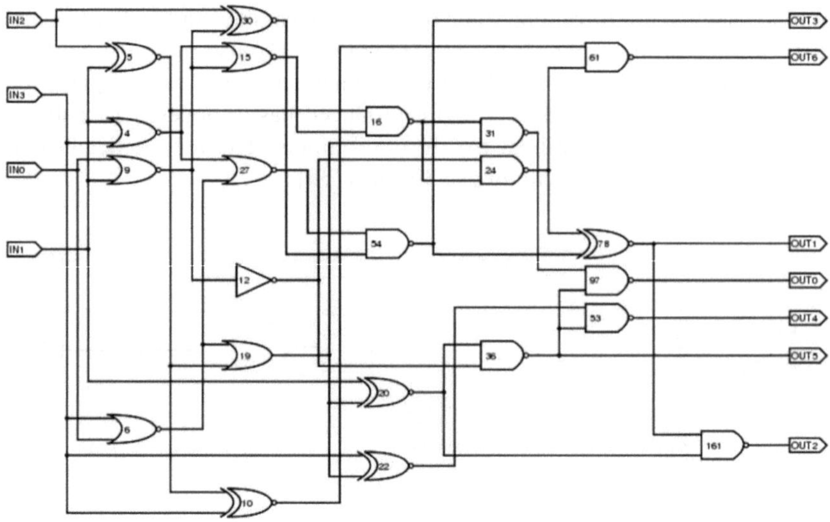

Fig. 5.5 Evolved seven-segment display driver with minimum gate count.

5.3.3 Summary

The goal of this research was not to create an algorithm that is more computationally efficient than CGP (or any other logic-building algorithm) but to demonstrate that the use of an multi-objective optimization could effectively improve the characteristics of a circuit, at the cost of more resources. The results demonstrate that by allowing the algorithm to continue to optimize results once a functionally correct design has been discovered, significant improvements in the extracted designs can be found. The MOCGP fitness function discovers designs which have lower gate counts and shorter path lengths than those of the designs discovered by standard CGP, which will translate to real-world improvements in circuit area and operating speed in a fabricated design.

5.4 Evolution of Polymorphic Circuits

CGP can be used to evolve compact implementations of circuits whose efficient construction using conventional methods is quite difficult or even impossible. The design of polymorphic circuits is a typical example because conventional circuit synthesis tools are not able to work with polymorphic gates.

5.4.1 Polymorphic Electronics

Polymorphic electronics was introduced by Stoica's group at the NASA Jet Propulsion Laboratory as a new concept for reconfiguration of electronic devices [61].

Polymorphic circuits consist of both ordinary and polymorphic gates. Polymorphic gates are unconventional logic components which can switch their logic function according to a changing environment. For example, Fig. 5.6 shows the behaviour of a polymorphic gate, controlled by the level of the power supply voltage (Vdd), which operates as NOR when Vdd = 3.3–3.8 V and as NAND when Vdd = 3.9–5.0 V [50]. Another gate was developed that operates as AND when the temperature is 27°C and as OR when the temperature is 125°C [61].

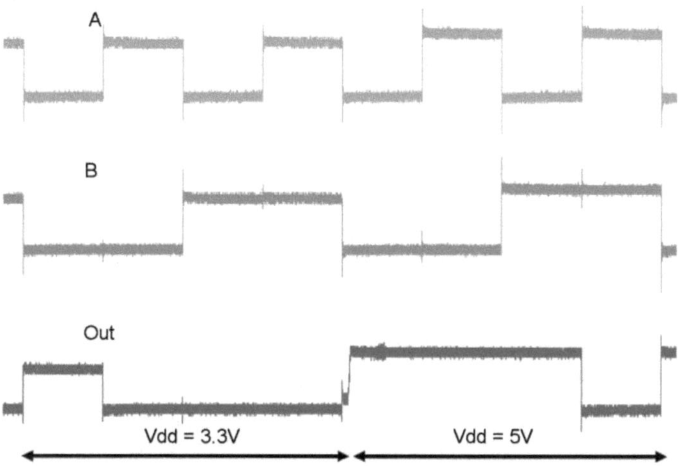

Fig. 5.6 Behaviour of a polymorphic NAND/NOR gate controlled by Vdd, according to [50].

The main motivation for the use of polymorphic gates is to obtain reconfigurable (and thus potentially adaptive) circuits at a very low cost and without the need to implement a reconfiguration infrastructure (switches, multiplexers, configuration registers etc.). Figure 5.7 shows an example of a polymorphic digital-circuit and its equivalent behaviour in both modes of the polymorphic NAND/NOR gate (i.e. $y = \overline{(a \oplus b) \wedge c}$ and $y = \overline{(a \oplus b) \vee c}$). Although polymorphic gates can be implemented relatively effectively using current CMOS technology, we can expect an expansion of polymorphic devices with further development of nano electronics and molecular electronics. Other current applications of polymorphic electronics include multifunctional circuits [80], self-checking adders [46], reduction of test vector volume [60] and adaptive filtering [49].

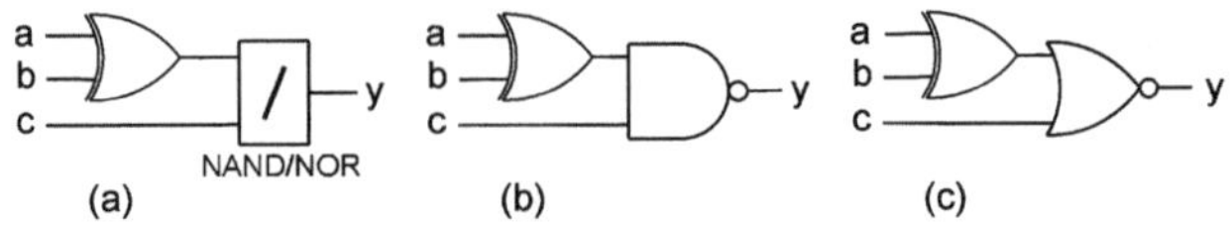

Fig. 5.7 (a) Example of a polymorphic circuit; (b) equivalent circuit in mode 1; (c) equivalent circuit in mode 2.

5.4.2 Gate-Level Evolution of Polymorphic Circuits

Here the objective was to design a circuit which operates as a multiplier (the function for which is denoted by f_1) when all polymorphic gates are set to the first mode and as a sorting network (the function for which is denoted by f_2) when all polymorphic gates are set to the second mode. It was assumed that the resulting circuit would contain ordinary gates and NAND/NOR polymorphic gates (other polymorphic gates are not currently available for implementation, as we will see in Sect. 5.4.4).

In order to use CGP for the design of polymorphic circuits and the optimization of the number of gates, we used the same approach as in Sect. 5.2. However, the fitness function had to be modified so that a candidate circuit could be evaluated in both modes. The new fitness value was defined as follows [48]:

$$fitness = B_1 + B_2 + (n_c.n_r - z), \tag{5.2}$$

where B_1 and B_2 are the numbers of correct output bits for f_1 and f_2, respectively, obtained as the response for all possible input combinations; z denotes the number of gates utilized in a particular candidate circuit, and $n_c.n_r$ is the total number of programmable gates available. The last term was considered only if the circuit behaviour was perfect in both modes; otherwise, $n_c.n_r - z = 0$. Table 5.6 summarizes the parameters and the results obtained for various instances of the problem [51].

Table 5.6 Parameters of evolved polymorphic multipliers/sorters. The gates in 'Gate set' are numbered as (1) NAND/NOR, (2) AND, (3) OR, (4) XOR, (5) NAND, (6) NOR, (7) NOT A, (8) NOT B, (9) MOV A and (10) MOV B, where MOV denotes the identity operation. The population size is 15

Multiplier/sorter	$2b \times 2b/4b$	$3b \times 2b/5b$	$3b \times 3b/6b$	$4b \times 3b/7b$
$n_c \times n_r$	10×12	100×1	120×1	16×16
l	1	100	120	16
Mutation (genes)	1	2	4	4
Gate set	1, 2, 9, 10	1–4, 9, 10	1–10	1, 2, 9, 10
Runs	10	10	10	10
Successful runs	10	10	9	3
Generations (average)	52,580	854,900	26,972,648	62,617,151
Min. number of gates	23	30	52	113

Similarly to the evolution of multipliers (Sect. 5.2), we can observe that the average number of generations grows exponentially with the number of inputs. No correct solution was obtained for eight-input circuits at the gate level. We can see that the problem is more complicated than the synthesis of conventional circuits. The computation time can be obtained as follows. In order to generate 1 million generations for a four-input multifunctional circuit, 80s are required by an Athlon64 3200+ processor. For a seven-input multi-functional circuit, 207 seconds are required by the same processor.

5.4.3 CGP as Optimizer

Another approach to the implementation of polymorphic circuits is to use independent conventional implementations for both of the required modes and multiplex their outputs by polymorphic multiplexers (see the illustration of polymorphic multiplexing in Fig. 5.8). In the case of a three-bit multiplier/six-bit sorter, the multiplier module will consist of 23 gates and the sorter module will consist of 24 gates, i.e. 47 gates will be used in total [51]. As six polymorphic multiplexers have to be connected to the outputs of the modules (each of them would cost five gates) the total cost of this multiplexing implementation is $47+30=77$ gates. It is evident that the 52-gate evolved circuit (see Table 5.6) is more area-efficient than circuits multiplexing the best known conventional implementations.

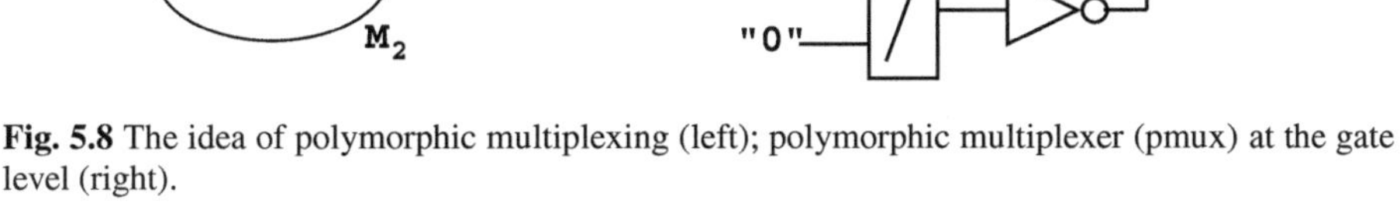

Fig. 5.8 The idea of polymorphic multiplexing (left); polymorphic multiplexer (pmux) at the gate level (right).

CGP can be also utilized to reduce the number of gates in a circuit initially created using polymorphic multiplexing or by any other conventional or unconventional method [16]. To do this, CGP was used with the same parameters as in previous experiments; however, the initial population consisted of a fully functional solution and its λ mutants. The goal of evolution was to keep the existing functionality and reduce the number of gates. Using this method, it was possible to reduce the number of gates of seven-input multiplier/sorter from 113 to 80 and even construct a 105-gate, eight-input multiplier/sorter. A significant reduction in gate count can be obtained if millions of generations can be performed, i.e. if sufficient time is allowed for optimization.

5.4.4 REPOMO32: CGP on a Chip

In order to demonstrate the applications of polymorphic electronics, the REPOMO32 (Reconfigurable Polymorphic Module) chip has been developed at Brno University of Technology [50]. Figure 5.9 shows the structure of the chip. It consists of 32 two-input configurable logic elements (CLEs) organized in an array of four rows and eight columns. The CLEs can be programmed to perform one of the following functions: AND, OR, XOR and polymorphic NAND/NOR (controlled by Vdd). When Vdd = 3.3V, the NAND/NOR gate exhibits the NOR function, and when Vdd = 5V, the gate exhibits the NAND function. In fact, REPOMO32 can be considered as a hardware implementation of CGP with parameters $n_i = 4, n_o = 8, n_a = 2, n_f = 4, l = 2$, where the outputs are unalterably connected to the outputs of the fourth and eighth columns of nodes.

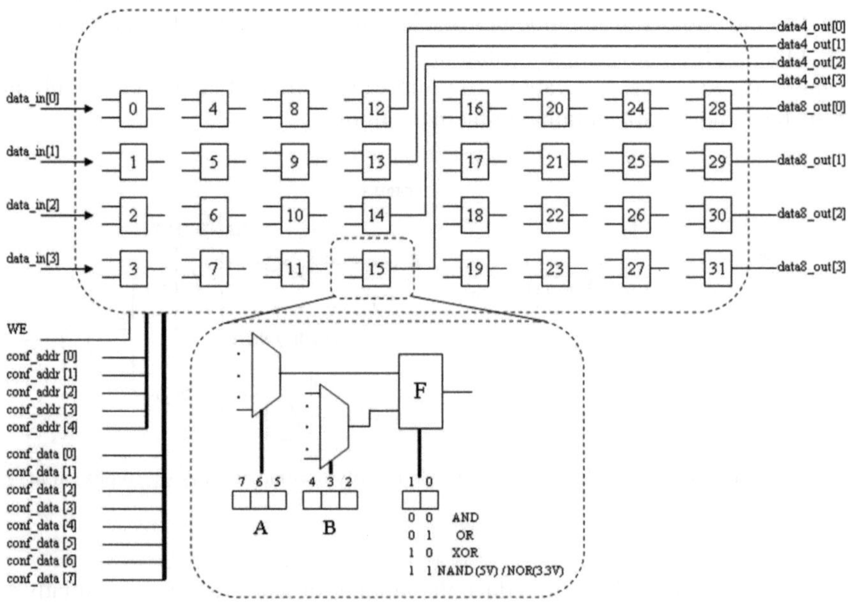

Fig. 5.9 Architecture of REPOMO32 chip.

REPOMO32's logic behaviour is defined by its configuration bits and the level of Vdd. The configuration bits control a set of multiplexers, which are responsible for interconnecting the CLEs and selecting their logic functions. In total, eight bits define the configuration of a single CLE. The configuration of the chip is stored in 32 eight-bit latch registers. The reconfiguration of a single CLE is performed by supplying the CLE's address (conf_addr) and the configuration data (conf_data), followed by activating the WE signal. The chip can be completely reconfigured in 32 configuration steps. The data4_out outputs are connected directly to the CLEs in the fourth column, and the data8_out outputs are connected directly to the CLEs in the last column. There are no synchronization registers in REPOMO32. The chip has 28 pins and occupies an area of 2900 times 1970 μm. It was fabricated using AMIS CMOS 0.7 μm technology. Maximum operating frequency is 38 MHz for Vdd = 3.3 V and 56 MHz for Vdd = 5 V.

In order to automate the process of reconfiguration, setting the input values and reading the output values, the REPOMO32 chip was connected to a configuration/evaluation controller implemented in a Field-Programmable Gate Array (Xilinx FPGA Spartan 3 XC3S50) available on a FITkit development board (Fig. 5.10). The FPGA generates the configurations and controls the REPOMO32. The FITkit board also controls the Vdd signal for the REPOMO32. In order to 'simulate' real-world changes in Vdd in a trustworthy way and to provide application-specific control of Vdd, it is necessary to generate not only a two-level Vdd (i.e. 3.3 V and 5 V), but also to switch Vdd between several voltage levels. This was achieved by a programmable power supply voltage.

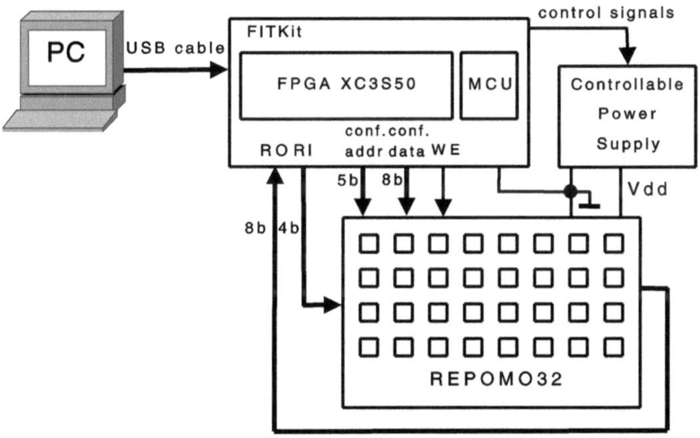

Fig. 5.10 Evaluation test bed for REPOMO32.

Intrinsic evolutionary design of polymorphic circuits can be performed using CGP implemented in software. This was done, and candidate circuits were then evaluated using the REPOMO32 and the fitness value was calculated in the FPGA. In order to evaluate a candidate solution, every chromosome was sent to the FPGA which reconfigured the REPOMO32 on the basis of the chromosome content. The FPGA also generated the input vectors for the REPOMO32, collected the responses from the REPOMO32, calculated the fitness value and reported the fitness value to the PC. In the current implementation, the evaluation of a single test vector requires 1.3 μs. A single candidate circuit can be evaluated in $2.2^4 \times 1.3$ μs (if both modes of the NAND/NOR gates are considered for all possible input combinations). This is sufficient time for the PC to prepare a new candidate circuit (configuration) and send it to the FPGA (which can then be re-sent to the REPOMO32 chip).

CGP operating with 4 x 8 programmable elements and a five-member population was used to evolve the three-bit parity/majority circuit. A mutation operator randomly modifies two genes of the chromosome. When perfect behaviour was achieved in both modes (i.e. parity for Vdd = 3.3 V and majority for Vdd = 5 V), the process of minimizing of the number of CLEs was initiated. On average, 875 generations were needed to find a working solution (calculated using 10 independent runs) [50]. Figure 5.11 shows two examples of evolved solutions. Their responses are shown in Fig. 5.12 for both Vdd levels. The parity/majority circuits shown in Fig. 5.11 are logically correct. However, it can be seen from Fig. 5.12 that circuit B does not exhibit as many glitches at the output as does circuit A, i.e. circuit B would be the better candidate for a practical use. Note that this observation could not be made only on the basis of a software fitness evaluation of the kind that is used in CGP. Hence the measurement on real hardware is important.

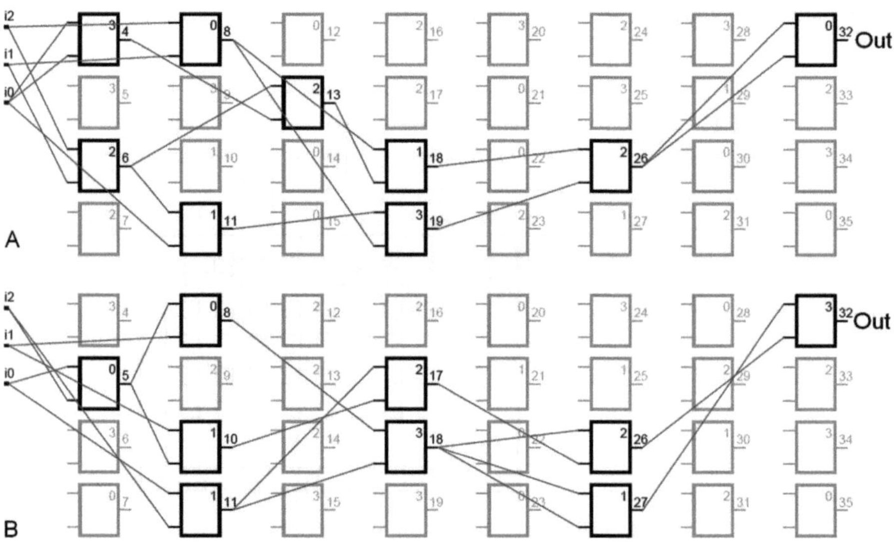

Fig. 5.11 Two evolved polymorphic three-bit parity/majority circuits. The function set is AND(0), OR(1), XOR(2) and NAND/NOR(3).

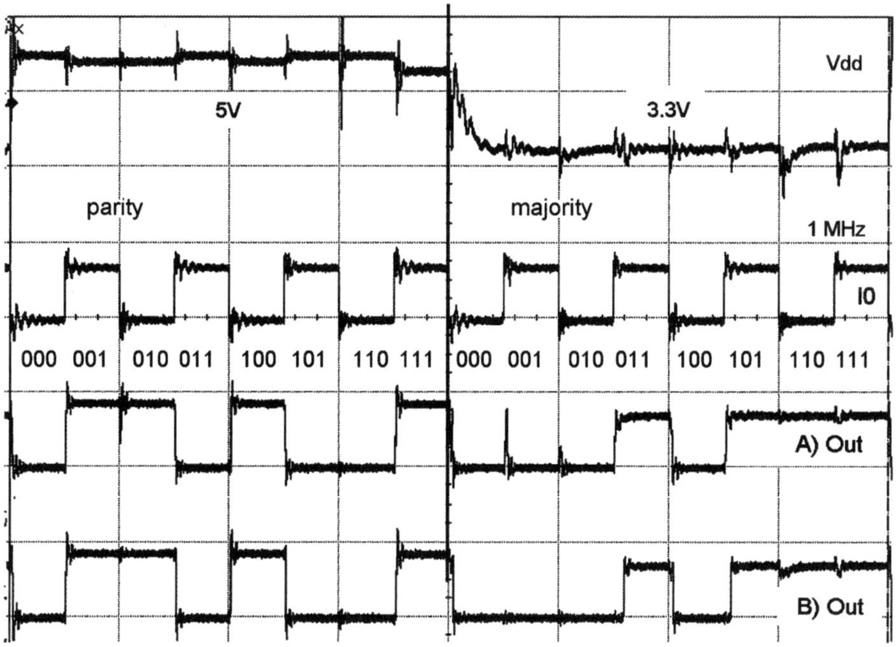

Fig. 5.12 Behaviour of two parity/majority circuits (at 1 MHz). Only the least significant input bit (I0) is shown.

5.5 Evolution of Multiple-Constant Multipliers

The experiments reported in Sect. 5.2 clearly demonstrated the scalability limits of gate-level evolutionary circuit design. Hence gate-level evolutionary design of, for example, 32-bit arithmetic circuits seems to be intractable. A question is whether one can escape, at least for some problems, from this conclusion. In order to evolve large circuits and simultaneously perform a perfect evaluation, we have to identify design problems for which the evaluation of a candidate solution requires either constant or linear time with respect to the number of inputs or components. In this section, we will show that for some specific cases, CGP can generate new implementations of complex arithmetic circuits. In those cases CGP operates at the functional level. The quality of candidate circuits cannot be evaluated for all possible assignments to the inputs, simply because there are too many of them.

5.5.1 Multiplierless Multiplication

A multiple-constant multiplier (MCM) is a digital-circuit which multiplies its single input x by N constants, and so generates N output products. As this circuit can be constructed from adders, subtractors and shifters, it is very useful for low-power implementations of finite-inpulse-response (FIR) filters, in which the input signal has to be multiplied by different but constant values [36]. Figure 5.13 shows one of the possible implementations of an FIR filter. Note that there are many publications dealing with the evolutionary design of multiplierless filters in which the fitness calculation is based on sampling the frequency characteristics and calculating the difference with respect to the required frequency characteristics [14, 30] and constant-coefficient multipliers (in which a symbolic verification algorithm is used [4]).

Finding the optimal solution of the MCM problem, i.e. the one with the smallest number of components (in particular, additions and subtractions), is known to be NP-complete. A very efficient heuristic algorithm for the MCM problem was recently published [72]. This is a graph-based algorithm which can handle problem sizes as large as 100 32-bit constants. This algorithm can be considered as the state-of-the-art method for the MCM design problem.

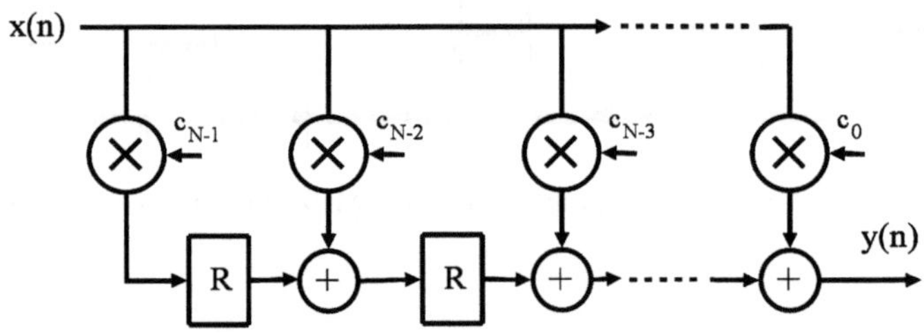

Fig. 5.13 An FIR filter implementation which utilizes a multiple-constant multiplier.

Although MCMs are relatively complex circuits composed of high-level components, their interesting feature from the point of view of evolutionary design is that a candidate MCM can be evaluated perfectly by applying only one input vector (e.g. $x = 1$), independently of the number of inputs and the bit width of the data path. The reason is that all components utilized are linear and thus MCMs implement linear transforms. This feature is unique because it enables a significant reduction of the fitness evaluation time. If a candidate MCM produces a correct result for a single input vector, then it will multiply correctly for any legal input value.

5.5.2 Results of CGP

CGP was applied to synthesize MCMs which generated N output values $c_1x, \dots, c_Nx$, where $c_1, \dots, c_N$ are given constants and x is the only input variable [69]. The fitness function was evaluated only for $x = 1$ and was defined as the sum of the absolute differences between the resulting and target values for each output. After perfect functionality, was achieved, the number of components was optimized. CGP operated with parameters $n_i = 1, n_o = N, n_a = 2$ and $n_f = 4$. The function set Γ included the addition, subtraction, shifts and the identity function (see Fig. 5.14). These functions and all connections were defined over b bits, where $b = 16$ in this case.

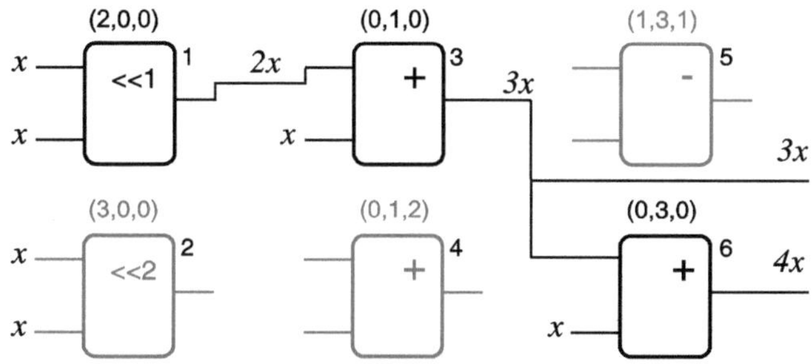

Fig. 5.14 Example of a candidate MCM. The CGP parameters are as follows: $n_i = 1$, $n_o = 2$, $l = 3$, $n_c = 3$, $n_r = 2$, Γ = {add (0), sub (1), 1b-shift (2), 2b-shift (3)}. The input x is encoded as 0.

Table 5.7 summarizes the results for several different instances of MCMs with 3, 5, 10 and 20 16-bit constant coefficients. All experiments were repeated 200 times with a population of eight individuals, and five genes mutated in the chromosome. Table 5.7 shows that the proposed evolution-based approach is able to generate multipliers that are competitive with the results obtained using the state-of-the-art heuristic approach [72]. We can see that CGP can reduce the total number of components and the delay of the MCMs.

Figure 5.15 compares one of the best evolved solutions with the solution provided by the heuristic approach [72] for three selected constants. This evolved solution contains two shifters fewer and exhibits a shorter delay than the solution provided by the heuristic approach. CGP is able to handle MCMs with tens of outputs. For the experiments reported in [69], the number of components was reduced by 7.3% and the delay by 8% with respect to the results of [72]. Note that 'M' indicates units of a million and when placed between digits denotes the decimal point. To carry out 10 million generations requires 230s. on a Celeron 2.4 GHz processor. This is quite a reasonable time when the average number of generations required to find a solution is 2.2 million.

5.6 CMOS-Level Circuit Evolution

Moore's law states that every two years, the number of transistors on an integrated circuit doubles [40]. This is due to the shrinking of devices through advances in technology. However, as these devices approach the atomic level, intrinsic variations are becoming more abundant, leading to a lower production yield and higher failure

Table 5.7 Results of evolutionary design of MCMs with various coefficients

Setting		Average results			Best result			
$n_c \times n_r$	max generation	Generations	#add/sub	Success rate	Delay	Add/sub	Shifts	Operations
3 constants: 2925, 23111, 13781								
Heuristics [72]					8	8	8	16
5×6	20M	1M62	14	68.5	5	9	8	17
6×6	20M	1M27	14	86.5	6	8	8	16
7×4	40M	2M15	13	99.0	7	8	6	14 (Fig. 5.15)
5 constants: 83, 221, 71, 387, 13								
Heuristics [72]					5	6	6	12
4×6	20M	461k	10	99.5	4	7	6	13
5×6	20M	207k	11	99.5	5	6	6	12
6×6	20M	114k	11	100.0	6	6	5	11
10 constants: 117, 1123, 743, 221, 1069, 7605, 987, 16689, 3033, 29								
Heuristics [72]					8	14	13	27
10×4	40M	4M8	23	99.0	7	15	12	27
7×6	20M	4M7	23	95.5	6	17	11	28
9×4	40M	9M5	22	91.0	9	17	9	26
20 constants: 1, 3, 5, 7, 11, 13, 17, 19, 23, 29, 31, 37, 41, 43, 47, 53, 59, 61, 67, 71								
Heuristics [72]					4	19	8	27
4×10	40M	457k	23	100	4	19	4	23
5×10	40M	347k	23	100	4	19	4	23
6×5	40M	772k	21	100	5	19	3	22

rates in conventional designs. This has been recognized by the ITRS[1] as one of the main challenges facing the semiconductor industry.

5.6.1 Intrinsic Parameter Fluctuations

The intrinsic variations are those caused by atomic-level differences between devices that can be considered identical in their layout, construction and environment. These differences occur in the doping profiles, in the thickness of various layers and at boundaries between different device areas. Further intrinsic variation also exists in the interconnections between devices and at any bonded junction within a circuit. The principle intrinsic variations are summarized below:

Random dopant fluctuations. The variations caused by the dopant atoms within the device. Whilst the implantation and annealing processes are tightly controlled, there is always a stochastic element to the number of atoms and their precise position within the silicon lattice. This uncertainty can lead to a substantial variation in the threshold voltage, subthreshold slope and drive current for the

[1] International Technology Roadmap for Semiconductors.

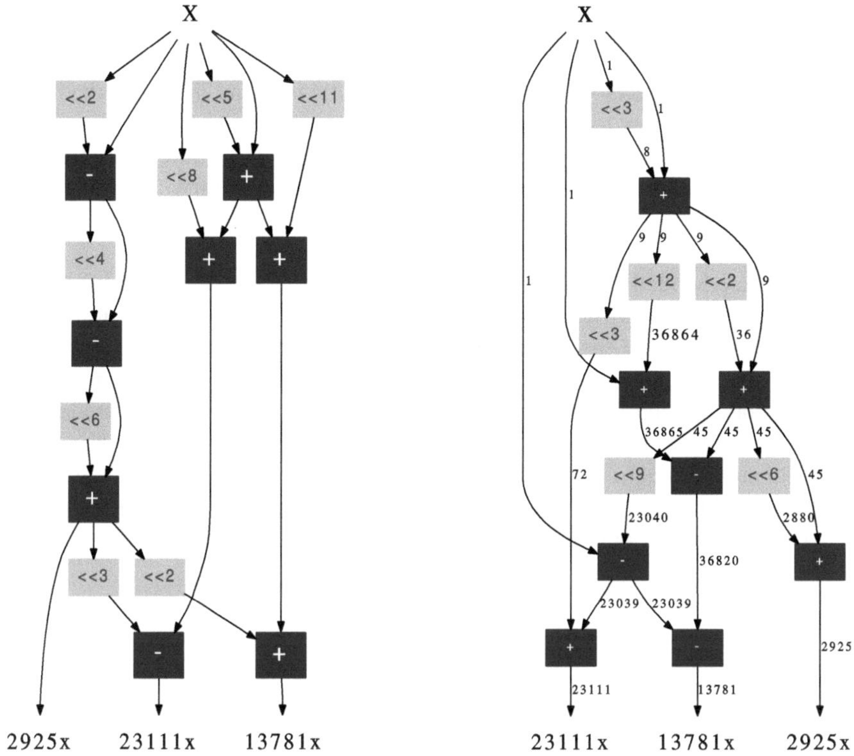

Fig. 5.15 MCM with three coefficients (2925, 23111, 13781): according to [72] (left), and the best evolved solution [69] (right).

device, with the doping near the surface and channel of the device having the largest effect on the V_T variation [5].

Line edge roughness. The horizontal deviation of a fabricated feature boundary from its ideal form, caused by imperfections in manufacturing during the mask, photoresist and etching processes and also by the discrete nature of the molecules used within the photo-resist layer. The resulting roughness on the edges of the blocks etched into the wafer leads to a stochastic variability from device to device [6].

Surface roughness. The vertical deviation of the actual surface compared with its ideal form. The shrinking of surface layers, particularly the oxide layer, causes variations in parasitic capacitances which can add to V_T variations [41].

Polysilicon grain boundaries. These are caused by the random arrangement of grains within the gate material. With the scaling down of the gate oxide thickness, implanted ions can penetrate through the polysilicon and insulator into the device channel, resulting in intrinsic variations on the lower surface of the polysilicon [13].

5.6.1.1 Modelling Intrinsic Variability

To reproduce the effects of intrinsic parameter fluctuations in future devices accurately, it is necessary to use statistical 3D simulation methods with fine-grained discretization [5, 6]. The simulations produce a series of I–V curves for a transistor, from which BSIM4 model parameters can be extracted. A library of BSIM4 models can then be used with NGSPICE to simulate the effect of intrinsic variation on a circuit.

The Device Modelling Group at the University of Glasgow have produced a library of 200 different NMOS and PMOS BSIM4 models based on a 35nm × 35nm Toshiba device. To use these models in a netlist the group have developed a tool named `Randomspice`, which creates multiple instances of a netlist and replaces the transistors in each instance of the netlist with randomly selected models from the library, ready to be evaluated by NGSPICE. The models can also be arranged in parallel subcircuits to allow different device widths (which must be multiples of 35nm, e.g. 140nm) to be simulated. The software also allows the creation of a single netlist which uses only uniform 35nm transistor models without any parameter fluctuations.

5.6.2 Modifying CGP for CMOS Design

This section describes part of a system called MOTIVATED[2] [74, 28, 75, 29, 27], which is capable of designing novel CMOS topologies using a modified form of CGP and the multi-objective algorithm NSGA-II. MOTIVATED uses variability-induced statistical models obtained from `Randomspice` to evaluate all of the evolved designs using NGSPICE simulations. However, owing to the computational overhead and the limitations on the resources available, only uniform models are used during the evolutionary process. The effect of intrinsic variability on the evolved designs is analysed post-evolution.

5.6.2.1 Representation

The representation used is a modification of Clegg et al.'s CGP representation [10]. The modified representation separates the directed-graph topology of the genotype from the transistor circuit topology of the phenotype,. Thereby allowing loops and multiple connections to occur within the phenotype, whilst maintaining the feed-forward nature of the CGP representation. This is achieved by encoding the function of each node (i.e. transistor) using three floating-point values. These three values encode the drain, gate and source connections of the transistor. Each of the CGP node's two inputs is also encoded using a single floating-point value. The CGP genotype

[2] Multi Objective Toolkit for Intrinsic Variation Aware Transistor-level Evolutionary Design.

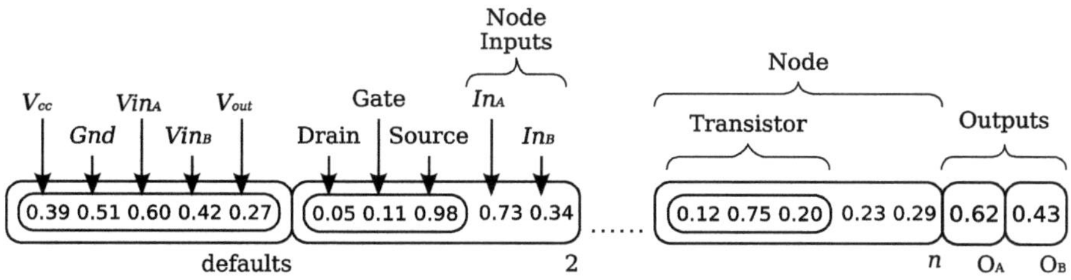

Fig. 5.16 The modified CGP representation. The 'defaults' node encodes the connections to the default terminals of a transistor circuit with two inputs and one output. The nodes o_A and o_B encode the CGP output connections. The remaining nodes, 2 to n, each consist of five genes: three genes to encode a transistor and two genes to encode the CGP node inputs. The node index label is underneath each node.

also has two output nodes, each of which is encoded using a single floating-point value. The purpose of each output node is to act as a root for a directed graph in the genotype. Therefore two graphs can be extracted from the genotype: one that encodes the PMOS transistors, and another that encodes the NMOS transistors. This allows any node in the genotype to represent both a PMOS and an NMOS transistor, thereby partly exploiting the complimentary nature of CMOS. Also, to ensure that the default terminals (V_{cc}, gnd, V_{in} and V_{out}) are present in the phenotype, a connection to each of them is also encoded. An example of the modified representation is given in Fig. 5.16.

The benefits of the representation described are that the smallest solution encoded in the genotype contains a single PMOS and a single NMOS transistor, which will always form a structure that resembles CMOS (i.e. an inverter). Also, the solutions encoded in the genotype always guarantee a fully connected design that includes all of the active components and does not include any dangling connections or unconnected terminals, thereby ensuring that the evolved designs should always be evaluated in NGSPICE. This is an important factor when one is designing a representation for evolving analogue designs, as there is a high probability of randomly generated designs failing to be evaluated in NGSPICE, especially in the initial population.

5.6.2.2 Decoding Process

In order to map the CGP genotype to the phenotype netlist so that it can be evaluated, a decoding procedure must be undertaken, which consists of three consecutive stages:

Genotype to active nodes. The first stage of the decoding process recursively extracts the active nodes from the genotype (similarly to decoding a standard CGP genotype). Starting independently from each program output, the directed graph

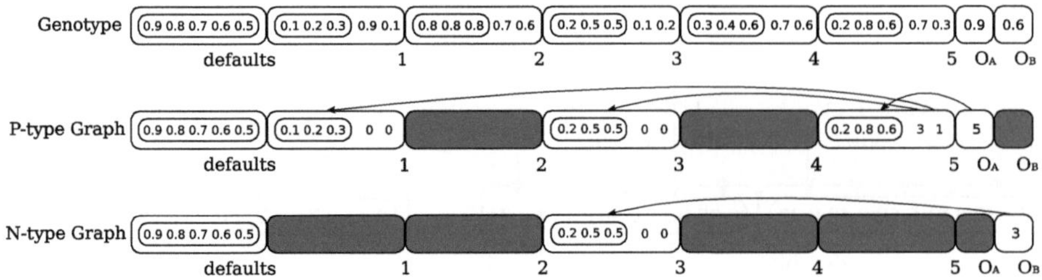

Fig. 5.17 The first stage of the decoding procedure.

is decoded using the following equation, where g_i is the CGP connection gene's value, and n_j is the index label assigned to each node and program input:

$$n_{next} = \lfloor g_i \times n_j \rfloor \quad \text{where } 0 \leq i \leq g_{total} \text{ and } 0 \leq j \leq n_{total}. \tag{5.3}$$

The decoded output determines which node in the genotype is visited next. The inputs of the node are then decoded in the same way as the output, and determines the node or program input to be visited next. The decoding process continues until the program inputs are reached. At this point, the directed graph formed consists of active nodes, and the first stage of the decoding process is complete. An example of the first stage of the decoding process is shown in Fig. 5.17.

Active nodes to transistor netlist. The second stage of the decoding procedure extracts the transistor connections from the function gene for the active nodes determined in stage 1, and the default terminals, which are also included in the genotype. In order to decode the drain, gate and source connections of each transistor, the possible connection options must first be calculated for each transistor terminal for both the PMOS and the NMOS transistors. Rather than allowing an unconstrained approach, where each transistor terminal may connect to any other transistor or default terminal, as used in [74], constraints are placed on the connectivity of each transistor and default terminal to ensure that the evolved designs follow design rules similar those used in conventional CMOS designs. This modification was necessary to ensure that the evolved design's operation was easier to understand and trust by designers, and was also easier to fabricate than an unconventional design, thereby making the evolved designs more likely to be feasible in industrial production. The constraints used for the transistor connections are shown in Table 5.8a. Similarly, a set of constraints is also required for decoding the default terminals to ensure that the existing CMOS-like design rules for the transistor connections are not broken. These constraints are shown in Table 5.8b.

To decode the floating-point values for each transistor terminal, each value is multiplied by the total number of constraints available for the terminal type, and the floor of the values is used. A similar process is used for decoding the default terminals. Once all transistors and default terminals are decoded, stage 2 of the decoding procedure is complete. An example of stage 2 of the decoding procedure is shown in Fig. 5.18.

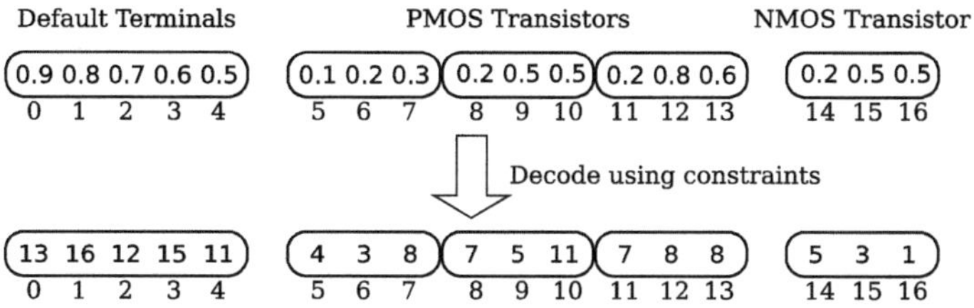

Fig. 5.18 Stage 2 of the decoding procedure.

Table 5.8 Decoding constraints for the transistor (a) and default (b) terminal connections

(a)

Terminal	PMOS	NMOS
Drain	V_{out}	V_{out}
	PMOS source	NMOS source
	NMOS drain	PMOS drain
Gate	Vin_A	Vin_A
	Vin_B	Vin_B
	PMOS drain	NMOS drain
Source	V_{cc}	*Gnd*
	PMOS drain	NMOS drain

(b)

Default	Transistor
V_{cc}	PMOS source
Gnd	NMOS source
Vin_A	PMOS gate, NMOS gate
Vin_B	PMOS gate, NMOS gate
V_{out}	PMOS drain, NMOS drain

Transistor netlist to a valid NGSPICE *netlist.* The third and final stage of the decoding procedure ensures that all of the default terminals are present in the final netlist and restructures the terminal connections so they are in terms of the default terminals or intermediate nodes. To ensure that the default terminals are present in the netlist, the decoded value for each default terminal obtained from stage 2 of the decoding procedure is used to indicate the position where the default terminal is substituted into the transistor connections. An example of this process is shown in Fig. 5.19a. After the substitutions have taken place, a relabelling algorithm is used to convert the bidirectional connection graph into a unidirectional connection graph, where all of the transistor connections are connected to either a default terminal or an intermediate node. An example is shown in Fig. 5.19b. At this point, the decoding procedure is complete, and all of the connections are in a valid format for SPICE and are inserted into a template netlist, ready for evaluation.

5.6.2.3 Multi-objective Evolutionary Strategy

From the results of previous work by Walker et al. [74] it became apparent that it was not possible to evolve novel CMOS topologies that were feasible in industrial production using a fitness function based on matching the voltage output of the target circuit. Even when constraints were put in place that penalized individuals which evolved designs reliant on feedback to the inputs, the CGP algorithm still did not have enough knowledge of the design space (including such things as power and delay) to evolve circuits showing some sign of a CMOS structure. The only option that would allow a number of different design criteria to be taken into account by the CGP algorithm when deciding the fitness of an individual would be to incorporate a multi-objective evolutionary strategy.

The method chosen was based on the popular NSGA-II algorithm [12]. NSGA-II combines a non-dominated sorting algorithm with a crowding distance operator, which ranks the individuals in the population into a number of Pareto fronts, based on each individual's scoring on a number of objectives. This process allows trade-offs between the objectives, so that individuals are preserved which do well on some objectives whilst performing worse on other objectives. For example, an individual which has low power but is slow and an individual which has high power but is fast would both be preserved. As evolution continues, the individuals in the population will converge to a single Pareto front, known as the *Pareto optimal*. Therefore, NSGA-II will produce a number of solutions along the Pareto optimal front for all objectives, which allows the user to choose the solution that best meets their requirements.

Traditionally in CGP, a parent and its offspring may have different genotypes, owing to mutation, which decode to identical phenotypes with the same fitness score. The $(1+4)$ evolutionary strategy used in CGP would promote the offspring over the parent in such a case, in order to promote neutral drift within the search space. In order to preserve this feature of CGP, the non-dominated sorting algorithm of NSGA-II was modified so that when a tie occurs in the fitness score between a parent and its offspring on all objectives, the offspring is classed as dominating the parent, and is therefore ranked higher than the parent. This allows evolution to continue searching for better solutions within the neutral space and may prevent early stagnation of the algorithm.

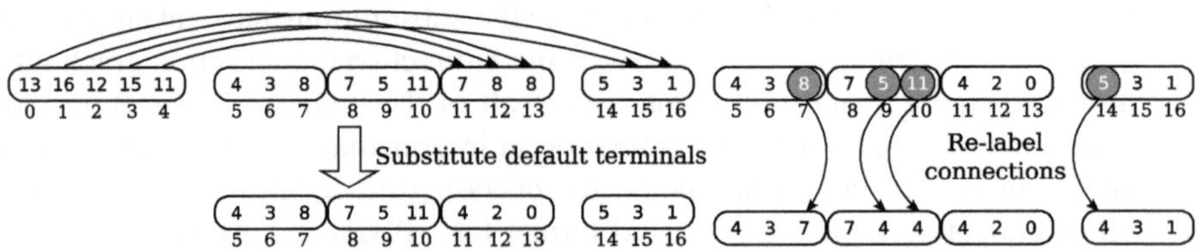

(a) Substitution of the default terminals. **(b)** Relabelling the connections.

Fig. 5.19 Stage 3 of the decoding procedure.

As the topologies are evolved from scratch, the first priority of the CGP algorithm is to find a functionally correct solution to the problem, which can then be optimized for all of the objectives. If all of the objectives are allowed to be optimized before a functionally correct solution is found, potential solutions will exist that perform well on certain objectives, such as power or delay, but will not function correctly. Therefore, priority is given to a single objective which performs a simple functionality test (discussed in Sect. 5.6.2.4). While CGP is trying to find a functionally correct solution, the evolutionary strategy used is similar to that of the $(\mu+\lambda)$ evolutionary strategy.

5.6.2.4 Objectives

In order to evolve topologies that would be feasible in industry and not be reliant on exploiting short circuits and feedback loops, a number of simple objectives were devised that would characterize the evolved topologies. Some of these objectives had aspects similar to the measures used by human designers for characterizing conventional designs, such as consideration of delay, power, and size. The nine objectives were:

Functionality test. The midpoint of each clock cycle in the evolved design was compared with the corresponding point in the target design to see whether the correct high or low threshold had been reached. For each cycle in which the correct threshold had not been reached by the evolved design, the functionality value was incremented. Therefore, an evolved design which is functionally correct received a score of 0.

Start-mid-cycle rail distance. Whenever an output state change (from low to high or vice versa) did not occur between two consecutive clock cycles, the distance was measured between the voltage output of the evolved design and the corresponding rail (V_{CC} or *gnd*) for all sample points between the start and the mid-point of the cycle. The objective was cumulative, so the final value was the summation of the distances in all cycles where a state change did not occur. This objective tried to maintain a constant output voltage by keeping it as close to the rails as possible whilst the inputs of the evolved design were switching.

Mid-end-cycle rail distance. For all cycles, the cumulative distance was measured between the voltage output of the evolved design and the rail (V_{CC} or *gnd*) corresponding to the target state (high or low), over all sample points between the mid-point and the end of each cycle. This objective minimized this distance and pulled the voltage output of the evolved design to the rails, whether it had switched state or not.

Threshold delay. The threshold delay objective was derived from the standard method of calculating delay, apart from one small change. Delay is normally calculated from the point where the input signal reaches 50% to the point where the voltage output of the test circuit reaches 50%. In the study described in this section,, the threshold delay was calculated from the start of the cycle (as the input signal went from low to high within one sample at the start of each cycle) until the point where the voltage output of the evolved circuit crossed the relevant threshold. It was calculated only for the cycles when a state change occured, and therefore it also encouraged the voltage output to be pulled to the threshold of the desired state, as a voltage output which did not reach the appropriate rail would have the worst delay for that cycle.

Supply voltage deviation. In order to help penalize short circuits and also minimize the voltage drop across the supply, the cumulative absolute distance of the supply voltage from the ideal supply voltage was calculated for all sample points during the SPICE simulation.

Input voltage deviation. This objective was similar to the supply voltage deviation objective, except that the cumulative absolute distance of the two input signals from the two ideal input signals was calculated. This helped to penalize feedback loops and circuits which tried to draw voltage from the inputs.

Supply current. In order to minimize the power drawn from the supply by the evolved design, the weighted cumulative absolute current from the supply was calculated across all sample points in the simulation. The objective was weighted so that the weight increased linearly from the start to the end of each cycle. This penalized evolved designs which drew a continuous current and encouraged evolved designs that drew current only at the switching points.

Input Current. Once again, this measure is calculated in a similar way to the supply current in the previous section, except that it calculates the weighted cumulative absolute current drawn from the two inputs. This penalizes evolved designs which draw heavily on the inputs for power.

Size. As the widths of the transistors were fixed, the size of the evolved design was calculated simply as the number of transistors which it contained. This objective allowed the designs to be minimized without sacrificing functionality.

5.6.3 Experiments

The constrained multi-objective CGP algorithm was used to evolve XOR and XNOR CMOS designs. The NGSPICE test bench used to evaluate the evolved designs, which contained the circuitry for the voltage, input and output stages is illustrated in Fig. 5.20a. The inverters in the input stage acted as a buffer, and helped penalize evolved circuits reliant on feedback loops to the input voltage pulses, whilst the potential divider in the supply stage penalized evolved designs that exploited shortcircuits between V_{cc} and ground. The inverter in the output stage ensured that a realistic load was applied to the test circuit. The voltmeters and ammeters captured data from both of the inputs and from V_{cc} and the output, which was then used in calculating the objectives discussed in Sect. 5.6.2.4. The two inputs to the test bench were created using a pair of synchronous-clock pulse sources. One source was held logic low for two cycles and then logic high for two cycles; the other source was held logic low for three cycles and then logic high for two cycles. An NGSPICE transient analysis was then performed on the circuit for a period of 21 clock cycles.

Table 5.9 The parameters for constrained multi-objective CGP (a) and NGSPICE (b)

(a) CGP

Parameter	Value
Number of parents	50
Number of offspring	2
Genotype length (genes)	250
Mutation rate (% of genes)	3

(b) NGSPICE

Parameter	Value
V_{cc} (V)	0.8
Clock (GHz)	2
Sample rate (per cycle)	250
PMOS width (nm)	350
NMOS width (nm)	175

The pulses chosen for the inputs ensured that every possible state transition was evaluated during the analysis, as illustrated in Fig. 5.20b.

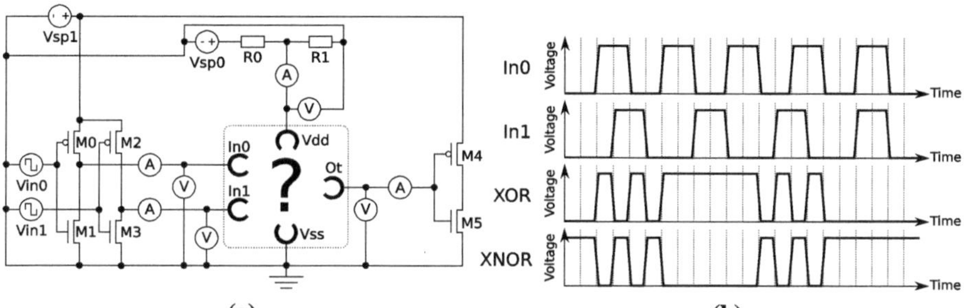

Fig. 5.20 The NGSPICE test bench (a) and the input and output pulses (b) for evolving CMOS XOR and XNOR designs.

The parameters used by the constrained multi-objective CGP are shown in Table 5.9. The population contained 150 individuals, 50 of which were selected as parents. Each parent generated two offspring to produce the new population. The mutation rate is given as a percentage of the genes per individual. The width of the PMOS and NMOS transistors remained fixed throughout the experiments.

5.6.3.1 Results

The CGP algorithm was run for 20,000 generations to evolve XOR and XNOR CMOS topologies. All experiments were conducted on a single PC with a 2.4 GHz quad-core CPU and 4 GB RAM, and took approximately 2 days each. The best evolved XOR and XNOR CMOS topologies are shown in Fig. 5.21.

The evolved designs are structurally very similar to conventional designs from a standard cell library (SCL) [43]. The evolved 10-transistor XOR design contains two more PMOS transistors and one fewer NMOS transistor than the SCL nine-transistor XOR design, and the evolved eight-transistor XNOR design contains one less PMOS transistor than the SCL nine-transistor XNOR design. This implies that the constraints used in the CGP decoding procedure produce evolved designs resembling CMOS rather than unconventional structures, and could therefore be feasible to manufacture industrially.

Examining the characteristics of the evolved and SCL designs also shows similarities. The evolved designs draw approximately 33% and 50% more current from the supply than do the SCL designs, but in both cases current is drawn only whilst the MOSFETs are switching. Also, neither the evolved nor the SCL designs draw large amounts of current from the inputs. Both of these features can be attributed to the use of the multi-objective evolutionary strategy, which is capable of monitoring the voltage and current drawn from both the supply and the inputs, as well as the functionality of the circuit. Therefore, it is possible to find evolved designs which produce the correct voltage output and also meet other requirements of the design.

By combining the constrained CGP decoding procedure with the multi-objective evolutionary strategy, most of the problems associated with the unconventional evolved designs in Walker et al.'s earlier work [74] have been rectified, and the designs now produced appear to be industrially feasible.

5.6.3.2 Variability Analysis of the Evolved Designs

In order to assess the variability tolerance of the evolved XOR and XNOR designs compared with the SCL XOR and XNOR designs, both were examined using variability-enhanced models instead of uniform models. Each design underwent 50 NGSPICE transient analysis simulations in order to statistically assess the effect

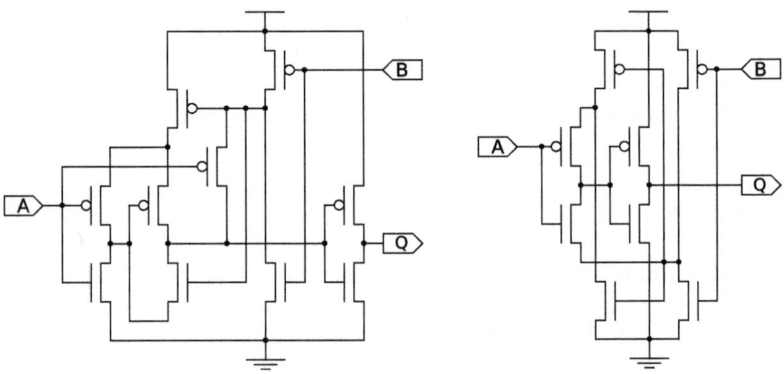

(a) (b)

Fig. 5.21 Evolved designs for a 10-transistor XOR circuit (a) and an eight-transistor XNOR circuit (b).

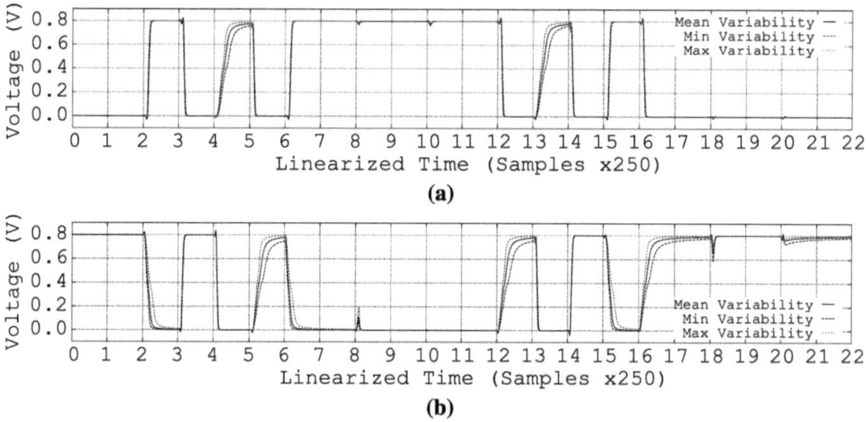

Fig. 5.22 Effect of variability on the voltage output of the evolved 10-transistor XOR (a) and the evolved eight-transistor XNOR (b).

of intrinsic variability. Figure 5.22 shows the results for the voltage output of the evolved designs.

Although all designs show some variability, which will affect the delay and power consumption, the SCL designs show approximately five to six times less variability than do the evolved designs. Further analysis of the evolved designs shows that during the transitions in the voltage output where variability occurs most, certain MOSFETs in the centre of the design experience body-biasing issues, which affects the threshold voltage of the MOSFET. As the body connections of all PMOS devices are connected to V_{cc} and all NMOS devices are connected to *Gnd*, body-biasing occurs when the source terminal of a PMOS device is driven to *Gnd* or the source terminal of an NMOS device is driven to V_{cc}. The point at which this occurs in the evolved XOR is when $V_{in}A$ is low and $V_{in}B$ is high. It also occurs in the evolved XNOR when $V_{in}A$ is high and $V_{in}B$ is low or when both inputs are low. However, the body-biased MOSFETs are still able to drive the gate terminals of the final MOSFETs in the circuit, which explains why the voltage outputs of the evolved designs are still similar to those of the SCL designs.

One potential solution is to connect the bodies of all MOSFETs directly to their respective source connections. This design methodology is often used when designing specialized CMOS amplifiers. The effect of variability on the evolved designs using this design methodology is compared against the standard evolved and SCL designs in Fig. 5.23. It can clearly be seen that connecting the body to the source of each MOSFET improves the worst-case delay and power consumption of both evolved designs, and also approximately halves the effect of variability. However, the SCL design still has a lower delay and power consumption, and is approximately two to three times less affected by variability. If the body and source connections of all MOSFETs were connected during evolution, better designs might be evolved. This will be investigated in future work.

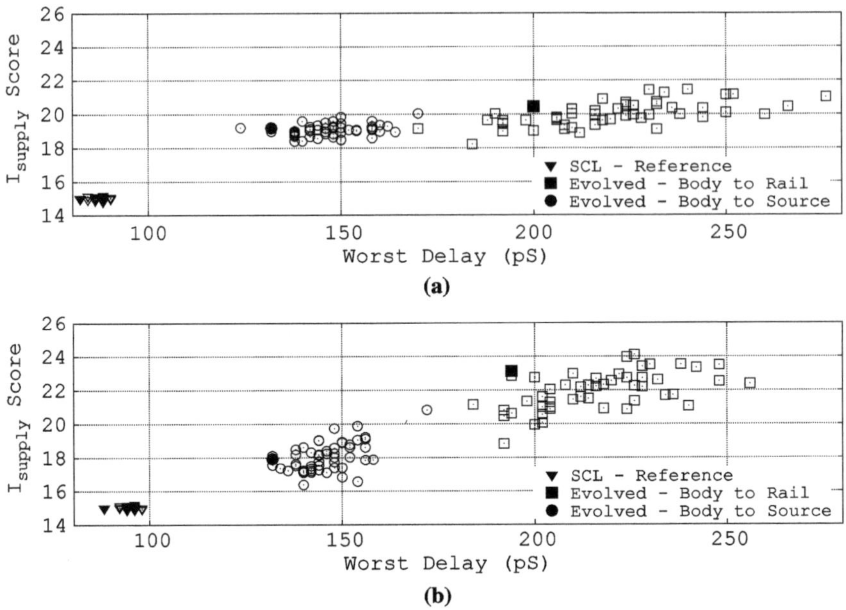

Fig. 5.23 Effect of variability on the worst-case delay and I_{supply} score of the XOR (a) and XNOR (b) for the SCL and evolved designs, and also the evolved design with body and source connected.

5.6.4 Conclusions and Future Work

This section has described a modified version of CGP with a multi-objective evolutionary strategy which is capable of evolving variability-tolerant CMOS designs that could be industrially feasible. The method has successfully evolved XOR and XNOR designs which were similar in structure and characteristics to conventional CMOS designs. The results showed a vast improvement on the previous results of Walker et al. [74], and clearly show the need to incorporate multiple aspects of the CMOS design space in order to evolve designs similar to conventional designs. However, when the variability of the evolved designs was analysed, they were found to be more susceptible to variability than conventional SCL designs. This was due to body-biasing issues somewhere in the centre of the evolved designs, which were not anticipated or picked up by the objectives used. Therefore, the multi-objective evolutionary strategy could not penalize individuals for such features. The evolved designs were shown to improve when the body and source terminals of each MOSFET were connected together in the design. This will be incorporated into the evolutionary process in future work and investigated further. Other potential solutions that will be investigated are adding objectives to the multi-objective evolutionary strategy to check for body-biasing in the individuals, and incorporating some form of current-flow analysis [58] within the evolutionary process. In order to evolve designs that are tolerant to variability, it appears that the variability analysis needs to be incorporated into the evolutionary process. However, owing to the extra computational time required, this is only feasible using a high-performance computing cluster or grid service.

5.7 Evolution of Classification Hardware Using a Modular Approach

Many classification tasks, especially in embedded and real-time settings, require not only sufficient classification accuracy but also high performance, high energy efficiency and low resource usage. These objectives can be reached by a direct hardware implementation. Evolvable hardware approaches (EHW) create direct hardware implementations and have been applied successfully to classification tasks, such as the recognition of characters [25], road signs [65], faces [21] and electromyographic signals [31, 18, 9], and the diagnosis of the Parkinson's disease [56] (see also Chap. 11) .

Here, we present an EHW approach based on the EGCP hardware representation model to design classifiers for electromyographic (EMG) signals. EMG signal classification is a prerequisite for the control of prosthetic hands [9, 20]. We discuss our classifier architecture, the EMG signal domain, the hardware representation model, the evolutionary optimization technique and, finally, our experimental results. The results are promising and show that EHW-based EMG signal classifiers reach a classification accuracy close to that of state-of-the-art classification techniques. Furthermore, the EHW approach lends itself to online adaptability and fault recovery, making the technology a promising candidate for future self-adaptive classifiers.

5.7.1 Classifier Architecture

The hardware architecture is probably the most important design issue for an EHW classifier. Figure 5.24 shows the structure of our classifier architecture. The basic element is the category classifier (CC), implementing an evolved pattern-matching rule. A category detection module (CDM) groups several category classifiers together and counts the number of satisfied rules. A global-maximum detector determines the category with the most hits, which is the classification result. In the case of a draw, the category with the lowest numerical index is selected.

Rather than directly evolving a huge ECGP model, all successful EHW classifiers impose a top-level structure such as the one shown in Fig. 5.24. An earlier approach used a structure that was similar to a programmable-logic-array and consisted of AND gates followed by OR gates [25]. The Increased Complexity Evolution (ICE) architecture [64] features a modular top-level structure that splits into several category detection subsystems. A similarly modular approach is the Functional Unit Row (FUR) architecture [19].

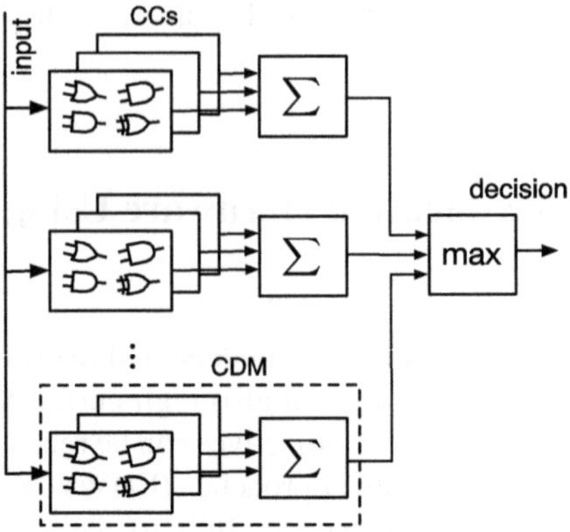

Fig. 5.24 EHW classifier architecture. Each category detection module (CDM) contains a set of category classifiers (CCs) that implement evolved pattern-matching rules. The CDM with the highest number of satisfied rules defines the global decision.

Our classifier architecture can be seen as a generalized version of the previous modular architectures. All of these modular architectures aim at three goals. First, the modularization of the classifier into CDMs and CCs reduces the complexity of the evolved circuits, which in turn leads to faster convergence of the optimization process. Second, the architecture scales with the number of categories by adding more CDMs. Third, classification accuracy can be increased by using multiple CCs and taking a majority vote decision at the end.

5.7.2 EMG Signal Domain

For EMG data acquisition, we used a measurement system comprising four components: EMG electrodes (Tyco Arbo*, Ag/AgCl, 35 mm), amplifiers (Biovison [8]), A/D converters (National Instruments [42]) and a standard computer. Our system continuously monitored four sensor channels with 14 bit resolution at a sampling rate of 6 kHz. Two important requirements for such a measurement system are reduction of noise in the analogue signal domain and a reproducible biomechanical experimental set-up.

To reduce noise, we employed an optical bridge (Sonowin [57]) to galvanically decouple the signal amplifiers and the A/D converters from the computer that accumulated the data. A separate battery provided a stable power supply to the amplifiers and A/D converters. The amplifiers were placed as near as 10 cm to the skin-attached electrodes in order to minimize parasitic inductance.

We placed four electrode pairs on the top, bottom, medial and lateral sides of the forearm as shown in Fig. 5.25, with a reference at the wrist. The exact electrode

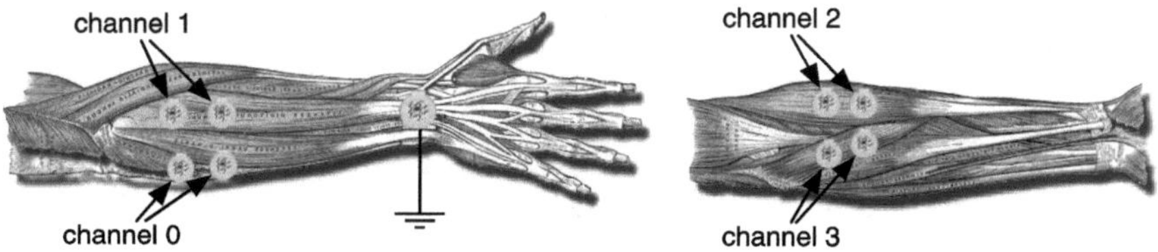

Fig. 5.25 Sensor placement (muscle anatomy taken from [22]).

positions were determined specifically for each test subject to obtain strong signals. After this initial calibration, the electrode positions were marked so that we were able to re-establish the experimental set-up on different days.

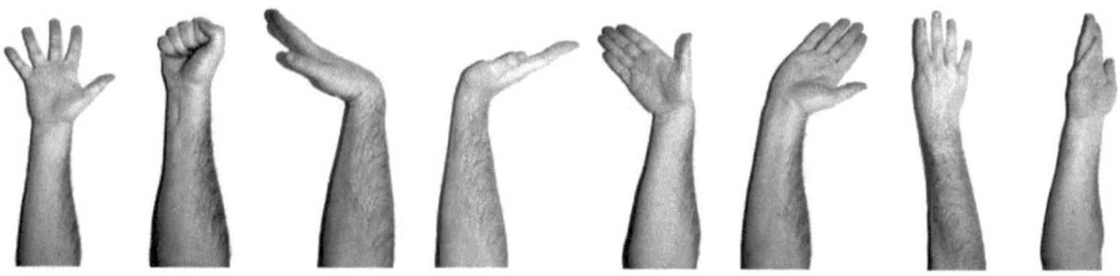

Fig. 5.26 Movements: (a) open, (b) close, (c) flexion, (d) extension, (e) ulnar deviation, (f) radial deviation, (g) pronation and (h) supination.

In a single experimental run, the test subject had to perform 20 iterations of a sequence of eight different movements. These movements were open, close, flexion, extension, ulnar deviation, radial deviation, pronation, and supination, and are depicted in Fig. 5.26. A single movement consisted of two phases: a 9s relaxation and an 11s contraction part. The EMG signal for the contraction part was divided into a 3s phase at the onset of the contraction containing the transient components of the EMG signal, and an 8s steady-state phase which corresponded to a constant-force contraction. A part of this steady-state phase was used for classification. Figure 5.27 presents an example of a complete EMG signal.

Signal preprocessing and feature extraction was done completely in the digital domain. Following the approach presented by Kajitani et al. [31], we extracted the features in four steps:

1. For every channel $k, k = 1, \ldots, 4$, and movement $p, p = 1, \ldots, 8$, we calculated the sensor DC offset o_{kp} as the mean value of all signal samples between the third and the fifth second of the relaxation phase of the signal.
2. The steady-state signal d_{ikp} was compensated for DC offset and smoothed by a root mean square (RMS) method with a window size $w_s = 600$. The first 1.9s (11,400 samples at 6 kHz) of the compensated smoothed signal d'_{jkp} were calculated from

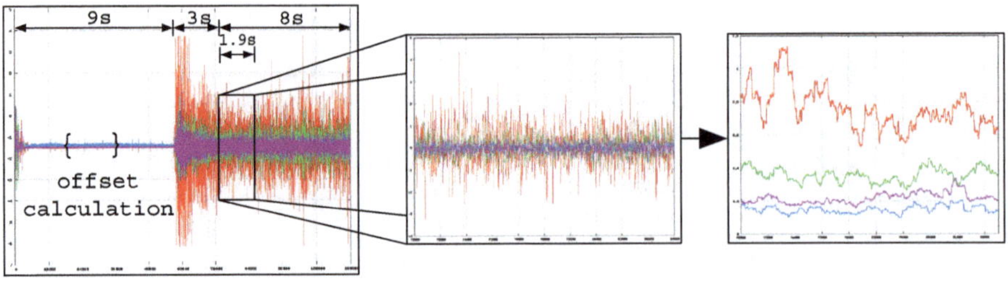

Fig. 5.27 EMG signal preprocessing. The left figure shows the raw signals for all four channels, consisting of a 9s relaxation phase, a 3s transient phase with intensified activity and an 8s steady state-contraction phase. The centre figure shows the DC-offset-compensated first 1.9s of the steady-state phase, and the right figure the RMS smoothed signals from which the features were extracted.

$$d'_{jkp} = \left[\frac{1}{w_s} \sum_{i=j}^{j+w_s-1} (d_{ikp} - o_{kp})^2 \right]^{\frac{1}{2}},$$

where $j = 1, \ldots, 11{,}400$.

3. We then applied a logarithm-transformed moving average to the compensated, smoothed signal, using a window size of $w_f = 6{,}000$ samples and a shift of $s_f = 600$ samples. The non-normalized feature thus consisted of 10 values and was defined as

$$f_{l_m kp} = -\log \left(\frac{1}{w_f} \sum_{j=l_m}^{l_m+w_f-1} d'_{jkp} \right),$$

where $l_m = 1 + (m-1) \cdot s_f$ and $m = 1, \ldots, 10$.

4. Finally, we normalize the features for each channel separately:

$$g_{lkp} = \frac{f_{lkp} - \min_{l,p}(f_{lkp})}{\max_{l,p}(f_{lkp}) - \min_{l,p}(f_{lkp})}.$$

Taking all $k = 4$ channels into account, the feature vector for a single movement consisted of 10×4 values. These 40 values were previously linearly quantized by a 1-out-of-16 encoder and fed into the classifiers. This configured the number of inputs (n_i) to the category classifiers of the CDM to be 640 bits.

5.7.3 Classifier Hardware Representation Model

We encoded the category classifiers of the CDMs by ECGP chromosomes. With six category classifiers per CDM and eight categories, the complete classifier architecture was configured by a multi-chromosome genotype with 48 ECGP chromosomes. Each chromosome comprised one row ($n_r = 1$) of 50 to 250 ($n_c = 50, \ldots, 250$) functional units without restriction on the length of the feed-forward wires ($l = n_c$). The functional units were either look-up tables (LUTs) with four inputs and one output ($n_i = 4, n_o = 1$) or modules with up to ten LUTs. We did not restrict the function set for the LUTs. We applied the techniques for automatic module creation and reuse (with ADFs) described in Chap. 3. However, finding category-specific patterns in input vectors modulated by a large amount of randomness results in circuits with less structure than evolved circuits for arithmetic benchmarks. Based on this observation, we focused solely on age-based module creation [34]. Table 5.10 lists the parameters chosen for the ECGP model, as described in Sect. 3.6.1.2.

Table 5.10 Parameters for the evolution of the ECGP EHW classifier

$n_i/n_o/n_r/n_c$	640/1/1/50–250
n_a/n_f	4 / $\mathbb{B}^4$
#fitness evaluations per generation	4
Mutation probability	1.0
Mutation rate	0.03
One point mutation probability	0.6
Compress/expand probability	0.1/0.2
Module point mutation probability	0.04
Add/remove module input probability	0.01/0.02
Add/remove module output probability	0.01/0.02
Maximum module size	10

5.7.4 Fitness Assignment and Evolutionary Algorithm

There are several ways to evolve the CDMs and CCs in the proposed classification architecture. Related work has used incremental and two-step evolution to partition the task of evolutionary optimization into smaller subtasks. In [21], single CDMs were evolved incrementally and independently, before being assembled into a complete classifier. Two-step evolution was presented in [64]. In the first step, the CMDs were evolved independently. In the second step the CDMs were combined and the CDM selectors of the overall architecture were evolved, considering the fitness of the complete system.

We used direct evolution of the complete classifier. Through a series of experiments, we have observed that partitioning the evolutionary optimization process leads to classifiers with acceptable classification performance rather quickly, but letting the evolutionary algorithm exploit the complete search space delivers the highest classification performance.

For classifier evolution, we defined the fitness as the reciprocal of the squared classification error. With categories C_p ($p = 1,\ldots,P$) and a training vector set X ($X = \dot{\cup}_{i=1}^{P} X_i$), where X_i is the set of training vectors for category i, the fitness of a classifier c was defined as

$$f(c) = \left[1 + \frac{1}{|X|} \sum_{i=1}^{P} \left[\sum_{x \in X_i} |i - c(x)| \right]^2 \right]^{-1}.$$

We employed a $(1+4)$ evolutionary strategy to evolve the classifier (see Sect. 2.6.3). This evolutionary scheme creates four offsprings by mutation and selects the fittest child to become the parent for the next generation. If no child has a fitness superior to the parent's fitness, the parent proceeds to the next generation.

5.7.5 Experiments and Results

We compared the performance of our classification architecture with the following popular state-of-the-art classification algorithms: k-th nearest neighbour (kNN), decision trees (DT), artificial neural networks (ANN), and support vector machines (SVM). We used k-fold cross-validation to determine the classification rates for all classifiers. k-fold cross-validation segments an overall data set into k subsets of approximately equal size. In k runs, one subset is used for testing, and the others are used for the training of the classifiers.

We report on two experiments here. The objective of the first experiment (*Day 1–3*) was the investigation of the asymptotic classification performance. The collected EMG data, consisting overall of 60 repetitions of eight movements recorded on three consecutive days, was evaluated using the leave-one-out validation scheme, which set k to the number of feature vectors. The second experiment (*2 of 3*) was defined from the point of view of applications: an amputee would rely on data from past days to train a prosthesis for use on the next day. Thus, we used threefold cross-validation with data sets defined by the three recording days.

To compare the classification performance of the different approaches, we used the classification accuracy expressed by the error rate. Table 5.11a presents the training error rates, which show the classifiers' approximation abilities, and Table 5.11b presents the test error rates, showing the classifiers' generalization abilities. Table 5.12 presents the test error rates for the individual movements. In both tables, EHW denotes our classification, architecture and the best-performing classifier is marked in bold.

Since the EHW classifier was evolved from random genomes, each classifier implemented a different combinational function and the classification rates vary slightly. For the *Day 1–3* experiment, the leave-one-out technique required us to evolve a rather high number of classifiers, which averaged out the differences in the initial genomes. The *2 of 3* experiment, however, generated only three classifiers. To achieve sound error rates, we evolved 10×3 classifiers and averaged the results. The training error for EHW was 5% in all experiments, as we used this threshold as the termination criterion.

The comparison of experiments *Day 1–3* and *2 of 3* in Table 5.11 shows that almost all algorithms achieved better training but worse test results for the *2 of 3*

Table 5.11 The training errors (approximation) are summarized in (a), and the test errors (generalization) are summarized in (b)

(a)

	Day 1–3	*2 of 3*
*k*NN	3.80 %	3.74 %
DT	2.68 %	2.69 %
ANN	**0.22%**	**0.11%**
SVM	3.58 %	3.03 %
EHW	5.00 %	5.00 %

(b)

	Day 1–3	*2 of 3*
*k*NN	4.25 %	**5.61 %**
DT	8.29 %	13.26 %
ANN	**2.70 %**	6.02 %
SVM	3.80 %	6.51 %
EHW	9.00 %	10.6 %

Table 5.12 2 of 3: individual movement errors (generalization)

	Close	Extension	Flexion	Open	Pronation	Radial deviation	Supination	Ulnar deviation
KNN	**3.92%**	**0.0%**	**0.0%**	5.17%	6.78%	19.23%	**3.57%**	6.78%
DT	5.88%	6.78%	5.66%	13.79%	16.95%	17.31%	39.29%	15.25%
ANN	9.80%	1.69%	**0.0%**	8.62%	**5.08%**	15.38%	5.36%	**3.39%**
SVM	5.88%	**0.0%**	**0.0%**	8.62%	8.47%	19.23%	**3.57%**	6.78%
EHW	5.90%	12.20%	**0.0%**	**3.50%**	12.20%	**12.80%**	22.20%	5.80%

experiment. This is due to the fact that for the *2 of 3* experiment the training set was much smaller allowing tighter approximation, while the verification set was larger than in the *Day 1–3* experiment.

The best-performing algorithms in both experiments were *k*NN, ANN and SVM with only marginal performance differences, followed by EHW and DT. Interestingly, the good performance of the simple *k*NN technique points to the fact that our EMG signal classification problem is not too hard. Table 5.12 shows, however, that the individual movements were not consistently classified best by *k*NN, ANN and SVM. For the 'flexion', 'open' and 'radial deviation' movements, the EHW classifier was on par with or even outperformed *k*NN, ANN and SVM.

The main result of our EMG signal classification experiments is that for prosthesis control, EHW classifiers achieve accuracies sufficiently close to that of state-of-the-art classification algorithms [24]. Depending on the specific set of movements to be classified other algorithms might deliver slightly increased accuracy. The appeal of EHW classifiers, however, is rooted in their compactness, fast computation and suitability for self-adaptation.

5.8 EvoCaches: Application-Specific Adaptation of Cache Mappings

In this section we present the EvoCache ('evolvable cache'), a novel approach for implementing application-specific caches. The key innovation of the EvoCache is that it makes the function that maps memory addresses from the CPU address space to the cache indices programmable. It supports arbitrary Boolean mapping functions that are implemented in a small reconfigurable logic fabric. To find suitable cache mapping functions it relies on Cartesian genetic programming for circuit representation and evolutionary strategies for fast optimization. We have evaluated the use of EvoCaches in an embedded processor for two specific applications (JPEG and BZIP2 compression) with respect to execution time, cache miss rate and energy consumption. We show not only that the evolvable-hardware approach to optimizing cache functions significantly improves cache performance for the training data used during optimization, but also that the evolved mapping functions generalize very well. Compared with a conventional cache architecture, an EvoCache applied to test data achieved a reduction in execution time of up to 14.31% for JPEG (10.98% for BZIP2), and a reduction in energy consumption by 16.43% for JPEG (10.70% for BZIP2). We also discuss the integration of EvoCaches into an operating system and show that the area and delay overheads introduced by EvoCaches are acceptable.

5.8.1 The EvoCache Concept

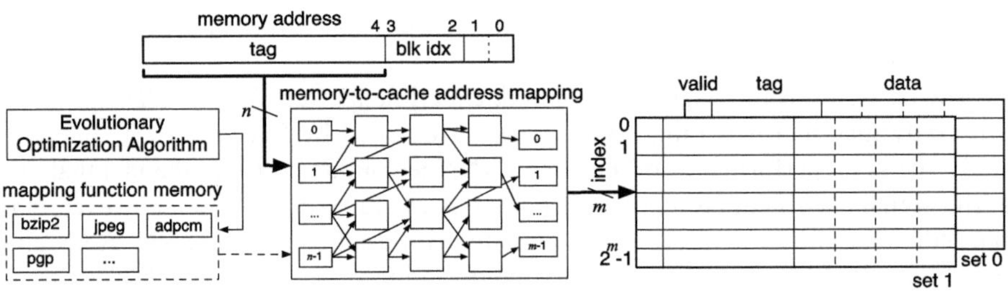

Fig. 5.28 The EvoCache architecture provides a configurable mapping from CPU memory addresses to cache indices. The optimization process reconfigures node functions and the wiring between the nodes. The nodes represent Boolean functions with n_a inputs. The figure shows an example of a two-way set-associative cache [35].

The key idea of the EvoCache approach is presented in Fig. 5.28. A very small reconfigurable logic fabric implements a hashing function that maps part of a memory address to a cache line index. The hashing function is optimized to achieve a low overall execution time for a specific application. The algorithmic methods for optimization originate in the evolvable hardware domain, which aims at automated circuit design and optimization by combining evolutionary algorithms and reconfigurable hardware technology.

Our architecture provides a mapping-function memory that can store several configurations for the reconfigurable logic fabric, which allows rapid switching to different memory-to-cache address mappings. To prevent aliasing, i.e. storing several potentially dirty copies of the same physical address at different indices in the cache, the cache is flushed when a new mapping is activated.

The EvoCache approach is orthogonal to other work aimed at selecting and/or reconfiguring the cache organization in an application-specific way (e.g. [2, 45, 81]). Although Fig. 5.28 displays an address mapping for a byte-addressable architecture to a two-way associative cache with a block size of four words, the EvoCache principle is applicable to all possible configurations and levels of caches. Compared with classical modulo mappings or mappings based on bit permutations [59] and XOR functions [67], EvoCaches utilize more complex, evolved hashing functions, allowing them to reduce an application's overall execution time and energy requirement, as we will show in the remainder of this section.

Including EvoCaches into a processor architecture will also increase the logic area, the hit time and the overall number of memory cells for the cache. The increase in logic area is due to the reconfigurable fabric itself, which is assumed to be small, as the fabric comprises only a handful of look-up tables. Additionally, we require a mapping-function memory to store the configurations for the logic fabric. The size of a configuration is architecture-dependent. The architecture used for the case study described in this section had a configuration size of 151 bytes. The increase in the cache size is due to the fact that the flexibility in the hashing function requires us to store the full address, excluding block and byte offsets, as tags in the cache. The additional overhead incurred depends on the actual cache configuration. For example, a conventional four-way set-associative cache holding 16 kilobytes of data with a block size of two words for a byte-addressable architecture with 32-bit addresses comes with an overhead of 25.56%, where the overhead includes the valid bit and the tag for each cache block. Switching to an EvoCache of same data size and organization increases the overhead to 34.88%. We think that this overhead is bearable, since most processor designs today are not restricted by silicon area but by performance and performance per unit energy. The increase in hit time is more critical. The additional delay depends strongly on the depth of the LUT network. This depth can be restricted in the optimization process to satisfy timing constraints. Moreover, for many embedded processors with clock frequencies well below 1 GHz, the pressure on the timing is moderate. High-performance processors, on the other hand, have several levels of cache, where only the first level is optimized for hit time. Here, the EvoCache approach can still be applied to higher-level caches.

Integrating EvoCaches with a standard operating-system environment requires only a few modifications. To keep the information about the cache mapping as close as possible to the application's binary, we chose to store it as an optional section in the binary itself. Since all commonly used binary formats, such as COFF, ELF and MachO, support storing multiple code and data areas (sections) in the binary, this feature can be easily added without requiring a new binary format. The cache-mapping information can be added by a standard linker. Since this information is small (typically a few hundred bits), the size of the binary is only slightly increased.

To activate the cache mapping when an application is started, the application loader needs to be extended. After loading the application's text and data sections, the loader configures the mapping-function memory according to the information stored in the binary. The operating system also stores the cache mapping as part of the context of a process. In the case of multi-tasking operating systems, the operating system changes the cache mapping at every context switch to a user task.

The proposed change in the binary format integrates support for EvoCaches in a backward-compatible way. First, the additional section containing the cache mapping will be ignored when the application is executed on a system without EvoCaches. Second, systems with EvoCaches can still execute standard binaries. If the loader detects that no cache-mapping information is present, it will initialize the classical modulo cache mapping.

In the following subsections, we describe the determination of a suitable cache mapping function for an application and a specific input data set with an evolutionary optimizer, and then evaluate the performance on different input data sets to verify the generalization capability of EvoCaches.

5.8.2 System Simulation and Metrics

This subsection describes the configuration of the hardware representation model, the evolutionary optimization algorithm, and the method used to evaluate the fitness of candidate circuits.

We configured the CGP model to use LUTs with four inputs ($n_a = 4$) as node functions. The functional set f for the nodes was not constrained, i.e. $f = \mathbb{B}^{16}$. To reduce the search space and thus increase the efficacy of the evolutionary optimizer, we configured the CGP model to have only one row ($n_r = 1$) but $n_c = 32$ columns. The levels-back parameter was set to $l = 31$. The circuit's inputs were fed from $n_i = 27$ primary inputs taken from the memory address. The $n_o = 15$ bit outputs of the circuit encoded the cache line index. The circuit depth is an important parameter for EvoCaches, as it is proportional to the delay of the resulting hashing function, which adds to the cache hit time. While constraining the circuit depth during optimization can be easily done, the experiments were conducted with an unconstrained circuit depth. Instead, in Sect. 5.8.3 we report on the depths and sizes of the evolved circuits.

We used a $1+4$ ES as the optimization scheme, where in every generation one parent creates four children through mutation. One of the fittest children proceeds to the next generation. The parent is promoted to the next generation if it excels over all children. The mutation operator modifies a single gene during child creation, i.e. the function of a single logic node or the wiring of one of its inputs is affected.

5.8.2.1 Fitness Evaluation and SimpleScalar Integration

For the experiments, we leveraged our MOVES EHW toolbox [33], which comprises various hardware representation models and evolutionary optimizers. Additionally, the toolbox can generate a set of jobs for fitness evaluation and distribute them on a computer cluster.

The set-up of the tool is presented in Fig. 5.29. The MOVES toolbox includes the CGP model and the ES. Whenever a new candidate circuit is generated, it is passed to the SimpleScalar [7] processor simulator for fitness evaluation. SimpleScalar reads the description of the circuit and simulates the execution of a specified benchmark and specified input data on a processor with a given cache configuration in a cycle-accurate manner. We chose SimpleScalar for system simulation as it is easily extensible and it models a variant of the widely used MIPS instruction set architecture. Two modifications to the original SimpleScalar tool were necessary. First, its command line interface was extended to include the activation and specification of up to four mapping functions. These circuit specifications are read in and stored in a data structure. Second, for the actual mapping between addresses and the cache line index, SimpleScalar needs to determine the logic result for the mapping function. To this end, the circuit evaluation routine already available in the MOVES toolbox was extracted into a library (moves.lib) and linked with SimpleScalar. In each simulation run, SimpleScalar determined an application's overall run time and fed it back as a fitness value into the evolutionary optimizer.

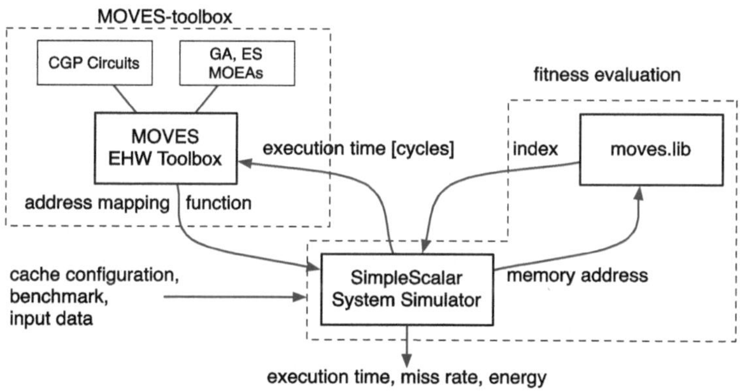

Fig. 5.29 EvoCache tool set-up: SimpleScalar is invoked by the MOVES toolbox and returns the overall execution time in clock cycles as a fitness measure.

5.8.2.2 Miss Rate and Energy

Besides a cycle-accurate run time, SimpleScalar determined the miss rates for the various levels of cache. Our interest in the miss rates was motivated by the fact that related work used miss rates to measure the fitness of a specific cache configuration. However, for more sophisticated processor architectures, metrics based solely on miss rates might be less conclusive than execution time. The downside of using the cycle-accurate execution time as the main metric is the long simulation time. We constrained the simulation time to three to five minutes for a single fitness evaluation, which resulted in an overall runtime of roughly one week for a single complete ES run. These constraints on the simulation time resulted in limiting the input data size for the benchmarked applications to some 100 kilobytes which posed sufficient pressure on the cache architecture of an embedded processor as modelled in our work. However, a modern general-purpose processor's cache architecture would not be stressed sufficiently, and thus would require the simulation of application runs on larger data sizes.

As an estimate of the energy we used a variant of the energy model presented in [81] which splits the energy demand into a static and a dynamic part. We modelled an embedded processor with up to two levels of cache and an external memory. For each of the caches, i.e. split level-one caches L1:I and L1:D and a unified level-two cache L2:U, and for the external memory, the static or stand-by energy per cycle was given by $E_{L1:I,s}, E_{L1:D,s}, E_{L2:U,s}$ and $E_{M,s}$. With c as the number of clock cycles required for program execution, the static energy was

$$E_{static} = (E_{L1:I,s} + E_{L1:D,s} + E_{L2:U,s} + E_{M,s}) \cdot c.$$

The dynamic energies per access were given by $E_{L1:I,d}, E_{L1:D,d}, E_{L2:U,d}$ and $E_{M,d}$ and the numbers of accesses by $a_{L1:I}, a_{L1:D}, a_{L2:U}$ and a_M. Thus, the dynamic energy results was given by

$$E_{dynamic} = E_{L1:I,d} \cdot a_{L1:I} + E_{L1:D,d} \cdot a_{L1:D} + E_{L2:U,d} \cdot a_{L2:U} + E_{M,d} \cdot a_M.$$

The actual values in nanojoules for the static energy per cycle and dynamic energy per access were derived from the CACTI cache model [55] for the 90 nm technology node. For the external memory, these values were derived from the datasheet for a standard V58C2256 DDR SDRAM module. The overall number of clock cycles and the number of accesses were determined by the SimpleScalar simulator. Finally, the CPU energy E_{cpu} was computed by assuming a CPU with an average power consumption of 0.45 mW per MHz at a clock frequency of 200 MHz implemented in 90 nm technology [78]. The overall energy for an application run thus added up to

$$E = E_{cpu} + E_{static} + E_{dynamic}.$$

5.8.3 Experiments and Results

To evaluate the EvoCache concept, we configured a processor and its memory hierarchy in a configuration similar to that of current ARM processors [78]. The configuration is shown in Fig. 5.30 and includes a split first-level cache and a unified second-level cache. The L1 caches were two-way associative with a hit latency of one cycle, 64 sets and a block size of 16 bytes. The L2 cache had an associativity of four ways with a hit latency of six cycles, 128 sets and a block size of 32 bytes. The memory bus between the L2 cache and the external memory was eight bytes wide. The external memory showed an access time of 18 cycles and a two-cycle delay for consecutive data transfers in burst mode. Hence, the miss penalty for the L2 cache amounted to 24 cycles. Using this configuration, a conventional cache system for a byte-addressable architecture with 32-bit addresses would have a 22-bit tag and a six-bit index for the L1 caches and a 20-bit tag and a seven-bit index for the L2 cache. For an EvoCache, the original tags and indices merge into a single tag of 28 and 27 bits for the L1 and L2 caches, respectively. We evolved mapping functions for two optimization scenarios. In the first optimization scenario, only the first-level caches (LI:I and L1:D) were EvoCaches with evolved mapping functions, while in the second scenario, all three caches received evolved mapping functions. Thus, a single chromosome describing the system's mapping functions consisted of two CGP chromosomes in the first optimization scenario and of three CGP chromosomes in the second optimization scenario.

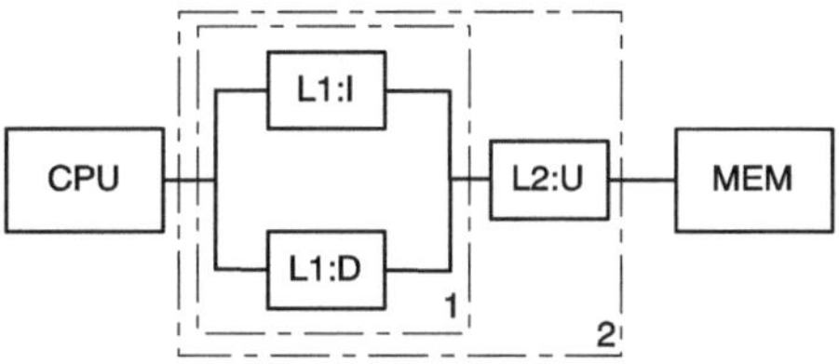

Fig. 5.30 Two memory hierarchy configurations considered for the optimization of the address-mapping function: (1) optimization of split first-level caches (L1:I,L1:D), (2) optimization of an additional unified second-level cache (L1:I, L1:D, L2:U).

For evaluation we simulated the execution of two benchmarks, `BZIP2` (version 1.0.4) and `JPEG` (version 6a), each with different sets of input data. `BZIP2` is a recent data compressor based on the Burrows–Wheeler transformation [53] and has been reported to cause a large amount of cache misses. The picture-encoding application `JPEG` [76] is a commonly used benchmark for performance analysis.

For each combination of benchmark and optimization scenario, we proceeded as follows. First, we evolved a mapping function for a given input data set, referred to as the training data. This optimization step was repeated 16 times. To study the potential of EvoCaches, we analyzed the development of the fitness of the best and the worst individual in each generation and the average over all 16 runs for two reference systems. These were a cacheless system with a one-cycle memory access time, which is unrealistic but serves as a point of reference, and a two-level cache with classical modulo address-mapping functions.

Second, we determined the generalization behaviour by evaluating the best evolved mapping functions on different sets of input data, referred to as the test data. These results are actually more important than the results achieved for the training data, as they reflect a practical use case of EvoCaches.

For BZIP2, the training data set consisted of the HTML code from Wikipedia's page 'Genetic Programming' [1]. The test data consisted of 30 data sets partitioned into HTML data, Linux binaries and human-readable text files. For JPEG, the training data set originated from the standard picture contained in the JPEG source code distribution. We used 10 data sets from [66, 11] as the test data.

5.8.3.1 Training EvoCaches

Table 5.13 summarizes the training results. The numbers in the 'absolute' columns were calculated as the reciprocal of the overall execution time relative to the cacheless reference system with a one-cycle access time. That is, since BZIP2 required 13,131,325 cycles to process the training data on the cacheless reference system and 34,417,080 cycles on the classical modulo cache system (a 2.62 times slowdown), the modulo cache was rated at 0.3815. Analogously the JPEG benchmark was executed on the one-cycle access memory system in 47,723,253 cycles, and in 80,148,296 cycles on the classical modulo cache system. This is a slowdown of 1.67 times and the JPEG benchmark was rated at 0.5954. The ratio in the 'relative' columns noted ratio is the speed-up relative to the modulo cache system. We can observe that the average performance achieved by optimizing both cache levels is actually higher than the performance achieved by optimizing only the level-one caches. The best individual, with 17.1% improvement in run time is found, however, by optimizing the level-one caches.

Table 5.13 Performance of the average and the best individuals on training data for BZIP2 after 2500 generations and for JPEG after 1400 generations

		BZIP2		JPEG	
		Absolute	Relative	Absolute	Relative
modulo cache		0.3815	–	0.5954	–
L1:I,D	Average	0.4038	5.8%	0.6623	11.2%
	Maximum	0.4086	7.1%	0.6975	17.1%
L1:I,D, L2:U	Average	0.4037	5.8%	0.6718	12.8%
	Maximum	0.4174	9.4%	0.6962	16.9%

5.8.3.2 Testing EvoCaches

Table 5.14 EvoCache generalization performance for BZIP2 trained on the Wikipedia 'Genetic Programming' HTML page. The test data was partitioned into compressed Linux binaries in ELF format (bash, cpio, dbus-daemon, awk, sh, gawk, tar, tcsh, vim and zsh), Web pages in HTML format (Ancient Egypt [W], Ancient Greece [W], Ancient Rome [W], Germany [W], heise.de, Andrey Kolmogorov [W], sailinganarchy.com, spiegel.de, wired.com and slashdot.org) and text files (rfc 2068, 2246, 845, 1000, 1001, 1002, 1005, 1008, 1009 and 2658). The data sets indicated by [W] were collected from wikipedia.org

		BZIP2						JPEG	
		ELF		HTML		TXT			
		L1	L12	L1	L12	L1	L12	L1	L12
Execution time	Best	4.92%	5.90%	5.31%	6.98%	7.30%	10.98%	14.31%	12.96%
	Average	4.36%	5.60%	3.86%	4.94%	4.49%	6.66%	12.73%	10.78%
	worst	3.36%	4.87%	2.47%	3.46%	-4.15%	1.95%	11.48%	9.12%
Miss rate	Best	6.11%	9.00%	5.94%	8.92%	8.45%	11.38%	41.25%	40.35%
	Average	5.59%	8.51%	4.15%	6.41%	4.82%	8.26%	37.40%	37.19%
	worst	4.13%	7.94%	1.64%	4.88%	-9.31%	2.63%	31.64%	30.46%
Energy requirement	Best	4.64%	5.49%	5.61%	7.31%	6.88%	10.70%	16.43%	14.46%
	Average	4.13%	5.23%	4.53%	5.29%	4.93%	6.98%	14.19%	11.93%
	Worst	3.27%	4.57%	3.21%	3.77%	2.83%	2.16%	12.53%	10.49%

To verify the generalization performance of EvoCaches, we evaluated the execution times, the miss rates and the energy requirements for BZIP2 and JPEG with the various optimizations scenarios. The test data for BZIP2 comprised 10 data sets taken from Linux binaries (ELF benchmark), 10 data sets taken from HTML dumps of popular web sites (HTML benchmark), and ten data sets taken from RFCs (TXT benchmark). The detailed results are shown in Table 5.14. In this table, the optimization scenario L1 denotes the optimization of level-one caches, and L12 the optimization of both levels of caches. The numbers for each individual benchmark and optimization scenario were averaged over the corresponding 10 data sets and measured relative to the performance of a conventional system with modulo address mappings. Positive percentages indicate an improvement in execution time, a reduction in miss rate and a reduction in energy. The miss rates for all caches were added to obtain the miss rate metric.

The following observations can be made for the BZIP2 benchmark when we analyze the results in Table 5.14:

- EvoCaches generalize well and deliver substantial performance improvements for all the test data. The improvements in execution time are by up to 10.98% and the reductions in energy are by up to 10.70%.
- Having EvoCaches in both levels of cache (L1:I, L1:D and L2:U) leads to higher performance gains than having EvoCaches only in level one.

To test EvoCaches on the JPEG benchmark, we selected 10 images from [66, 11]. The detailed results are shown in Table 5.14 and can be summarized as follows:

- EvoCaches again generalize well, with even larger improvements in execution time (up to 14.31%) and reductions in energy (up to 16.43%).
- The average performance when L1 caches only are optimized is about 2% higher than when optimizing both cache levels are optimized. This corresponds with the observation made when the EvoCaches were trained for JPEG, where the best training performance was reached by optimizing L1 caches only. Consequently, the individual with the best test performance acheived that performance even though it was not being optimized additionally for the L2 cache.
- While the reductions in the miss rates are rather high, the reductions in execution time are lower. This demonstrates that for multiple levels of cache (or sophisticated processor architectures), the total miss rate is not necessarily a suitable metric for quantitatively determining a performance improvement.

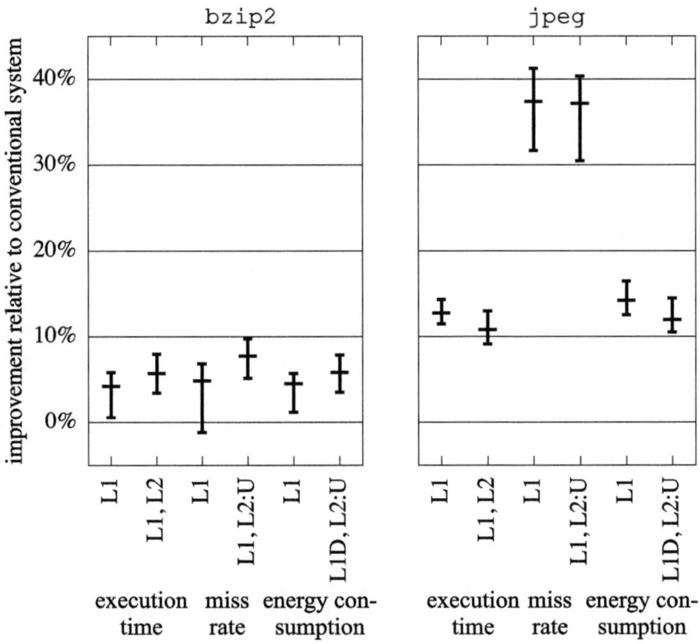

Fig. 5.31 Summary of the EvoCache generalization performance for BZIP2 and JPEG. The data is for randomly initialized mapping functions. The best, worst and average values are indicated for every optimization scenario and metric.

The results of our experiments are summarized in Fig. 5.31. The figure shows, for both of the benchmarks (BZIP2 and JPEG) and all of the optimization scenarios, the relative improvement for EvoCaches in execution time, miss rate and energy requirement over a modulo address mapping function.

The area parameters (number of 4 input LUTs (4-LUTs)) and the delay (depth of the circuit) parameters for the resulting reconfigurable logic circuits are presented

Table 5.15 Area and delay parameters for the evolved reconfigurable address-mapping functions

		BZIP2		JPEG	
		Delay	Size	Delay	Size
L1:I,D	Average	4.63	16.94	4.38	15.13
	Maximum	7	22	7	22
	Best	4	18	6	19
L1:I,D	Average	4.19	15.81	4.44	15.44
L2:U	Maximum	8	24	6	22
	Best	3	14	4	17

in Table 5.15. In addition to the average and maximum values, the values for the fittest circuit that was used for testing are also listed. These circuits show depths between three and six LUTs. It should be noted that the circuits resulted from an evolutionary design process and have thus not been optimized for area or delay. Delay minimization could possibly further reduce the circuit's propagation time and thus the cache hit time.

5.8.4 Conclusion

In this section, we have presented EvoCaches, which rely on two main ideas. First, the function mapping an address to a cache line index is implemented by a small reconfigurable logic fabric. Second, the function is optimized by an evolutionary algorithm with the goal of achieving the minimum overall execution time for a specific application. We have defined various optimization scenarios, optimizing split level-one caches and, additionally, a unified level-two cache, and conducted experiments with the `BZIP2` and `JPEG` benchmarks. After evolving the mapping functions, we tested the best solutions on independent data sets and evaluated the overall execution times, miss rates and energy requirements. Compared with conventional caches, we have observed run time improvements of up to 10.98% for `BZIP2` and up to 14.31% for `JPEG`, and energy reductions of up to 10.70% for `BZIP2` and up to 16.43% for `JPEG`.

5.9 Acknowledgements

This work was supported by the German Research Foundation under project number PL 471/1-3 within the 'Organic Computing' priority programme. James Alfred Walker would like to thank all partners on the EPSRC-funded Nano-CMOS project (ref. EP/E001610/1), especially the Device Modelling Group at the University of Glasgow for providing the variability-enhanced models and the `Randomspice` application. Lukas Sekanina was partly supported by the Czech Science Foundation under Contract No. P103/10/1517 and by the research programme MSM 0021630528.

References

1. Genetic Programming. http://en.wikipedia.org/wiki/Genetic_programming
2. Albonesi, D.H.: Selective Cache Ways: On-demand Cache Resource Allocation. In: Proc. ACM/IEEE International Symposium on Microarchitecture, pp. 248–259. IEEE Computer Society (1999)
3. Ali, B., Almaini, A.E.A., Kalganova, T.: Evolutionary Algorithms and Their Use in the Design of Sequential Logic Circuits. Genetic Programming and Evolvable Machines **5**(1), 11–29 (2004)
4. Aoki, T., Homma, N., Higuchi, T.: Evolutionary Synthesis of Arithmetic Circuit Structures. Artificial Intelligence Review **20**(3–4), 199–232 (2003)
5. Asenov, A.: Random Dopant Induced Threshold Voltage Lowering and Fluctuations in sub 50 nm MOSFETs: A Statistical 3D 'Atomistic' Simulation Study. Nanotechnology **10**, 153–158 (1999)
6. Asenov, A.: Variability in the Next Generation CMOS Technologies and Impact on Design. In: International Conference of CMOS Variability (2007)
7. Austin, T., Larson, E., Ernst, D.: SimpleScalar: An Infrastructure for Computer System Modeling. Computer **35**(2), 59–67 (2002)
8. Biovision: EMG Amplifier. www.biovison.eu
9. Boschmann, A., Kaufmann, P., Platzner, M., Winkler, M.: Towards Multi-movement Hand Prostheses: Combining Adaptive Classification with High Precision Sockets. In: Proc. European Conference on Technically Assisted Rehabilitation (2009)
10. Clegg, J., Walker, J.A., Miller, J.F.: A New Crossover Technique for Cartesian Genetic Programming. In: Proc. of the Genetic and Evolutionary Computation Conference, pp. 1580–1587 (2007)
11. Classic Test Still Images. http://hlevkin.com
12. Deb, K., Pratap, A., Agarwal, S., Meyarivan, T.: A Fast and Elitist Multi-Objective Genetic Algorithm: NSGA-II. IEEE Transactions on Evolutionary Computation **6**, 181–197 (2002)
13. Eccleston, W.: The Effect of Polysilicon Grain Boundaries on MOS Based Devices. Microelectronic Engineering **48**, 105–108 (1999)
14. Erba, M., Rossi, R., Liberali, V., Tettamanzi, A.: An Evolutionary Approach to Automatic Generation of VHDL Code for Low-Power Digital Filters. In: Proc. European Conference on Genetic Programming, vol. 2038, pp. 36–50. Springer (2001)
15. Gajda, Z., Sekanina, L.: Reducing the Number of Transistors in Digital Circuits Using Gate-Level Evolutionary Design. In: Proc. Genetic and Evolutionary Computation Conference, pp. 245–252. ACM Press (2007)
16. Gajda, Z., Sekanina, L.: Gate-Level Optimization of Polymorphic Circuits Using Cartesian Genetic Programming. In: Proc. IEEE Congress on Evolutionary Computation, pp. 1599–1604. IEEE (2009)

17. Gajda, Z., Sekanina, L.: An Efficient Selection Strategy for Digital Circuit Evolution. In: Proc. International Conference on Evolvable Systems, *LNCS*, vol. 6274, pp. 13–24. Springer (2010)
18. Glette, K., Gruber, T., Kaufmann, P., Torresen, J., Sick, B., Platzner, M.: Comparing Evolvable Hardware to Conventional Classifiers for Electromyographic Prosthetic Hand Control. In: Proc. NASA/ESA Conference on Adaptive Hardware and Systems, pp. 32–39. IEEE Computer Society (2008)
19. Glette, K., Torresen, J.: A Flexible On-Chip Evolution System Implemented on a Xilinx Virtex-II Pro Device. In: Proc. International Conference on Evolvable Systems, *LNCS*, vol. 3637, pp. 66–75. Springer Berlin / Heidelberg (2005)
20. Glette, K., Torresen, J., Kaufmann, P., Platzner, M.: A Comparison of Evolvable Hardware Architectures for Classification Tasks. In: Proc. International Conference on Evolvable Systems, *LNCS*, pp. 22–33. Springer (2008)
21. Glette, K., Torresen, J., Yasunaga, M.: An Online EHW Pattern Recognition System Applied to Face Image Recognition. In: Applications of Evolutionary Computing, *LNCS*, vol. 4448, pp. 271–280. Springer (2007)
22. Gray, H.: Anatomy of the Human Body (1918). Retrieved from Wikimedia Commons
23. Greenwood, G., Tyrrell, A.M.: Introduction to Evolvable Hardware. IEEE Press (2007)
24. Hargrove, L., Losier, Y., Lock, B., Englehart, K., Hudgins, B.: A Real-Time Pattern Recognition Based Myoelectric Control Usability Study Implemented in a Virtual Environment. In: Engineering in Medicine and Biology Society, pp. 4842–4845. IEEE Press (2007)
25. Higuchi, T., Iwata, M., Kajitani, I., Iba, H., Hirao, Y., Manderick, B., Furuya, T.: Evolvable Hardware and its Applications to Pattern Recognition and Fault-Tolerant Systems. In: Towards Evolvable Hardware: The evolutionary Engineering Approach, *LNCS*, vol. 1062, pp. 118–135. Springer (1996)
26. Higuchi, T., Liu, Y., Yao, X.: Evolvable Hardware. Springer (2006)
27. Hilder, J.A., Walker, J.A., Tyrrell, A.M.: Designing Variability Tolerant Logic using Evolutionary Algorithms. In: Proc. International Conference on Ph.D. Research in Microelectronics & Electronics (PRIME) (2009)
28. Hilder, J.A., Walker, J.A., Tyrrell, A.M.: Optimisation of Variability Tolerant Logic Cells using Multiple Voltage Supplies. In: Proc. IEEE Workshop on Evolvable and Adaptive Hardware, Proc. IEEE Symposium Series on Computational Intelligence, pp. 17–24. IEEE (2009)
29. Hilder, J.A., Walker, J.A., Tyrrell, A.M.: Optimising Variability Tolerant Standard Cell Libraries. In: Proc. IEEE Congress on Evolutionary Computation, pp. 2273–2280. IEEE (2009)
30. Hounsell, B.I., Arslan, T., Thomson, R.: Evolutionary Design and Adaptation of High Performance Digital Filters within an Embedded Reconfigurable Fault Tolerant Hardware Platform. Soft Computing **8**(5), 307–317 (2004)
31. Kajitani, I., Sekita, I., Otsu, N., Higuchi, T.: Improvements to the Action Decision Rate for a Multi-Function Prosthetic Hand. In: Proc. International Symposium on Measurement, Analysis and Modeling of Human Functions, pp. 84–89 (2001)
32. Kalganova, T., Miller, J.F.: Evolving More Efficient Digital Circuits by Allowing Circuit Layout Evolution and Multi-Objective Fitness. In: Proc. NASA/DoD Workshop on Evolvable Hardware, pp. 54–63. IEEE Computer Society (1999)
33. Kaufmann, P., Platzner, M.: MOVES: A Modular Framework for Hardware Evolution. In: Proc. NASA/ESA Conference on Adaptive Hardware and Systems, pp. 447–454. IEEE (2007)
34. Kaufmann, P., Platzner, M.: Advanced Techniques for the Creation and Propagation of Modules in Cartesian Genetic Programming. In: Proc. Genetic and Evolutionary Computation Conference (GECCO'08), pp. 1219–1226. ACM Press (2008)
35. Kaufmann, P., Plessl, C., Platzner, M.: EvoCaches: Application-specific Adaptation of Cache Mappings. In: Proc. NASA/ESA Conference on Adaptive Hardware and Systems, pp. 11–18. IEEE Computer Society (2009)
36. Meyer-Baese, U.: Digital Signal Processing with Field Programmable Gate Arrays. Springer (2004)
37. Miller, J.F.: Digital Filter Design at Gate-level Using Evolutionary Algorithms. In: Proc. Genetic and Evolutionary Computation Conference, pp. 1127–1134. Morgan Kaufmann (1999)

38. Miller, J.F.: Evolution of Digital Filters Using a Gate Array Model. In: Proc. Workshop on Evolutionary Image Analysis and Signal Processing, *LNCS*, vol. 1596, pp. 17–30. Springer (1999)
39. Miller, J.F., Job, D., Vassilev, V.K.: Principles in the Evolutionary Design of Digital Circuits – Part I. Genetic Programming and Evolvable Machines **1**(1), 8–35 (2000)
40. Moore, G.E.: Cramming More Components onto Integrated Circuits. Electronics **38** (1965)
41. Moroz, V.: Design for Manufacturability: OPC and Stress Variations. In: International Conference on CMOS Variability (2007)
42. National Instruments: USB-6009. www.ni.com
43. Petley, G.: VLSI and ASIC Technology Standard Cell Library Design. www.vlsitechnology.org
44. Poli, R., Page, J.: Solving High-Order Boolean Parity Problems with Smooth Uniform Crossover, Sub-Machine Code GP and Demes. Genetic Programming and Evolvable Machines **1**(1–2), 37–56 (2000)
45. Ranganathan, P., Adve, S., Jouppi, N.P.: Reconfigurable Caches and Their Application to Media Processing. Proc. International Symposium on Computer Architecture **28**(2), 214–224 (2000)
46. Ruzicka, R., Sekanina, L., Prokop, R.: Physical Demonstration of Polymorphic Self-checking Circuits. In: Proc. IEEE International On-Line Testing Symposium, pp. 31–36. IEEE (2008)
47. Sekanina, L.: Evolvable Components: From Theory to Hardware Implementations. Natural Computing. Springer (2004)
48. Sekanina, L.: Evolutionary Design of Gate-Level Polymorphic Digital Circuits. In: Applications of Evolutionary Computing, *LNCS*, vol. 3449, pp. 185–194. Springer (2005)
49. Sekanina, L., Ruzicka, R., Gajda, Z.: Polymorphic FIR Filters with Backup Mode Enabling Power Savings. In: Proc. NASA/ESA Conference on Adaptive Hardware and Systems, pp. 43–50. IEEE (2009)
50. Sekanina, L., Ruzicka, R., Vasicek, Z., Prokop, R., Fujcik, L.: REPOMO32 – New Reconfigurable Polymorphic Integrated Circuit for Adaptive Hardware. In: Proc. of IEEE Symposium Series on Computational Intelligence - Workshop on Evolvable and Adaptive Hardware, pp. 39–46. IEEE Computational Intelligence Society (2009)
51. Sekanina, L., Starecek, L., Kotasek, Z., Gajda, Z.: Polymorphic Gates in Design and Test of Digital Circuits. International Journal of Unconventional Computing **4**(2), 125–142 (2008)
52. Sekanina, L., Vasicek, Z.: On the Practical Limits of the Evolutionary Digital Filter Design at the Gate Level. In: Applications of Evolutionary Computing, 3907, pp. 344–355. Springer (2006)
53. Seward, J.: bzip2: A Freely Available, Patent Free, High-quality Data Compressor (2009). URL www.bzip.org/
54. Shanthi, A.P., Parthasarathi, R.: Practical and Scalable Evolution of Digital Circuits. Applied Soft Computing **9**(2), 618–624 (2009)
55. Shivakumar, P., Jouppi, N.P.: CACTI 3.0: An Integrated Cache Timing, Power, and Area Model. Tech. rep., COMPAQ Western Research Lab, Palo Alto, California 94301 USA (1999)
56. Smith, S.L., Greensted, A.J., Timmis, J.: Hardware Acceleration of an Immune Network Inspired Evolutionary Algorithm for Medical Diagnosis. In: Proc. International Conference on Evolvable Systems, *LNCS*, vol. 5216, pp. 34–46. Springer Berlin / Heidelberg (2008)
57. Sonowin: USI-01 USB Isolator. www.sonowin.de
58. Sripramong, T., Toumazou, C.: The Invention of CMOS Amplifiers Using Genetic Programming and Current-Flow Analysis. IEEE Trans. on CAD of Integrated Circuits and Systems **21**, 1237–1252 (2002)
59. Stanca, M., Vassiliadis, S., Cotofana, S., Corporaal, H.: Hashed Addressed Caches for Embedded Pointer Based Codes. In: Proc. Int. Conf. on Parallel Processing, pp. 965–968. Springer (2000)
60. Starecek, L., Sekanina, L., Kotasek, Z.: Reduction of Test Vectors Volume by Means of Gate-Level Reconfiguration. In: Proc. IEEE Design and Diagnostics of Electronic Circuits and Systems Workshop, pp. 255–258. IEEE Computer Society (2008)

61. Stoica, A., Zebulum, R.S., Keymeulen, D.: Polymorphic Electronics. In: Proc. International Conference on Evolvable Systems, *LNCS*, vol. 2210, pp. 291–302. Springer (2001)
62. Stomeo, E., Kalganova, T., Lambert, C.: Generalized Disjunction Decomposition for Evolvable Hardware. IEEE Transaction Systems, Man and Cybernetics, Part B **36**(5), 1024–1043 (2006)
63. Sutherland, I., Sproull, B., Harris, D.: Logical Effort – Designing Fast CMOS Circuits. Morgan Kaufmann (1999)
64. Torresen, J.: Increased Complexity Evolution Applied to Evolvable Hardware. In: Smart Engineering System Design: Neural Networks, Fuzzy Logic, Evolutionary Programming, Data Mining, and Complex Systems (ANNIE), pp. 429–436. ASME Press (1999)
65. Torresen, J.: Scalable Evolvable Hardware Applied to Road Image Recognition. In: Proc. NASA/DoD workshop on Evolvable Hardware, pp. 245–252. IEEE Computer Society (2000)
66. The USC-SIPI Image Database. URL sipi.usc.edu/database
67. Vandierendonck, H., Bosschere, K.D.: Constructing Optimal XOR-Functions to Minimize Cache Conflict Misses. In: Proc. Int. Conf. on Architecture of Computing Systems (ARCS), pp. 261–272. Springer (2008)
68. Vasicek, Z., Sekanina, L.: Formal Verification of Candidate Solutions for Post-Synthesis EvoluTionary Optimization in Evolvable Hardware. Genetic Programming and Evolvable Machines **12**(3), 305–327 (2011)
69. Vasicek, Z., Zadnik, M., Sekanina, L., Tobola, J.: On Evolutionary Synthesis of Linear Transforms in FPGA. In: Proc. International Conference on Evolvable Systems, *LNCS*, vol. 5216, pp. 141–152. Springer (2008)
70. Vassilev, V., Job, D., Miller, J.F.: Towards the Automatic Design of More Efficient Digital Circuits. In: J. Lohn, A. Stoica, D. Keymeulen, S. Colombano (eds.) Proc. NASA/DoD Workshop on Evolvable Hardware, pp. 151–160. IEEE Computer Society (2000)
71. Vassilev, V.K., Miller, J.F.: Embedding Landscape Neutrality to Build a Bridge from the Conventional to a More Efficient Three-Bit Multiplier Circuit. In: Proc. Genetic and Evolutionary Computation Conference, p. 539 (2000)
72. Voronenko, Y., Puschel, M.: Multiplierless Multiple Constant Multiplication. ACM Transactions on Algorithms **3**(2) (2007)
73. Wakerly, J.F.: Digital Design: Principles and Practices, 3rd edn. Prentice Hall, New Jersey, US (2000)
74. Walker, J.A., Hilder, J.A., Tyrrell, A.M.: Evolving Variability-Tolerant CMOS Designs. In: Proc. International Conference on Evolvable Systems, pp. 308–319. Springer (2008)
75. Walker, J.A., Hilder, J.A., Tyrrell, A.M.: Towards Evolving Industry-Feasible Intrinsic Variability Tolerant CMOS Designs. In: Proc. IEEE Congress on Evolutionary Computation, pp. 1591–1598. IEEE (2009)
76. Wallace, G.K.: The JPEG Still Picture Compression Standard. Communications of the ACM **34**(4), 30–44 (1991)
77. Wang, J., Lee, C.H.: Evolutionary Design of Combinational Logic Circuits Using VRA Processor. IEICE Electronics Express **6**, 141–147 (2009)
78. ARM10E Processor Family. http://www.arm.com/products/CPUs/families/ARM10EFamily.html
79. Zebulum, R., Pacheco, M., Vellasco, M.: Evolutionary Electronics – Automatic Design of Electronic Circuits and Systems by Genetic Algorithms. The CRC Press International Series on Computational Intelligence (2002)
80. Zebulum, R.S., Stoica, A.: Four-Function Logic Gate Controlled by Analog Voltage. NASA Tech Briefs **30**(3), 8 (2006)
81. Zhang, C., Vahid, F., Lysecky, R.: A Self-tuning Cache Architecture for Embedded Systems. Trans. on Embedded Computing Systems **3**(2), 407–425 (2004)

Chapter 6
Image Processing and CGP

Lukas Sekanina, Simon L. Harding, Wolfgang Banzhaf and Taras Kowaliw

6.1 Introduction

Computerized image processing has been studied intensively for many years [32]. Because of the inherent complexity of the problem, stochastic optimization algorithms, including evolutionary algorithms, have been applied to improve existing image processing methods and develop new algorithms [4].

Section 6.2 deals with the automatic design of low-level image filters, ones comparable in quality to conventional filters. Moreover, when the evolved filters are constructed as single-purpose circuits in a field-programmable gate array (FPGA), the implementation cost is usually lower than the cost of conventional solutions. We will describe the basic approach to filter evolution and its extensions, such as the use of a bank of evolved filters, and filtering using extended kernels.

Evolutionary design of more advanced image operators such as dilation/erosion filters is presented in Sect. 6.3. As these filters are composed of relatively complex elementary functions (sine, square root etc.), they are primarily intended for advanced image processing software tools.

An image classification task will be described in Sect. 6.4, where CGP graphs are used to define transformations in the case of a medical image problem. We will show that CGP image transformations may be used to significantly improve classification accuracy with respect to a collection of predefined image features.

6.2 Evolutionary Design of Image Filters for FPGAs

Image preprocessing (which includes image filtering, edge detection, histogram equalization, brightness and contrast adjustment, and other operations on images) is the first stage of many image processing applications. The quality of preprocessing significantly influences the accuracy, reliability, robustness and performance of

J.F. Miller (ed.), *Cartesian Genetic Programming*, Natural Computing Series,
DOI 10.1007/978-3-642-17310-3_6, © Springer-Verlag Berlin Heidelberg 2011

subsequent image processing steps such as segmentation, classification and recognition. In order to perform the required preprocessing (such as image filtering, edge detection) a problem-specific filter has to be created. Traditionally, engineers use a library of predefined filters and operators and manually tune promising variants of these filters for a given application. In the process of tuning, various properties of the filters might be optimized, in particular their coefficients and structure [3, 7, 24]. There are other important parameters to be optimized, such as the area, delay and power consumption, when the goal is to implement the filter as a digital circuit.

Historically, linear filters have been the most popular filters for image processing. Their popularity is due to the existence of robust mathematical models which can be used for their analysis and design. However, there exist many areas in which nonlinear filters provide significantly better results [6]. The advantage of nonlinear filters lies in their ability to preserve edges and suppress noise without loss of detail. As there is no suitable general theory for the design of nonlinear operators, evolutionary design techniques have been utilized to accomplish this task in recent years.

The first utilization of CGP for image filter design was due to Sekanina [26]. It was shown later that CGP can automatically design image filters that are competitive with filters designed conventionally in terms of filtering quality and the cost of implementation on a chip. In this section, we will briefly survey how CGP can be used to evolve human-competitive image filters and edge detectors. In order to compare the results produced by CGP, we will briefly describe the most popular conventional methods that are utilized to suppress selected types of nonlinear noise. We will take into account greyscale images only; however, the concept can be naturally extended to colour images.

6.2.1 Sliding-Window Function

Software and hardware implementations of image filters operate mostly in the spatial domain. As spatial filters operate with pixel values in the neighbourhood of the centre pixel (this neighbourhood is called the window or kernel), it is necessary to implement a local neighbourhood function (sometimes referred to as sliding-window function). This function is applied separately to all pixel locations. This function is also invariable for all locations (i.e. spatially invariant). Figure 6.1 shows the concept of the sliding window function and the utilization of CGP for the evolutionary design of such a function.

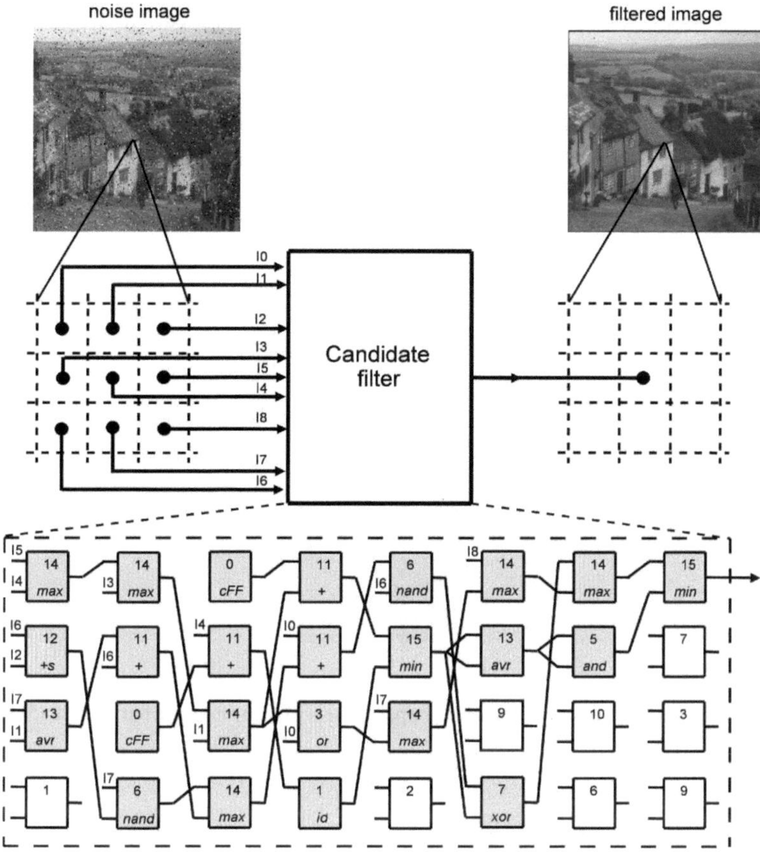

Fig. 6.1 Candidate local neighborhood function represented in CGP.

6.2.2 *Types of Noise*

Impulse noise is a frequently encountered type of noise. In most cases, impulse noise is caused by malfunctioning pixels in camera sensors, faulty memory locations in hardware or errors in data transmission. We distinguish two common types of impulse noise: the salt-and-pepper noise (also referred to as intensity spikes or speckle) and random-valued shot noise. For images corrupted by salt-and-pepper noise (see Fig 6.2a), the noisy pixels can take only the maximum or minimum value. In the case of the random-valued shot noise, the noisy pixels have an arbitrary value. Impulse noise cab be characterized by one parameter the noise intensity p, which gives the ratio of the number of corrupted pixels to the total number of pixels.

Impulse burst noise typically occurs in remote sensing images such as satellite images (see Fig. 6.2b). The main reason for the occurrence of bursts is interference in the frequency-modulated carrying signal caused by signals from other data sources. This interference can occur several times during the transmission of a single image and corrupt several image pixels in one or more neighbouring rows. Impulse burst noise is a specific kind of noise which is difficult to filter out even if a non-linear filter is used. We will show in Sect. 6.2.8.3 that median filters are capable of suppressing impulse burst noise but, at the same time, they usually damage image detail too heavily.

(a) (b)

Fig. 6.2 Types of noise: (a) salt-and-pepper noise (5%), (b) impulse burst noise (5%).

6.2.3 Conventional Filters

Traditionally, impulse noise is suppressed by a median filter, the most popular non-linear filter. In this case the output of the local neighbourhood function is calculated from the median value of the kernel. However, the standard median filter gives poor performance for images corrupted by impulse noise of high intensity. A simple median filter utilizing a 3×3- or 5×5-pixel window is sufficient only when the noise intensity is less than approximately 10–20%. Various approaches have been proposed, to overcome this deficiency, including switching median filters [43], weighted median filters [2], weighted order-statistic filters [22] and adaptive median filters [15] (see e.g. [25] for a detailed survey of these methods). In addition to median filters, some other algorithms exist (e.g. [23]). Almost all filters of this type have already been implemented in hardware (see [40] for a survey of hardware implementations).

The adaptive median filter provides significantly better results than do standard median filters, especially for images corrupted by a high noise intensity [15]. The adaptive median filter operates with a kernel of $S_{max} \times S_{max}$ pixels. Let S_{xy} denote the processed (i.e. central) pixel. The kernel is processed by a set of sorting networks with $3 \times 3, 5 \times 5, \ldots, S_{max} \times S_{max}$ inputs. Each sorting network provides the minimum, maximum and median value of its window. The output value generated by the adaptive median filter is calculated on the basis of comparison of the outputs of the sorting networks as described in [15]. Figure 6.3 compares the results of applying some conventional filters to suppress 40% salt-and-pepper noise. Reasonable results can be obtained with an adaptive median filter; however, the implementation cost is too high for FPGAs (see Table 6.3).

Fig. 6.3 Images obtained by using conventional filters. (a) Original image. (b) Noisy image with 40% salt-and-pepper noise (peak signal-to-noise ratio (PSNR) 9.364 dB). (c) Filtered by median filter with a kernel size 3×3 (PSNR 18.293 dB). (d) Filtered by median filter with a kernel size 5×5 (PSNR 24.102 dB). (e) Filtered by adaptive median filter with a kernel size of up to 5×5 (PSNR 26.906 dB). (f) Filtered by adaptive median filter with a kernel size up to 7×7 (PSNR 27.315 dB).

In addition to various median-based filters, specific filters have been developed for impulse burst noise, such as training-based optimized soft morphological filters and variational approaches [6, 23, 19, 18]. Unfortunately, they are not suitable for hardware implementation.

6.2.4 Edge Detectors

Edge detection can be considered as image filtering too. The Sobel operator, which performs a 2D spatial-gradient measurement on an image, is one of the most popular edge detectors [32]. It utilizes two convolution kernels, which are defined as

$$p = \frac{1}{8}\begin{pmatrix} -1 & 0 & 1 \\ -2 & 0 & 2 \\ -1 & 0 & 1 \end{pmatrix}, \qquad q = \frac{1}{8}\begin{pmatrix} -1 & -2 & -1 \\ 0 & 0 & 0 \\ 1 & 2 & 1 \end{pmatrix},$$

By applying these two kernels, we can calculate the new pixel value as

$$NewPixelValue = c + |p| + |q|,$$

where c is a suitable constant, (e.g. $c = 128$).

6.2.5 Basic Approach to Filter Evolution

Every image filter operating with a 3×3-pixel kernel can be considered as a function (a digital circuit in the case of a hardware implementation) of nine eight-bit inputs and a single eight-bit output, which processes greyscale (eight-bits/pixel) images. As Fig. 6.1 shows, every pixel value of the filtered image is calculated using the corresponding pixel and its eight neighbours in the processed image.

A candidate filter was represented using $n_c \times n_r$ nodes, where the typical grid size was $n_c = 8$, $n_r = 4$. The settings of the other CGP parameters were $n_i = 9$, $n_o = 1$, $l = 1$, $n_a = 2$ and $\lambda = 8$. Each node represented a two-input function which receives two eight-bit values and produces an eight-bit output. An eight-bit node output was utilized to ensure a straightforward connectivity of the nodes in the hardware implementation. Table 6.1 lists the functions that were confirmed as useful for this task [27]. We note that these functions are also suitable for hardware implementation (i.e. there are no complex functions, such as multiplication or division).

Table 6.1 List of functions implemented in each programmable node

Code	Function	Description	Code	Function	Description
0	255	Constant	8	$x \gg 1$	Right shift by 1
1	x	Identity	9	$x \gg 2$	Right shift by 2
2	$255 - x$	Inversion	10	$swap(x, y)$	Swap nibbles
3	$x \vee y$	Bitwise OR	11	$x + y$	+ (addition)
4	$\bar{x} \vee y$	Bitwise $\bar{x}$ OR y	12	$x +^S y$	+ with saturation
5	$x \wedge y$	Bitwise AND	13	$(x + y) \gg 1$	Average
6	$\overline{x \wedge y}$	Bitwise NAND	14	$max(x, y)$	Maximum
7	$x \oplus y$	Bitwise XOR	15	$min(x, y)$	Minimum

CGP uses a single genetic operator – mutation – which modified 5% of the chromosome. Slany and Sekanina have analyzed various genetic operators for this task; however, the mutation-only search was confirmed to be the most successful [31]. The initial population was randomly generated. The evolution was usually terminated when a predefined number of generations was exhausted.

In order to evolve an image filter capable of removing a given type of noise, an original uncorrupted (training) image was needed to determine the fitness values of the candidate filters. The goal of CGP was to minimize the difference between the original image and the filtered image. The generality of the evolved filters (i.e. whether the filters could operate sufficiently well also for other images containing the same type of noise) was tested by means of a test (validation) set. The quality of filtering had to be expressed numerically. The peak signal-to-noise ratio (PSNR) is usually used in image processing for this purposes. The PSNR is defined was

$$\mathrm{PSNR} = 10\log_{10} \frac{255^2}{(1/MN)\sum_{i,j}(v(i,j) - w(i,j))^2}, \tag{6.1}$$

where $N \times M$ is the size of the image, v denotes the filtered image and w denotes the original image. Note that the higher the PSNR value the better the filter. However, computing the PSNR value is expensive, especially for hardware accelerators (see Sect. 7.3. Hence in most cases the fitness function was based on calculating the mean difference per pixel (MDPP)

$$MDPP = \frac{1}{MN} \sum_{i=1}^{M} \sum_{j=1}^{N} |v(i,j) - w(i,j)|, \tag{6.2}$$

and the goal of evolution was to minimize the MDPP value.

6.2.6 Bank of Evolved Filters

In order to create a salt-and-pepper noise filter which could provide similar filtering quality to the adaptive median filter and which would also suitable for hardware implementation, a bank of evolved filters (or, simply, a filter bank) was proposed in [36, 39]. A filter bank combines several simple image filters that have been designed by the basic approach using 3×3-pixel kernel. As Fig. 6.4 shows, the procedure has three steps: (1) reduction of the dynamic range of the noise, (2) processing using a bank of n filters and (3) deterministic selection of the result.

Step 1. It has been shown that evolved filters have problems with a high dynamic range of corrupted pixels (0–255). A straightforward solution to this problem is to create a component which inverts all pixels with the value 255, i.e. all spikes are transformed to have a uniform value. This operation is performed by the first stage of the filter bank.

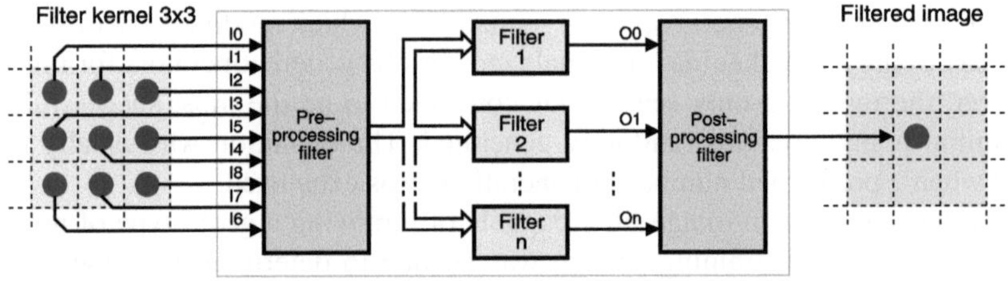

Fig. 6.4 Removal of salt-and-pepper noise using a filter bank.

Step 2. The Preprocessed image then enters a bank composed of n filters that operate in parallel. Because the evolutionary design of salt-and-pepper noise filters can be repeated many times, it is possible to obtain various different but similarly performing implementations of the filter. The bank of filters is then made up from the best-performing filters that have been evolved. In the present work, all these filters were designed by CGP using the same type of noise and training image and with the same aim – to remove 40% salt-and-pepper noise using 3×3-pixel kernel.

Step 3. Finally, the outputs from banks $1, \ldots, n$ have to be combined together. One possible solution is to use the n-input median function. Another type of aggregation function was proposed in [30].

6.2.7 Extended Kernel

Because of the nature of impulse burst noise, it is necessary to utilize a kernel larger than 3×3 pixels. Unfortunately, the image filter design method described in Sect. 6.2.5 is not scalable to larger filtering windows (e.g. if the number of inputs is increased to 25). In order to simultaneously support larger filtering windows and leave the problem reasonably difficult for evolution, CGP was extended by means of a selector that determined the pixels of a filtering window that would be used in evolved nine-input filters [35] (see Fig. 6.5). The selector was encoded as a string consisting of S bits, joined to the chromosome. If the bit of the selector string corresponding to a given pixel of the filter window had a logic value of 1, then the pixel was selected; otherwise, the pixel was not used as an input to the filtering logic. A special mutation operator was also developed for mutation of the selector [35].

6.2.8 Experimental Results

The applicability and performance of the CGP-based design method has been investigated in several papers dealing with evolution of filters (for Gaussian noise [26], impulse noise of lower intensity [29], salt-and-pepper noise of high intensity [36],

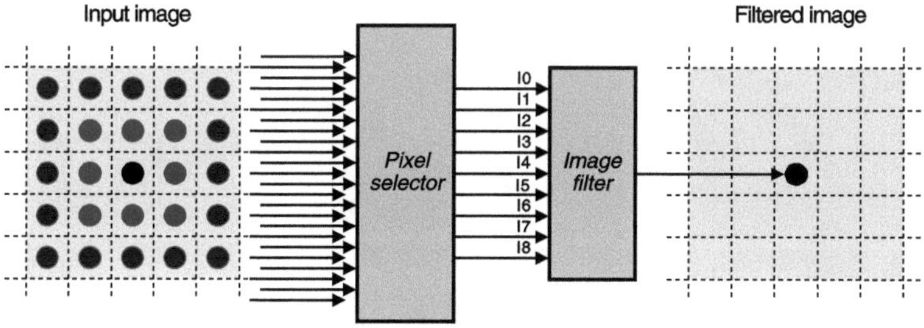

Fig. 6.5 The concept of filtering using a 5×5 filter kernel followed by a selector.

and impulse burst noise [35]), edge detectors [27] and edge detectors combined with impulse noise removal [38]. Extensive testing was carried out to analyze the optimal choice of the population size, mutation rate, size of the CGP array, size of the training image [28], set of node functions [27], pseudo-random number generator [38] and search algorithm [37]. We will briefly summarize the most important results for the basic method for a bank of evolved filters and for filtering with an extended kernel.

6.2.8.1 Filter Evolution Using a 3×3-Pixel Kernel

Table 6.2 summarizes the results that were obtained for the best filter for Gaussian noise, salt-and-pepper noise (5% intensity), random-valued shot noise and edge detection according to [28]. In these experiments, 500 independent runs (50,000 generations in each run) were performed for all noise types using a 256×256 pixel training image (the Lena image), a population size of 4, a mutation rate of 10 bits/chromosome and 4×8 programmable nodes. The third column of Table 6.2 indicates the generation in which the best filter was discovered. The last column gives the results for conventional filters. The MDPP is the mean difference per pixel (see Eqn. 6.2).

Table 6.2 Parameters of the best filters evolved using the Lena image

Problem	Best MDPP	Generations	Avg. MDPP	Std. dev.	Avg. generations	Conventional
Salt-and-pepper 5%	0.31	49808	0.95	0.51	29388	2.95 (median)
Random shot 5%	1.11	40616	1.65	0.31	26781	2.98 (median)
Gaussian noise	6.36	27179	6.73	0.24	31032	6.43 (mean)
Edge detection	1.16	49608	1.73	0.41	35074	–

Fig. 6.6 Original images (a, b, c), images with salt-and-pepper noise (d, e, f), images filtered using an evolved filter (g, h, i) and images filtered using a median filter (j, k, l). The evolved filter was trained via the Lena image (a, d).

The evolved filters were applied to other images to check whether the discovered filtering algorithms were general enough. The images in Fig. 6.6 demonstrate that the visual quality of the filtered images was sufficient and that the evolved filters worked correctly for the class of images considered. We can see from Fig. 6.6 that the median filter does not preserve details in the images and tends, in fact, to smudge the images. On the other hand, the evolved filters are good at preserving details. The performance and cost of this type of filter will be compared further with those of other filters in Sect. 6.2.8.2.

The following C program represents an implementation of the best-evolved edge detector, created according to Fig. 6.7. Note that `kernel` denotes the nine inputs of the image filter. Figure 6.8 compares the effect of the Sobel edge detector and an evolved edge detector on one of the validation images.

```
uint8 filter(uint8 kernel[9]) {
  uint i14,i17,i19,i22,i27,i29;

  i14 = min((kernel[1] + kernel[7]) >> 1, kernel[7]);
  i17 = i14 ^ kernel[7];
  i19 = min(i14 + (255 - kernel[1]), 255);
  i22 = 255 - i19;
  i27 = min(i22, (i17 + i19) >> 1);
  i29 = min(i22 + i27, 255);
  return (i27 + i29) & 0xff;
}
```

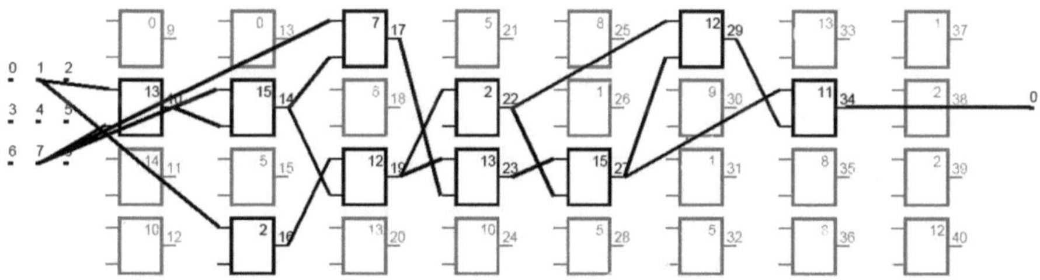

Fig. 6.7 The best edge detector from [37]. The node functions are numbered according to Table 6.1.

(a) input **(b)** Sobel operator **(c)** evolved filter

Fig. 6.8 Input image (a), and application of the Sobel edge detector (b) versus an evolved edge detector (c).

6.2.8.2 Bank of Evolved Filters

In order to construct a bank of filters, various filter implementations were evolved using the method described in Sect. 6.2.5. A 128×128-pixel training image, partially corrupted by 40% salt-and-pepper noise, was used in the fitness function. CGP operated with an eight-member population and 5% mutation. A single run was terminated after 200,000 generations. Three of the evolved filters were utilized in the bank filter and applied to suppress the 40% salt-and-pepper noise. The difference between the output of the filter bank and that of the adaptive median filter is almost invisible in Fig. 6.9.

Fig. 6.9 Comparison of images filtered using an adaptive median filter with a kernel size of up to 7×7 (a, b, c) and using a three-bank filter (d, e, f).

An FPGA platform was used for comparison of the implementation cost of median filters, adaptive median filters, evolved simple filters and filter banks [36]. The FPGA implementation was obtained in such a way that the CGP chromosome was transformed to VHDL code, which was then synthesized for a particular FPGA. Registers were automatically inserted in suitable places in order to allow the filter to be pipelined. The results of the synthesis (including the number of configurable logic blocks (CLBs)) are given for the relatively large Virtex II Pro XC2vp50-7 FPGA, which contains 23,616 slices (configurable elements of the FPGA), in Table 6.3. We can see that the filter bank requires a considerably smaller area on the chip than do the adaptive median filters, whose implementation was based on area-demanding sorting networks. One of the evolved filter banks is now protected using a utility model (a form of patent) by the Czech Industrial Property Office [30].

Table 6.3 Results of synthesis for various filters

Filter	Kernel size	CLBs	Max. frequency
Median filter	3×3	268	305 MHz
Median filter	5×5	1506	305 MHz
Median-column	5×1	69	310 MHz
Adaptive median	5×5	2024	303 MHz
Adaptive median	7×7	6567	298 MHz
FIB3	3×3	72	281 MHz
FIB5	5×5	108	242 MHz
single evolved	3×3	200	318 MHz
Three-bank filter	3×3	843	305 MHz

Figure 6.10 compares the results obtained from evolved filters (a single filter, and a three-bank filter) with conventional solutions (median filters (MFs) and adaptive median filters (AMFs)) using 25 images corrupted by salt-and-pepper noise of various intensities. We can observe that a single filter evolved using a 3×3-pixel kernel represents a better solution than the median filter if the noise intensity is low. For noise of high intensity, the bank of evolved filters is a better solution than the adaptive median filters. Nevertheless, neither the filter bank nor the adaptive median filters achieve the quality of the 'best SW algorithm' [5]. But that solution is not suitable for hardware implementation, as it does not utilize the concept of a local filtering window.

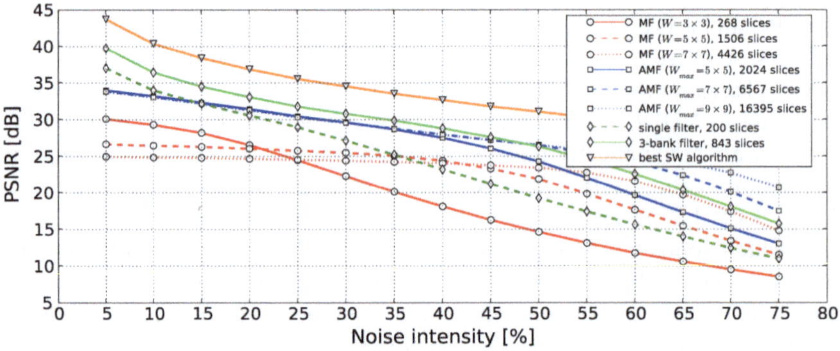

Fig. 6.10 Comparison of standard image filters (MFs, AMFs) and evolved filters (a single filter, and a three-bank filter) for salt-and-pepper noise of various intensities. The filtering quality is given as the mean PSNR value calculated for 25 images. The implementation cost is given in FPGA slices.

6.2.8.3 Impulse Burst Noise

An extended version of CGP was utilized for the evolution of impulse burst noise filters with a 5×5-pixel filtering window [35]. In fact, CGP was used to evolve nine-input filters, whose inputs were identified using a selector. The CGP parameters were initialized as follows: $n_c \times n_r = 6 \times 6$; population size = 8; mutation rate = 5%; number of generations = 50,000; and number of independent runs = 150. The initial population was generated randomly.

In order to evaluate the quality of filtering for the selected approaches, the mean value of PSNR was calculated for 15 test images with selected noise levels (Fig. 6.11). We compared evolved filters (FIB3 and FIB5), standard median filters, adaptive median filters and the three-bank filter.

Figure 6.12 shows one of the filters (denoted by FIB5) evolved with a 5×5 filtering window. Table 6.3 gives the implementation cost of the evolved filters for various existing approaches. The FIB5 filter occupies 108 slices and can operate at 242 MHz. We can observe that the evolved filters are relatively small but perform well.

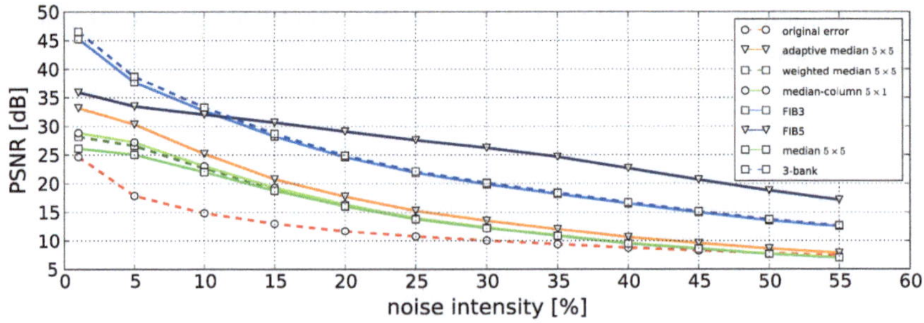

Fig. 6.11 Mean PSNR calculated using 15 images for different levels of impulse burst noise intensity.

Figure 6.11 shows the filtering capabilities of various filters for an image corrupted by 5% impulse burst noise. The results demonstrate that the evolved filter FIB5 provides the best results in most cases, especially for higher noise intensities. It is interesting that the best filter discovered by CGP (see Fig. 6.12) utilizes only the pixels of the central column of the filter window. These pixels are clearly the most important ones for suppressing this type of noise.

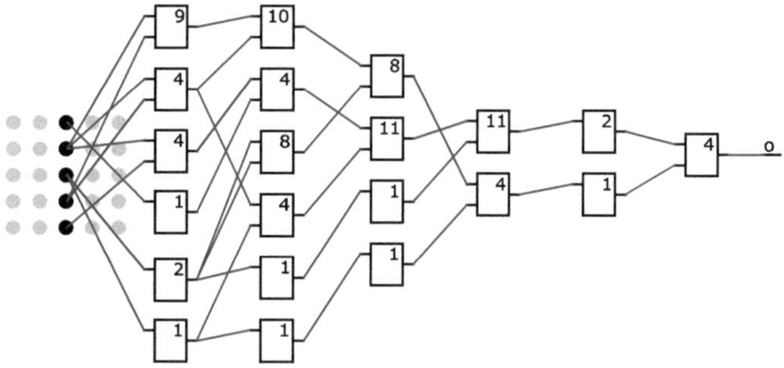

Fig. 6.12 The structure of the best evolved filter, FIB5 for impulse burst noise. The node functions are encoded as follows: (1), identity; (2), inversion; (4), minimum; (8), addition with saturation; (9), average; (10), if $(x > 127)$ then y else x; and (11), the absolute value of the difference.

(a) Original image **(b)** Corrupted image **(c)** FIB3 **(d)** FIB5

(e) Adaptive median 5×5 **(f)** Three-bank **(g)** Median 5×5 **(h)** Median-column

Fig. 6.13 One of the validation images corrupted by 5% impulse burst noise and filtered by various filters.

The experiments were conducted on a cluster consisting of 100 PCs (Pentium IV, 2.4 GHz, 1 GB RAM) using the Sun Grid Engine (SGE) which enables up to 100 independent experiments to be run in parallel. The evolution time of a single run was approximately 6 hours, until the CGP algorithm reached 50,000 generations.

6.2.9 Summary

The experimental results clearly demonstrate that CGP can automatically produce image filters that are competitive with conventional filters in terms of filtering quality. We observed that the images filtered by evolved filters exhibited more detail (and thus a higher visual quality) than images filtered by conventional filters (e.g. median filters). The implementation cost in an FPGA is also encouraging.

We will see in Sect. 7.3.2 that image filter design can be significantly accelerated when CGP is implemented using suitable hardware.

6.3 Evolving Advanced Image Filters

Using CGP, it is possible to evolve many types of advanced image filters. In previous work by Harding and Banzhaf [10, 11, 9, 12], the use of CGP to evolve a number of different common image filters was discussed. The task was framed as that of reverse-engineering some image filters such as the Sobel edge detection filter and the or dilation/erosion filter. By reverse engineering, we mean finding a mapping between an image and the output of a filter applied to it. The technique found may not be the same as that used by the process, but it produces similar results.

The approach used was similar to that described previously in this chapter, where kernel-based programs were evolved. An evolved CGP program takes a pixel and its neighbourhood, performs a calculation and returns a new value for the centre pixel. This set-up is illustrated in Fig. 6.14. The program is applied to each pixel in the image.

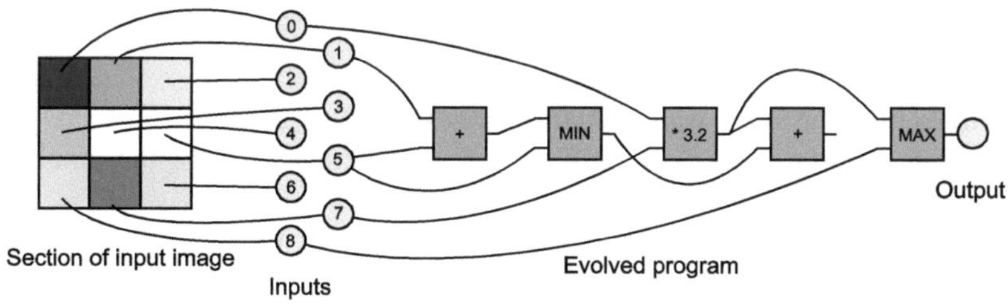

Fig. 6.14 In this example, the evolved program has nine inputs – these correspond to a section of an image. The output of the program determines the new colour of the centre pixel. Note that one node has no connections to its output. This means that this node is redundant, and will not be used during the computation.

Running an evolved program on a large number of pixels is a computationally intensive exercise. In Sect. 8.6, a methodology for accelerating this using graphics processing units (GPUs) is discussed. Here, we limit the discussion to the basic approach and to the evolved filters obtained.

Fig. 6.15 The training and validation image set. All images were presented simultaneously to the GPU. The first column of images was used to compute the validation fitness, and the remaining 12 for the training fitness. Each image is 256 × 256 pixels, with the entire set containing 1024 × 1024 pixels.

6.3.1 Fitness Function

The fitness function used was in many ways typical of those used in image processing. The quality of an evolved filter was the average error of the pixel values between the generated image and a target image.

With the acceleration in evaluation offered by a GPU, we were able to test individuals on a number of different images. Figure 6.15 shows the 16 different images used in the fitness evaluation. These were largely taken from the USC-SIPI image repository. To improve the quality of the results, 12 images were used for fitness evaluation, and four for a validation score. This allowed increased confidence that the evolved filters would generalize well.

Each image used was an eight-bit greyscale image, measuring 256×256 pixels.

The target images (i.e. the images after processing by a filter) were generated using the open source image-processing program GIMP [8].

The GPU implementation of the fitness function is discussed later, in Sect. 8.6.

6.3.2 Changes to the Standard CGP Genotype

The genotype and processing of the CGP genotype had a number of differences compared with the classical CGP implementation.

First, relative addresses were used instead of directly accessing connections. This approach is discussed in detail in Chap. 4.

In addition to the function type and connection information, each node also encoded a parameter as a floating-point number. This parameter was used by the evolved programs to encode a constant. The function set shown in Table 6.4 indicates where and how this parameter was used.

Another feature of the fitness function used here was that it allowed the image to be passed through the evolved filter a number of times. In addition to the encoded program, the genotype also contained an 'iteration counter' that specified how many times the filter should be applied to each pixel.

6.3.3 Evolutionary Algorithm, Parameters and Function Set

The algorithm used here was a simple evolutionary algorithm. The population had a size of 25. The mutation rate was set to be 5%, i.e. each gene in the genotype was mutated with probability 0.05. Crossover was not used. The iteration counter was also mutated with a 5% probability. The counter was mutated by adding a random number in the range -2 to 2. The counter was bounded between 1 and 5. Selection was done using a tournament selection of size 3. The five best individuals were promoted to the next generation without modification. The CGP graph was initial-

Table 6.4 CGP function set, defined on two inputs x and y and on the parameter c of the node

Function	Description
ITERATION	Returns the current iteration index
ADD	Adds the two inputs
SUB	Subtracts the second input from the first
MULT	Multiplies the two inputs
DIV	Divides the first input by the second, returning 1 if y is very close to zero
ADD CONST	Adds a constant (the node's parameter) to the first input
MULT CONST	Multiplies the first input by a constant (the node's parameter)
SUB CONST	Subtracts a constant (the node's parameter) from the first input
DIV CONST	Divides the first input by a constant (the node's parameter)
SQRT	Returns the square root of the first input
POW	Raises the first input to the power of the second input, returning 1 if the value is undefined.
SQUARE	Squares the first input
COS	Returns the cosine of the first input
SIN	Returns the sine of the first input
NOP	No operation – returns the first input
CONST	Returns a constant (the node's parameter)
ABS	Returns the absolute value of the first input
MIN	Returns the smaller of the two inputs
MAX	Returns the larger of the two inputs
CEIL	Rounds up the first input
FLOOR	Rounds down the first input
FRAC	Returns the fractional part of the number, $x - \lfloor x \rfloor$
LOG2	Log (base 2) of the first input
RECIPRICAL	Returns $1/\text{first input}$
RSQRT	Returns $1/\sqrt{\text{first input}}$

ized to contain 100 nodes (it is important to note that not all nodes were used in the generated program). Evolution was allowed to run for 50,000 evaluations.

Table 6.4 shows the available functions. These functions operate upon single-precision floating-point numbers. The function set was determined by mapping all the appropriate GPU programming API calls in the Microsoft Accelerator toolkit to functions.

6.3.4 Results

The results for evolving each filter are summarized in Table 6.5. In the following, the best validation result is shown, alongside the output of the target filter obtained from GIMP. Examples of the evolved programs are included to illustrate the types of operations that evolution found to replicate the target filters. Owing to space constraints, it is not possible to include such an analysis for every filter type. For the implementation of the original filters, there is extensive coverage in the literature and also in the source code for GIMP.

Results for the GPU acceleration can be found in Chap. 8.

Table 6.5 Results for each evolved filter. 'Best error' is the lowest error seen when testing against the validation images. 'Avg Validation Error' is the average of the best validation error. 'Avg. validation evaluations' is the average number of evaluations required to find the best validation error. 'Avg. training error' is the average of the lowest error found on the training images. 'Avg. training evaluations' is the average number of evaluations required to find the best training fitness. Each experiment was repeated for 20 trials

Filter	Best error	Avg. validation error	Avg. validation evaluations	Avg. training error	Avg. training evaluations
Dilate	0.57	0.71	2,422	0.67	3,919
Dilate2	5.84	6.51	11,361	6.10	39,603
Emboss	3.86	8.33	15,517	7.41	34,878
Erode	0.56	0.78	3,892	0.73	4,066
Erode2	5.70	6.72	26,747	6.64	40,559
Motion	2.08	2.32	29,464	2.24	43,722
Neon	1.32	2.69	15,836	2.41	35,146
Sobel	8.41	22.26	26,385	20.12	45,744
Sobel2	1.70	3.82	19,979	3.55	39,155
Unsharp	5.85	5.91	301	5.61	37,102

6.3.4.1 Dilation and Erosion

Figure 6.16 shows the result of evolving the Dilate filter. In Dilate2 (Fig. 6.17), the filter is applied twice. Figure 6.18 shows the result of evolving the Erode filter. In Erode2 (Fig. 6.19), the filter is applied twice.

It is possible to analyse the evolved programs to determine how the filters work. For the erosion filter, the best evolved program contained eight operations and required five iterations to run. The evolved expression is

$$OUTPUT = MAX(LOG_2(I_8), MIN(I_8 + (MIN(I_3, I_7) - MAX(I_7, I_8)), FLOOR(I_1))),$$

where I_1 to I_9 are the input pixels, with I_1 being in the top left of the kernel and I_9 in the lower right.

The best dilation program contained four instructions, and again required five iterations. The evolved program is

$$OUTPUT = MAX(MAX(I_9, MAX(I_1, I_5)), MAX(I_3, I_7)).$$

In contrast, the best evolved programs for applying erosion and dilation twice both contained 17 instructions (and again required five iterations). It is unclear why these programs should need to be so much more complicated.

6.3.4.2 Emboss, Sobel and Neon

The Emboss, Sobel and Neon filters are various types of edge detectors. It was chosen to evolve these three different types, as the outputs are very different. Emboss is a directional filter, whereas Sobel and Neon are not. It was found that all three types of filters could be accurately evolved.

Figure 6.20 shows the result of evolving the Emboss filter. When visually compared, the evolved Emboss filter is very similar to that used by GIMP (for this and the other sample images here, the most representative and visually useful subimage has been used). The evolved program for this filter contains 20 nodes:

$$\begin{aligned} Output = {} & ABS(MIN(-0.3857, \\ & POW(SQRT(I_2/((I_8+RSQRT(I_5))-0.863)), I_8))+ \\ & CEIL(MIN(((I_3-I_9)+(I_1-I_7)) \\ & -129.65, FRAC(I_5)))). \end{aligned} \tag{6.3}$$

Figure 6.21 shows the result of evolving the Neon filter.

GIMP has two versions of the Sobel filter. Figure 6.22 shows the result of evolving the normalized Sobel filter. In the case of Sobel2 (Fig. 6.23), the target was the output of the standard Sobel filter. The standard Sobel filter achieved poor error rates; however, the visual comparison is very good. It would appear that the evolved output is scaled differently, and hence the pixel intensities are different. If the two images are normalized, the error is reduced. However, the fitness function does not normalize automatically, but leaves this task to evolution.

The evolved Sobel filter is quite complicated:

$$\begin{aligned} A &= I_7 - I_3 \\ B &= I_9 - MAX(I_1, LOG_2(A)) \\ OUTPUT &= 2.0*(MIN(MAX(ABS(B)+FRAC(1)+ABS(2.0*A), \\ &\quad MAX(FLOOR(LOG_2(A)), 2.0*B)), \\ &\quad (CEIL(FRAC(I_5))*-0.760)+127.24)). \end{aligned}$$

The best evolved program for Sobel2 is considerably shorter:

$$\begin{aligned} Output = {} & MAX(ABS((I_9-I_1)*0.590), \\ & POW(I_3-I7, SQRT(0.774))/-2.245). \end{aligned}$$

Again, both programs required five iterations. This suggests there is some bias in the algorithm towards increasing the number of iterations to the maximum allowed.

6.3.4.3 Motion Blur

Figure 6.24 shows the result of evolving the Motion filter. The output of the evolved filter did not match the desired target very accurately. But although there is a degree of blurring, it is not as pronounced as in the target image. Motion blur is a relatively subtle effect, and as the target and input images are quite similar, it is likely that evolution will become trapped in a local minimum.

6.3.4.4 Unsharp

Figure 6.25 shows the result of evolving the Unsharp filter. Unsharp was the most difficult filter to evolve. It is suspected that this is due to the Gaussian blur that needs to be applied as part of the procedure. It is difficult to see how, with the current function set, such an operation could be evolved. In future work, this issue will need to be rectified.

Fig. 6.16 Dilate: evolved filter, GIMP filter and difference image.

Fig. 6.17 Dilate2: evolved filter, GIMP filter and difference image.

Fig. 6.18 Erode: evolved filter, GIMP filter and difference image.

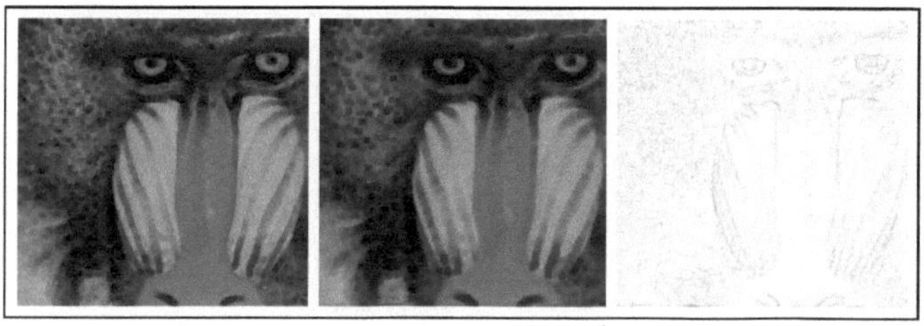

Fig. 6.19 Erode2: evolved filter, GIMP filter and difference image.

Fig. 6.20 Emboss: evolved filter, GIMP filter and difference image.

Fig. 6.21 Neon blur: evolved filter, GIMP filter and difference image.

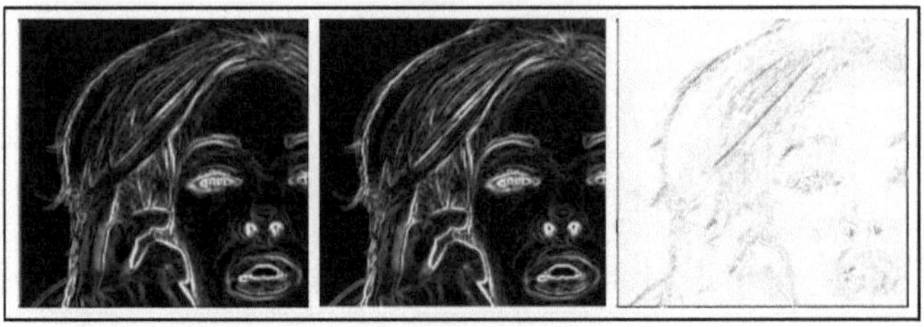

Fig. 6.22 Sobel: evolved filter, GIMP filter and difference image.

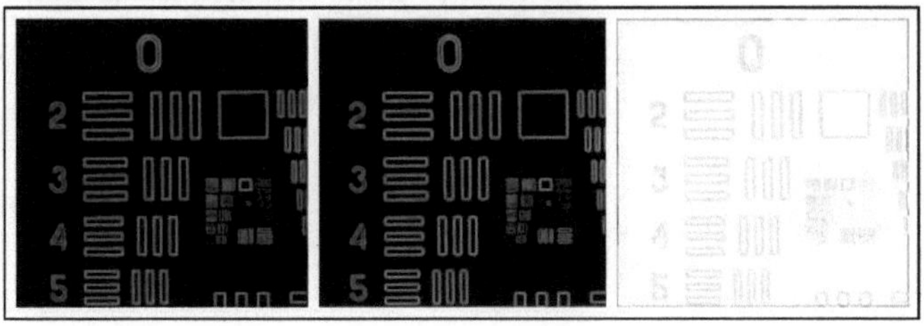

Fig. 6.23 Sobel2: evolved filter, GIMP filter and difference image.

Fig. 6.24 Motion blur filter: evolved filter, GIMP filter and difference image.

Fig. 6.25 Unsharp blur filter: evolved filter, GIMP filter and difference image.

6.4 The Automated Design of Features for Image Classification

In this section, we discuss the application of CGP to the automated discovery of features in an image classification task. We attempt to evolve transformations on the space of images, in the hope that particular transforms which emphasize distinguishing characteristics may be found. A set of moments describing the transformed image are extracted, and used for classification.

We have applied our work to the recognition of cells in a medical image database. Specifically, we attempted to recognize nuclear inclusions which indicate the presence (or absence) of OPMD, a form of muscular dystrophy. Recognition of these inclusions is a difficult task, usually requiring expert training of a human operator.

6.4.1 Motivation

Image classification (recognition) typically uses a set of features to reduce the dimensionality of the pattern space. A great deal of effort is spent on the design, selection and weighting of features. Often, practitioners begin with a set of 'standard' (generally applicable) features, and then use some machine-learning technique to find the most appropriate choices; these features often involve statistical moments, or entropy- or histogram-based measures (as in [17]). Domain-specific measures can be used as well, such as the cell-nucleus-specific measures used in the construction of the Wisconsin Breast Cancer Database [33].

The advantage of domain-specific features is, of course, potential increased relevance to the problem domain at hand, ideally increasing the efficacy of an image-processing system. The downside is the attention that the assembly of domain-specific features requires, where human operators – possibly domain experts – are often required for research, exploration and optimization.

The capacity to extract database-specific features automatically is enticing: one can conceivably have both application without expert intervention and the increased performance associated with database specificity. There are several existing techniques for the automated discovery of learned features. Common techniques include linear discriminant techniques and principal-component-analysis-based techniques, and there are also others [13]. There also exist several related genetic-programming-based applications in image processing, such as the automated detection of points of interest in an image [42], and the classification of image textures [21].

Our technique has followed similar motivation, but with some key differences: (a) we make nearly no assumptions regarding the distribution of class data; (b) our technique is specifically oriented towards image recognition, thus emphasizing local neighbourhoods of pixels in two dimensions; and (c) the results of our technique are easily analysed, both visually and analytically.

6.4.2 The Model

We treat an image space as a collection of pixels $p = (p_x, p_y) \in I$, and an image as a collection of intensities on that space, $f(p) \in [0,1]$, or simply $f(I)$.

We have seen how CGP kernel functions can be used to reverse-engineer image-processing filters. Here we use this notion, that of a sliding window function, and apply it to the design of transforms on the space of images. Since we know that this space of image transforms contains *some* useful image filters, we assume that more specialized or effective filters can be designed to highlight portions of an image database. Ultimately, we are searching for kernel functions that will help us in the classification of images.

Here we outline the definition of an evolvable feature extractor from a CGP graph. The CGP graph is used to define a transformation on the space of images. In this case, however, no explicit image target is given; instead, we hope to create a transform useful for accentuating the important features of the images. Next, these transformed images are converted to a set of numerical values. Finally, these numerical values are used to train or test a classifier. An overview of the entire process is illustrated in Fig. 6.26.

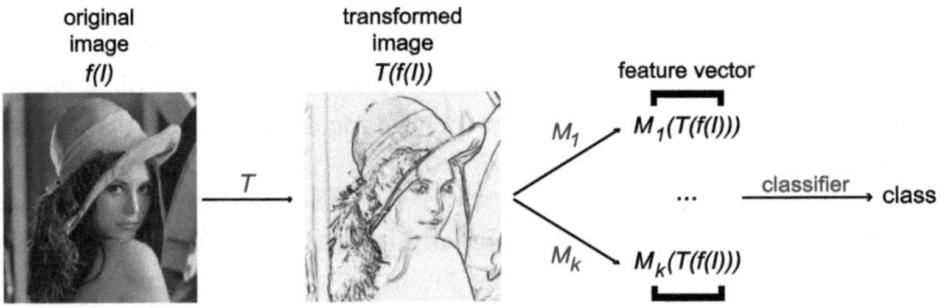

Fig. 6.26 An overview of the feature extraction procedure.

6.4.2.1 CGP Graphs as Image Transforms

We again use a simple form of CGP. The graph consists of a single row of connection nodes of length $n_c = L_n$. These are augmented by a large collection of inputs, labelled $I_0, \ldots, I_{n_i-1}$. Levels-back is chosen so that any connection node may connect to any previous node or input, and addressing is relative. There is a single output, automatically defined to be the final node in the representation. We use the function set {Max, Min, Add, Sub, Mult, Div, Const, Abs, Square, Pow}, i.e. a simple listing of functions known to beuseful in common convolution functions.

As before, the CGP graphs operate on a square of pixel intensities, $\bar{p} = \{p_0, p_1,$ ..., $p_{m^2}\}$, surrounding a central pixel p. Given a CGP graph T with $n_i = m^2$, we can apply the graph to $\bar{p}$ as follows:

$$T(f(\bar{p})) = T(f(p_0), f(p_1), \ldots, f(p_{n^2})). \tag{6.4}$$

Note that if $f(q)$ does not exist for some pixel q (beyond the edges of the image, say), then we return the value $f(q) = -1$. Furthermore, if $T(f(\bar{p})) \notin [0,1]$, we replace it by the closest boundary value, either 0 or 1.

Given some image $f(I)$, we can define a new image $T(f(I))$ as follows. Let I' be an image space of the same dimensions as I. For each pixel $p' \in I'$, let $f(p') = T(f(\bar{p}))$. Hence, every CGP graph T can be viewed as a transform on the space of images.

6.4.2.2 Feature Extraction

Thus far, we have discussed the generation of an image transform from a CGP graph, allowing us a means of mapping from the space of images to itself. However, our data still exists in a very high-dimensional space, unsuitable for a classifier. To perform the dimensionality reduction, we have to rely on some common statistical moments, chosen for simplicity and speed. The choice of a simple statistical measure is a purposeful attempt to constrain the results of the evolutionary process. That is, the use of GP is generally known to be capable of low bias and high variance [16], and a well-known but biasing step is likely to help us avoid overfitting.

There are many forms of statistical moments that could be used in this context. Local, orthogonal and multi-scale moments – such as wavelet-basis and Zernike polynomials, are common choices, but we are not considered desirable for the given application. We instead selected a collection of simple geometrically-based moments, some of which were transform-, rotation- and scale-invariant. These properties are desirable for cell images, since the exact location, zoom level and orientation of the cell image relative to the image boundaries is superfluous. We thus chose a collection of 16 moments for the reduction of images to numerical vectors, drawn from the first few geometric moments, central moments and Hu's moments [14]. For details, see [20].

6.4.2.3 A Cell Classification Problem

We worked with a database of images of cells, CellsDB, originally collected by the Centre hospitalier de l'Université de Montréal (CHUM), where the causes and associated symptoms of oculopharyngeal muscular dystrophy (OPMD) have been studied extensively [1]. Intranuclear inclusions (INIs) have been detected via both pathological studies and electron microscopy. These INIs are tubular, about 8.5 nm in external diameter and 3 nm in internal diameter and up to 0.25 μm in length, and converge to form tangles or palisades [34]. Detection of these inclusions is expected to lead to the detection of OPMD. CellsDB was collected and prepared by Tarundeep Dhot at the Centre for the Study of Brain Diseases at CHUM, and is pre-segmented and divided into two categories associated with the presence or absence of inclusions indicating OPMD: 'healthy' and 'sick'. An example of some cell images may be seen in Fig. 6.27.

Detecting OPMD-indicating INIs is a difficult task, requiring training for human classification. It is dependent upon the relative pixel intensities in a neighbourhood, and it is thus difficult to distinguish between noise and other intranuclear patterns. A further difficulty is that cell images come in a variety of scales: for this reason, we chose to mix images taken at 10×, 20×, and 40× zoom. We broke the database into two sets: 186 healthy and 200 sick cell images for training, and 200 healthy and 200 sick cell images for validation.

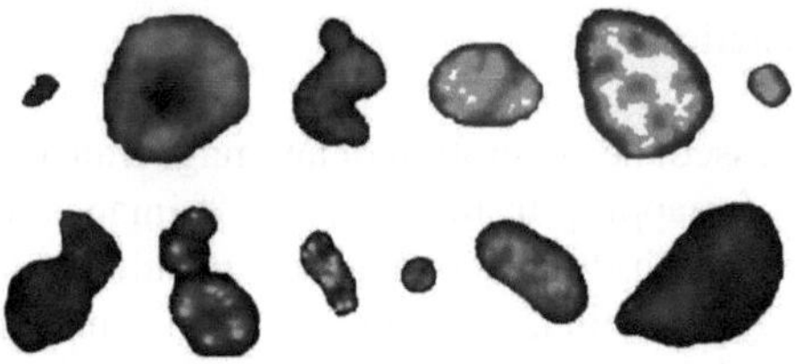

Fig. 6.27 Examples of healthy (top row) and sick (bottom row) cell images.

We chose to investigate the classifiability of this database using a collection of predefined moments, drawn from previous work in image classification. We implemented a collection of general features used in images processing: *image entropy*, *fractal dimension*, and our 16 statistical moments. We also selected a collection of cell-specific features, designed to operate on the nuclear boundary and interior of a cell image: *threshholded area*, *variance*, *mean radius*, *radius variance*, *perimeter* and *compactness*. These 24 features were calculated for the validation set of images, and evaluated using a collection of classifiers with 10-fold cross-validation. The results are summarized in Table 6.6. The best expected performance was an *SS* value of 0.580 – composed of a true-positive rate of 0.774 and a true-negative rate of 0.750 – under a 5-nearest neighbour (5-NN) classifier.

6.4.2.4 Evaluation of a Transform

A common and fast choice for the evaluation of a set of feature values is the use of a measure of inter-class and intra-class spread. Here, however, we used a classifier directly. Although slower, this effectively eliminated assumptions regarding the underlying distribution of data points, thus allowing us to consider a wider range of potential feature extractors.

We wished to award fitness to any particular collection of feature values based on its ability to distinguish between healthy and sick cells. To do so, we converted the database to a set of features, and then attempted to classify the cells using some given classifier and evaluation technique. The classifier returned a false-positive rate, *FPR*, for both classes, which we combined into a measure of *sensitivity speci-*

ficity:

$$SS = (1 - FPR(healthy))(1 - FPR(sick)) \tag{6.5}$$

$SS \in [0,1]$ iwould be maximized for perfect recognition. All classifiers and evaluation techniques were implemented via the Weka machine-learning system, version 3.5.7 [41].

For the training runs, a set of 16 feature values were computed for the transformed image, consisting of the 16 chosen moments. These were evaluated under a 50–50 train–test split.

For the validation runs, a set of 16 features were computed for the transformed image, and added to the original 24 predefined features. These 40 values were evaluated using a 1-NN classifier with 10-fold cross-validation.

The reason that a smaller set of features was used in the definition of the training evaluation was as follows: randomization of the database prior to classification is a useful feature to help prevent overfitting, but leads to a stochastic fitness function. Beginning with a set of features which is already adept at classification – as the predefined features are – makes the variance due to stochasticity greater than the fitness gains in early evolution, hence hindering the selection operator.

Table 6.6 shows the expected classification accuracies of the feature sets for completely randomly generated transforms. There is little difference in mean performance between the predefined features alone and the inclusion of a randomly generated transform, but a difference can be seen in the overall variance, suggesting that some new useful information is occasionally added.

Table 6.6 Comparison of classification on a standard database of unevolved features (∅) with a database augmented by a randomly generated transform. All figures are the mean (with the standard deviation in parentheses) of *SS* over 40 runs

Transform	**Decision tree**	**Ridor**	**1-NN**	**5-NN**
∅	0.495 (0.020)	0.461 (0.024)	0.517 (0.012)	0.580 (0.016)
Random transform	0.521 (0.080)	0.503 (0.059)	0.553 (0.070)	0.611 (0.067)

6.4.3 Transform Evolution

We conducted an informal parameter search to select a good running point for our evolutionary algorithm. Contrary to results for previous applications, in this context experimentation demonstrated to us that crossover was a useful genetic operator: we therefore included a single-point crossover in our work. We selected a population size of 200, a rate of mutation of 0.02, a probability of crossover of 0.6, single-member elitism, a graph size of $L_n = 100$ and a kernel size of 6×6. Each evolutionary algorithm was run for 50 generations, using a 1-NN classifier; 40 runs were undertaken in total.

Evolution was quite successful at increasing the fitness (training *SS*), which increased from a mean best fitness of 0.523 in the first generation to a mean best fitness of 0.620. In a few cases, evolution optimized training *SS* at the expense of validation *SS*, but a good general increase in the latter was seen in most runs. The mean best validation *SS* for the final generation was 0.676 (s.d. 0.052), and was maximized at a value of 0.766, a 32% improvement over the expected performance of the best classifier found for predefined features alone.

6.4.3.1 Best Discovered Transform

In this subsection, we explore the best evolved transform in more detail. The best individual of the final (50th) generation of the run with the highest validation fitness was selected. This best individual had a sensitivity specificity of 0.766, encompassing true-positive rates of 0.878 for healthy cells and 0.873 for sick cells.

The same evolved attribute values, when evaluated using a J48 decision tree instead of a 1-NN, gave a sensitivity-specificity of 0.801, encompassing true-positive rates of 0.912 for healthy cells and 0.878 for sick cells, or a 38% improvement over the expected performance of the best classifier found for predefined features alone. The performance for the best discovered transform, relative to the best expected performance for the unevolved features alone, is summarized in Table 6.7.

Table 6.7 Comparison of classification results for unevolved versus best evolved features

Classifier	Transform	SS	$TPR(H)$	$TPR(S)$
5-NN	$\emptyset$	0.580	0.774	0.750
1-NN	Best transform	0.766	0.878	0.873
Decision tree	Best transform	0.801	0.912	0.878

The best transform, once neutral code is removed, may be written as the following neighbourhood function:

$$\text{Output}(I_0,\ldots,I_{35}) = \text{Min}(I_{11} - I_{15}, \text{Pow}(I_9 - \text{Max}(I_6, I_{25}), I_8)).$$

The same transform is illustrated in CGP graph form in Fig. 6.28. This function mostly returns black (0), except when both $I_{11} - I_{15}$ and $(I_9 - \max\{I_6, I_{25}\})^{I_8}$ are (relatively) high. The first subsection ensures that the variance of the right-hand part of the neighbourhood is high (excluding inclusions that are too large). The second subsection ensures that I_9 is high, while both of I_6 and I_{25} are low or just I_8 is low. This ensures that that the left part of the image is dark, while there is some lightness in the right half, hence excluding light spots that extend to the right of the neighbourhood. Hence, we detect the left half of inclusions of the proper size and variance. Note that this function works for several different microscope zoom levels simultaneously.

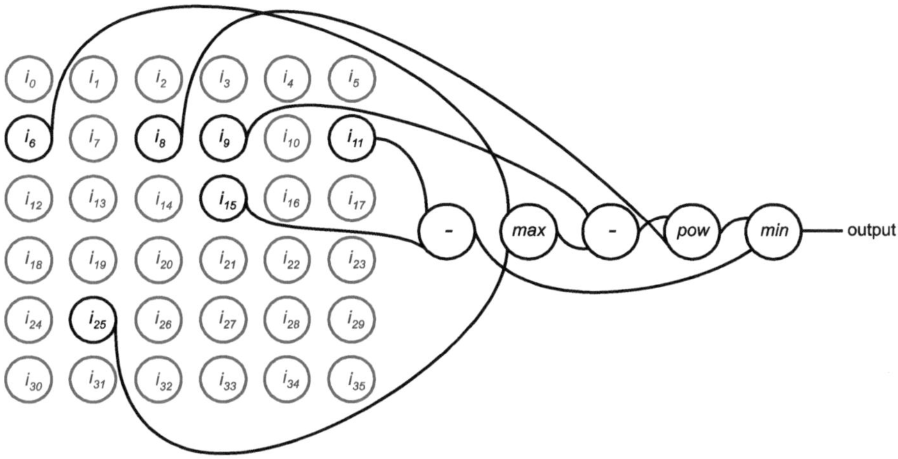

Fig. 6.28 A CGP graph view of the best discovered individual.

Although non-OPMD-indicating inclusions are still highlighted using this function, the OPMD-indicating inclusions are highlighted with more intensity, hence allowing greater recognition of the cells than with the original image features alone. Examples of the output are shown in Fig. 6.29.

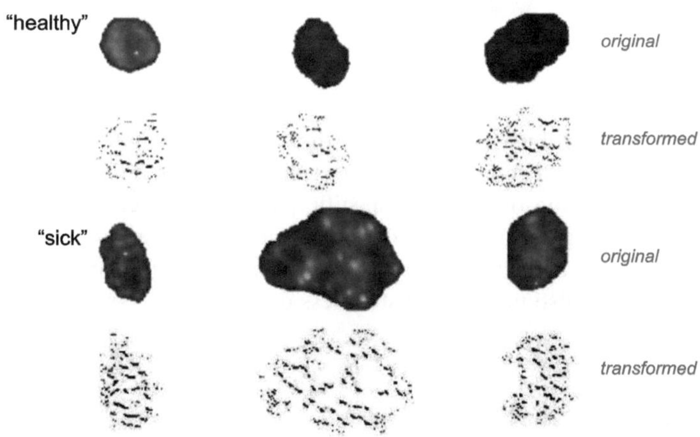

Fig. 6.29 Examples of the action of the best discovered transform on cell images. The transformed images have been subjected to contrast stretching and colour inversion to make the distinction between black and dark grey more visible.

Feature selection was run on the database of the original moments and the evolved moments, using an information gain attribute evaluator. The evaluator selected 30 of the features as significant, discarding the other 10. The top 10 ranked attributes, by information gain, were all moments of the evolved transformed image.

6.4.4 Future Directions

The use of a single image transform in the above work was based on our expectation of the pattern to be detected – we aimed to discover one type of nuclear INI, and hence we chose a single transform. Evolving several transforms simultaneously is also a possibility. Figure 6.30 shows a visualization of some transforms that were evolved when we selected for four transforms rather than one using the CellsDB database. In this case, a transform recovering the area of the interior of the cell nucleus is visible, and also partial edge detectors: this is an effective recovery of some of the more important information originally included in the predefined features. Clearly, some problem domains would require several transforms to distinguish the class data, and the best means of representing multiple image transforms and ensuring good cooperation between them is an open problem.

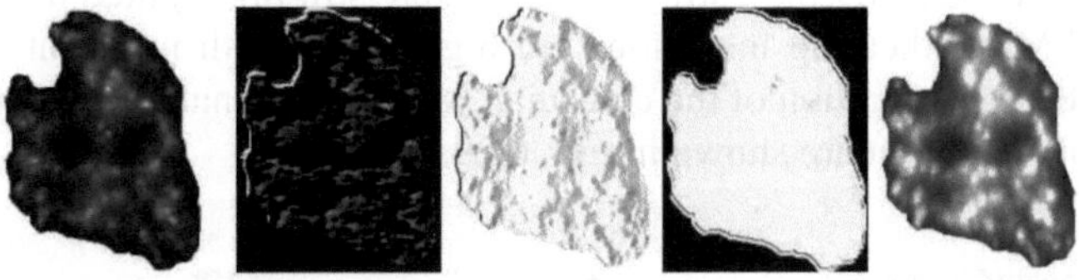

Fig. 6.30 The action of four simultaneously evolved transforms (*right*) on a cell image (*left*).

Evolutionary algorithms are known for their capacity to automatically discover parameters in learning techniques. Another valuable line of development would be the automated evolution of database-specific parameters, such as the selection of a dimension-reducing set of moments, the selection of an appropriate local pixel neighbourhood size, and the selection of an appropriate number of image transforms. These would also make the system less dependent on expert input.

6.5 Acknowledgements

Lukas Sekanina was partially supported by the Grant Agency of the Czech Republic under Contract No. P103/10/1517 and by the research programme MSM 0021630528. WB and SH gratefully acknowledge funding from Atlantic Canada's

HPC network ACENET, from the Canadian Foundation of Innovation, New Opportunities Grant No. 204503, and from NSERC under the Discovery Grant Program RGPIN 283304-07.

References

1. Brais, B., Bouchard, J.P., Xie, Y.G., Rochefort, D.L., Chretien, N., Tome, F.M., Lafreniere, R.G., Rommens, J.M., Uyama, E., Nohira, O.: Short GCG expansions in the PABP2 gene cause oculopharyngeal muscular dystrophy. Nature Genetics **18**, 164–167 (1998)
2. Brownrigg, D.: The weighted median filter. Commun. ACM **27**(8), 807–818 (1984)
3. Burian, A., Takala, J.: Evolved Gate Arrays for Image Restoration. In: Proc. Congress on Evolutionary Computing, pp. 1185–1192. IEEE Press (2004)
4. Cagnoni, S., Lutton, E., Olague, G.: Genetic and Evolutionary Computation for Image Processing and Analysis. EURASIP Book Series on Signal Processing and Communications, Volume 8. Hindawi Publishing Corporation (2007)
5. Chan, R.H., Ho, C.W., Nikolova, M.: Salt-and-Pepper Noise Removal by Median-type Noise Detectors and Edge-preserving Regularization. IEEE Transactions on Image Processing **14**, 1479–1485 (2005)
6. Dougherty, E.R., Astola, J.T. (eds.): Nonlinear Filters for Image Processing. SPIE/IEEE Series on Imaging Science & Engineering. SPIE/IEEE (1999)
7. Dumoulin, J., Foster, J.A., Frenzel, J.F., McGrew, S.: Special Purpose Image Convolution with Evolvable Hardware. In: Real-World Applications of Evolutionary Computing – Proc. Workshop on Evolutionary Computation in Image Analysis and Signal Processing, *LNCS*, vol. 1803, pp. 1–11. Springer (2000)
8. GNU: GNU image manipulation program (GIMP). www.gimp.org (2008). [Online; accessed 21-January-2008]
9. Harding, S.L.: Evolution of Image Filters on Graphics Processor Units Using Cartesian Genetic Programming. In: J. Wang (ed.) Proc. IEEE World Congress on Computational Intelligence. IEEE Press (2008)
10. Harding, S.L., Banzhaf, W.: Genetic Programming on GPUs for Image Processing. In: J. Lanchares, F. Fernandez, J. Risco-Martin (eds.) Proc. Workshop on Parallel and Bioinspired Algorithms, pp. 65–72. Complutense University of Madrid Press (2008)
11. Harding, S.L., Banzhaf, W.: Genetic programming on GPUs for image processing. International Journal of High Performance Systems Architecture **1**(4), 231–240 (2008)
12. Harding, S.L., Banzhaf, W.: Distributed Genetic Programming on GPUs using CUDA. In: I. Hidalgo, F. Fernandez, J. Lanchares (eds.) Proc. Workshop on Parallel Architectures and Bioinspired Algorithms, pp. 1–10 (2009)
13. Hastie, T., Tibshirani, R., Friedman, J.: The Elements of Statistical Learning: Data Mining, Inference, and Prediction, 2nd edn. Springer (2008)
14. Hu, M.K.: Visual pattern recognition by moment invariants. IRE Transactions on Information Theory **IT-8**, 179–187 (1962)
15. Hwang, H., Haddad, R.A.: Adaptive median filters: new algorithms and results. IEEE Transactions on Image Processing **4**(4), 499–502 (1995)
16. Keijzer, M., Babovic, V.: Genetic Programming, Ensemble Methods and the Bias/Variance Tradeoff – Introductory Investigations. In: Proc. European Conference on Genetic Programming, pp. 76–90. Springer (2000)
17. Kharma, N., Kowaliw, T., Clement, E., Jensen, C., Youssef, A., Yao, J.: Project CellNet: Evolving an Autonomous Pattern Recognizer. International Journal of Pattern Recognition and Artificial Intelligence **18**(6), 1039–1056 (2004)
18. Koivisto, P., Astola, J., Lukin, V., Melnik, V., Tsymbal, O.: Removing Impulse Bursts from Images by Training-Based Filtering. EURASIP Journal on Applied Signal Processing **2003**(3), 223–237 (2003)

19. Koivisto, P., Huttunen, H., Kuosmanen, P.: Training-based optimization of soft morphological filters. Journal of Electronic Imaging **5**(3), 300–322 (1996)
20. Kowaliw, T., Banzhaf, W., Kharma, N., Harding, S.: Evolving Novel Image Features Using Genetic Programming-Based Image Transforms. In: Proc. IEEE Congress on Evolutionary Computation. IEEE Press (2009)
21. Lam, B., Ciesielski, V.: Discovery of Human-Competitive Image Texture Feature Extraction Programs Using Genetic Programming. In: Proc. Genetic and Evolutionary Computation Conference, pp. 1114–1125. Springer (2004)
22. Marshall, S.: New direct design method for weighted order statistic filters. IEE proceedings. Vision, image and signal processing **151**(1), 1–8 (2004)
23. Nikolova, M.: A Variational Approach to Remove Outliers and Impulse Noise. J. Math. Imaging Vis. **20**(1–2), 99–120 (2004)
24. Porter, R.: Evolution on FPGAs for Feature Extraction. Ph.D. thesis, Queensland University of Technology, Brisbane, Australia (2001)
25. Schulte, S., Nachtegael, M., Witte, V.D., der Weken, D.V., Kerre, E.E.: Fuzzy Impulse Noise Reduction Methods for Color Images. In: Computational Intelligence, Theory and Applications International Conference 9th, Fuzzy Days in Dortmund, pp. 711–720. Springer (2006)
26. Sekanina, L.: Image filter design with evolvable hardware. In: Applications of Evolutionary Computing, *LNCS*, vol. 2279, pp. 255–266. Springer (2002)
27. Sekanina, L.: Evolvable components: From Theory to Hardware Implementations. Natural Computing. Springer (2004)
28. Sekanina, L., Martinek, T.: Evolving Image Operators Directly in Hardware. In: S. Cagnoni, E. Lutton, G. Olague (eds.) Genetic and Evolutionary Computation for Image Processing and Analysis, EURASIP Book Series on Signal Processing and Communications, Volume 8, pp. 93–112. Hindawi Publishing Corporation (2007)
29. Sekanina, L., Ruzicka, R.: Easily Testable Image Operators: The Class of Circuits Where Evolution Beats Engineers. In: Proc. NASA/DoD Conference on Evolvable Hardware, pp. 135–144. IEEE Computer Society (2003)
30. Sekanina, L., Vasicek, Z.: Nonlinear Image Filter (2009). Czech Utility model UV020017
31. Slany, K., Sekanina, L.: Fitness Landscape Analysis and Image Filter Evolution Using Functional-Level CGP. In: Proc. of European Conf. on Genetic Programming, *LNCS*, vol. 4445, pp. 311–320. Springer (2007)
32. Sonka, M., Hlavac, V., Boyle, R.: Image Processing: Analysis and Machine Vision. Thomson-Engineering (1999)
33. Street, W., Wolberg, W., Mangasarian, O.: Nuclear feature extraction for breast tumor diagnosis. In: IS&T/SPIE 1993 International Symposium on Electronic Imaging (1993)
34. Tome, F.M.S., Fradeau, M.: Nuclear changes in muscle disorders. Methods Achiev. Exp. Pathol. **12**, 261–296 (1986)
35. Vasicek, Z., Bidlo, M., Sekanina, L., Torresen, J., Glette, K., Furuholmen, M.: Evolution of Impulse Bursts Noise Filters. In: Proc. NASA/ESA Conference on Adaptive Hardware and Systems, pp. 27–34. IEEE Computer Society (2009)
36. Vasicek, Z., Sekanina, L.: An Area-Efficient Alternative to Adaptive Median Filtering in FPGAs. In: Proc. International Conference on Field Programmable Logic and Applications, pp. 216–221. IEEE Computer Society (2007)
37. Vasicek, Z., Sekanina, L.: An Evolvable Hardware System in Xilinx Virtex II Pro FPGA. International Journal of Innovative Computing and Applications **1**(1), 63–73 (2007)
38. Vasicek, Z., Sekanina, L.: Evaluation of a New Platform For Image Filter Evolution. In: Proc. NASA/ESA Conference on Adaptive Hardware and Systems, pp. 577–584. IEEE Computer Society (2007)
39. Vasicek, Z., Sekanina, L.: Reducing the Area on a Chip Using a Bank of Evolved Filters. In: Proc. International Conference on Evolvable Systems, *LNCS*, vol. 4684, pp. 222–232. Springer (2007)
40. Vasicek, Z., Sekanina, L.: Novel Hardware Implementation of Adaptive Median Filters. In: Proc. IEEE Design and Diagnostics of Electronic Circuits and Systems Workshop, pp. 110–115. IEEE Computer Society (2008)

41. Witten, H., Frank, E.: Data Mining: Practical Machine Learning Tools and Techniques, 2nd edn. Morgan Kaufmann (2005)
42. Zhang, M., Ciesielski, V.B., Andreae, P.: A Domain-Independent Window Approach to Multiclass Object Detection Using Genetic Programming. EURASIP Journal on Applied Signal Processing **2003**(8), 841–859 (2003)
43. Zhou, W., David, Z.: Progressive switching median filter for the removal of impulse noise from highly corrupted images. IEEE Trans on Circuits and Systems: Analog and Digital Signal Processing **46**(1), 78–80 (1999)

Chapter 7
CGP Acceleration Using Field-Programmable Gate Arrays

Lukas Sekanina and Zdenek Vasicek

7.1 Reconfigurable Chips

In comparison with fixed architectures such as processors and single-purpose chips, the structure and parameters of reconfigurable chips can be modified by writing configuration data to the configuration memory [4]. Most reconfigurable devices consist of configurable blocks whose functions and interconnections are controlled by configuration data. Also, the connection of configurable blocks to input–output pins can be configured. The configuration bits directly control the configuration subsystem of the reconfigurable chip, i.e. configurable switches, selection signals for multiplexers and the contents of programmable look-up tables. Therefore, a single chip can implement many different functions depending on its particular configuration. Creating a configuration for a reconfigurable device takes much less time than fabricating a new application-specific integrated circuit.

The reasons for using reconfigurable chips can be various. Firstly, the possibility of reconfiguration can extend the lifespan of a system. For example, when a new driver or peripheral device is added to a system, existing hardware could have a problem in communicating with it. However, if the system is implemented in a reconfigurable chip, the hardware can be updated by simply reprogramming the configuration memory. Secondly, an application which is too complex to fit on a single chip can still be carried out on a relatively small reconfigurable chip if the application can be divided into modules whose configurations alternate on the chip. Then, it is possible to reduce the power consumption, size or weight of the application. Thirdly, the reliability of the system can be increased by a suitable reconfiguration after a fault is detected and isolated. Finally, reconfigurable chips can be dynamically reconfigured in order to ensure adaptive behaviour in a changing environment.

J.F. Miller (ed.), *Cartesian Genetic Programming*, Natural Computing Series,
DOI 10.1007/978-3-642-17310-3_7, © Springer-Verlag Berlin Heidelberg 2011

The main disadvantage associated with the use of reconfigurable devices is that the circuits allowing the 'configurability' occupy a considerable area on the chip and make the whole system slower in comparison with application-specific integrated circuits fabricated using the same technology.

7.2 Field-Programmable Gate Arrays

FPGAs are the most significant and popular examples of reconfigurable chips. The first FPGAs were commercially introduced in the mid-1980s by Xilinx, Inc. They were typically used to facilitate rapid prototyping of new electronic products. Modern FPGAs can compete with application-specific integrated circuits in many domains, for example in advanced signal processing, medical imaging, computer vision, search engines, and embedded systems.

Figure 7.1 shows a typical architecture of a Xilinx FPGA.[1] This is a two-dimensional array of reconfigurable resources that include configurable logic blocks (CLBs), programmable interconnect blocks (PIBs) and reconfigurable I/O blocks (IOBs). A CLB consists of so-called slices; each of them contains function generators implemented using three-, four- or six-input look-up tables (depending on the particular family of FPGA), flip-flops and some additional logic. The configuration bit stream is stored in the configuration SRAM memory, which is also integrated on the chip.

FPGAs differ in the amount and type of resources available on the chip. The most advanced FPGAs contain more than 10,000 CLBs and integrate, in addition to CLBs, various embedded hard cores such as SRAM memories, fast multipliers, processors, gigabit interfaces, and PCI interfaces (see PowerPC processors in Fig. 7.1). Because the existence of these cores has been identified as important to designers in the past, it is reasonable to integrate them as hard cores on the chip instead of implementing them using CLBs and other resources.

The FPGA used in the work described in this chapter was the Virtex II Pro chip which contains 23,616 slices, 49,788 flip-flops, 852 I/O blocks and 232 block RAM modules [19]. Moreover, this FPGA integrates two IBM PowerPC 405 processor cores which are able to operate at 400 MHz. Each of these cores is equipped with a five-stage pipeline, a virtual-memory management unit, separate instruction-cache and data-cache units, three programmable timers, an on-chip memory controller (OCM), and a variety of interfaces, including a processor local bus (PLB) interface, a device control register (DCR) interface and a JTAG port interface.

FPGAs can be configured either externally or internally. In the case of external reconfiguration, the configuration bit stream is copied to the configuration memory from an external (flash) memory. Internal reconfiguration is available in Xilinx Virtex FPGAs via the Internal Access Configuration Port (ICAP), which allows the FPGA configurations to be read and modified by circuits created directly in the

[1] Although there are other FPGA vendors, we will deal only with Xilinx FPGAs in this chapter.

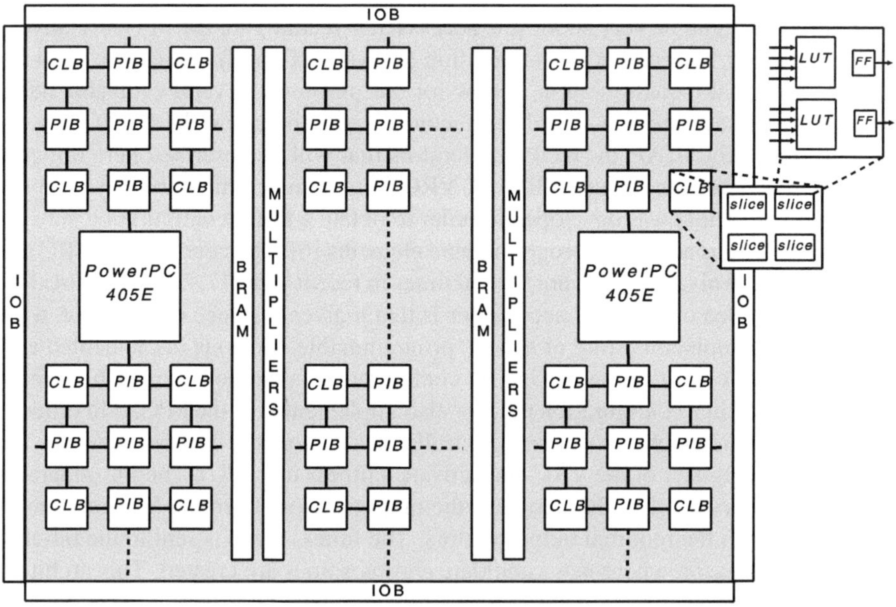

Fig. 7.1 FPGA Virtex II Pro architecture, which contains two PowerPC processors.

same FPGA. Some FPGAs support *dynamic partial reconfiguration*, which means that some parts of the FPGA can be reconfigured while the remaining parts of the FPGA perform computation.

As the circuit design process for FPGAs is very similar to programming, the resultant system can be obtained relatively quickly. The designer has to describe the circuit structure or behaviour using a hardware description language (such as VHDL or Verilog). Then, the source code is almost automatically transformed into the configuration bit stream for a particular FPGA. The transformation, which includes the synthesis, placement and routing, is performed by CAD tools. This process can be constrained using various requirements; for example, the maximum delay of the circuit can be specified. Also, it is possible to simulate intermediate results of the transformation, modify the original source code when needed and optimize the design.

7.3 Hardware Accelerators for CGP

In CGP, candidate configurations directly represent digital circuits that can intrinsically be evaluated in an FPGA. The goal of this chapter is to show that candidate configurations can be not only evaluated but also generated in an FPGA. As the evaluation time can be very short, it is necessary to reconfigure the FPGA relatively often. However, the current reconfiguration subsystem (ICAP) that can be used for dynamic internal reconfiguration is slow for our purposes. Several evolvable hardware systems have previously utilized reconfiguration based on the ICAP [10]. Instead of using the ICAP, the CGP accelerators that will be described here utilize a *virtual reconfigurable circuit* (VRC). A VRC is a second configurable layer on the top of an FPGA that was developed in order to obtain a fast reconfiguration subsystem and application-specific programmable elements [6]. The concept of a VRC has been used for evolvable hardware several times in recent years [7, 20, 3, 1, 5, 14, 18].

The basic idea of the CGP accelerator is that a given instance of CGP (i.e. a reconfigurable graph consisting of $n_r \times n_c$ programmable nodes) is implemented as a reconfigurable circuit in the FPGA. Its configuration is defined using a bit stream which is stored in a configuration register also implemented in the FPGA. In order to evaluate a candidate chromosome, a controller has to store the chromosome into the configuration register of the VRC and activate a fitness unit (FU). The FU generates the input vectors for the VRC, collects the output vectors from the VRC and compares them with the required output vectors. The fitness value is sent to the on-chip PowerPC processor, where new candidate chromosomes are created. This architecture was introduced in [12]. Alternatively, instead of being performed in the on-chip PowerPC processor, genetic operations can be performed either by a specific circuit [9, 8] or by a soft processor such as the Xilinx MicroBlaze core [2].

In order to utilize a reasonable amount of FPGA resources and maximize the throughput of the accelerator, it is not desirable to use the most favorable CGP settings applied almost exclusively in the work described in the previous chapters (i.e. $n_r = 1$ and $n_c = l$). The reason is that pipeline data processing requires $l = 1$ to calculate one output vector in a single clock cycle. Furthermore, setting $l > 1$ means that a programmable node can be connected to many other programmable nodes and, in order to select one of them, large and so area-expensive multiplexers have to be instantiated in the FPGA. These constraints do not exist in software implementations of CGP; however, it is necessary to take them into account in the FPGA accelerators for CGP.

7.3.1 Architecture Overview

The CGP accelerator consists of a Genetic Unit (GU), a processor and memory interface (PMI), a fitness unit (FU), a VRC and a control unit (CU), which is a communication interface to a personal computer (PC) (see Fig. 7.2). The PC is used just to define the parameters of the CGP and the target data (the truth table or training set). External SRAM memories are needed to store large training sets (e.g. training images for evolution of image filters), while on-chip BlockRAM (BRAM) memories are used to store small training sets. While the GU, PMI and CU can be reused by all applications, the VRC and fitness unit are usually application-specific.

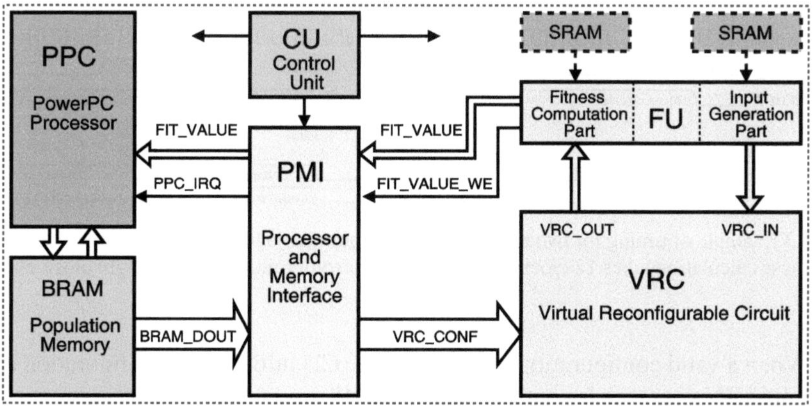

Fig. 7.2 Generic architecture of a CGP accelerator implemented in the FPGA Virtex 2 Pro [12].

All components (except the VRC) are connected to an internal bus called the LocalBus, which provides an effective communication interface between the FPGA and the PCI bus. In order to maximize the overall performance, the CU plays the role of master and controls the entire system. In particular, it starts and stops the evolution, determines the number of generations and other parameters of the search algorithm and generates control signals for the remaining components.

Upon a request, the PowerPC generates a new candidate individual; otherwise, it is idle in its main loop. The instruction memory of the PowerPC is implemented using on-chip BRAM memories and connected to the LocalBus in order to send and read programs to and from an external PC. However, since the program for the PowerPC is short, it can be stored completely in an instruction cache.

The population of candidate configurations is stored in on-chip BRAM memories. The population memory is divided into banks; each of them contains a single configuration bit stream of the VRC. An additional bit (associated with every bank) determines the validity of the data; only valid configurations can be evaluated. In order to overlap the evaluation of a candidate configuration with the generation of a new candidate configuration, at least two memory banks have to be utilized. While a circuit is being evaluated, a new candidate configuration is generated. A new configuration is used immediately after the evaluation of the previous circuit has been completed. If b banks are utilized, the PowerPC processor has b times more time to generate a new candidate circuit (i.e. the search algorithm and genetic operators can be more sophisticated).

The PMI component consists of two subcomponents, working concurrently. The first subcomponent, controlled by the CU, reconfigures the VRC using configurations stored in the population memory. The second subcomponent is responsible for sending the fitness value to the PowerPC processor. This process is controlled by the FU. The PMI component also provides an interface to the population memory via the LocalBus. The process of configuration of the VRC and evaluation of the candidate circuit works as follows [12]:

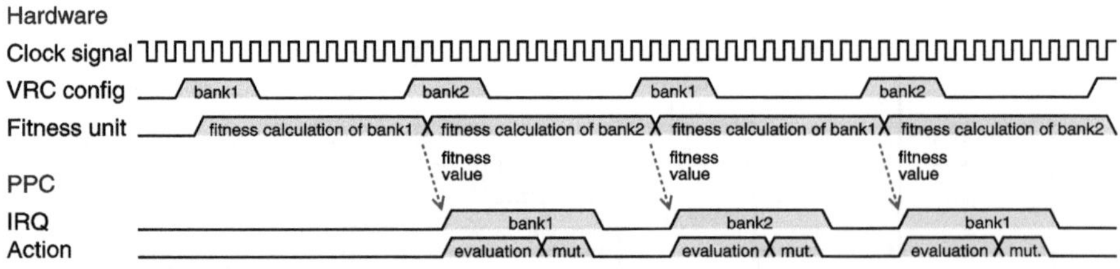

Fig. 7.3 Example of timing for two banks. The reconfiguration of the VRC takes four clock cycles, the fitness calculation takes 12 clock cycles and the interrupt routine requires eight clock cycles.

1. When a valid configuration is available, the CU initiates a reconfiguration of the VRC. This process is controlled by the PMI.
2. As soon as the first column of programmable nodes has been reconfigured, the CU initiates the fitness calculation process, performed by the FU.
3. When the last column of programmable nodes has been reconfigured, the corresponding memory bank is invalidated and the bank counter is incremented.
4. Three clock cycles before the end of the evaluation, the FU indicates the forthcoming end of the evaluation.
5. The CU initiates a new configuration of the VRC and repeats the sequence 1–4 again.
6. As soon as the fitness value is valid, it is sent (together with the corresponding bank number) to the PowerPC. An interrupt (IRQ) is generated to activate a service routine of the PowerPC. In this routine, a new candidate configuration is generated for the given bank. The PowerPC processor acknowledges the interrupt (IRQACK) and sets the validity bit.

Figure 7.3 shows the concurrent operations of several processes running in the hardware and the PowerPC processor (including the configuration of the VRC, the evaluation of a candidate circuit and the generation of candidate configurations). These processes are synchronized in such a way that no clock cycles are lost because of waiting for resources. Therefore, the time for evolution can be expressed as

$$t_{evol} = QP\frac{1}{f},$$

where Q is the number of evaluations, P is the number of training vectors and f is the operation frequency (the time required to fill the pipeline at the beginning of the computation is insignificant and so is not considered here).

The accelerator allows various search algorithms to be applied [12]. These algorithms utilize a population of candidate solutions and a single genetic operator, namely mutation, which inverts k bits of the chromosome (i.e. of the configuration). No crossover operator is used. The performance of several search algorithms, various mutation operators and pseudo-random number generators has been compared in [12, 13].

7.3.2 VRC for Symbolic Regression Problems

The CGP accelerators that will be described in this chapter differ from each other mainly in the organization of the VRC and in the fitness unit. Figure 7.4 shows a VRC implemented for the image filter design problem (see Sect. 6.2.5), which is a kind of a symbolic regression problem over a fixed-point number representation [12]. Every candidate image filter is considered as a digital circuit of nine eight-bit inputs and a single eight-bit output.

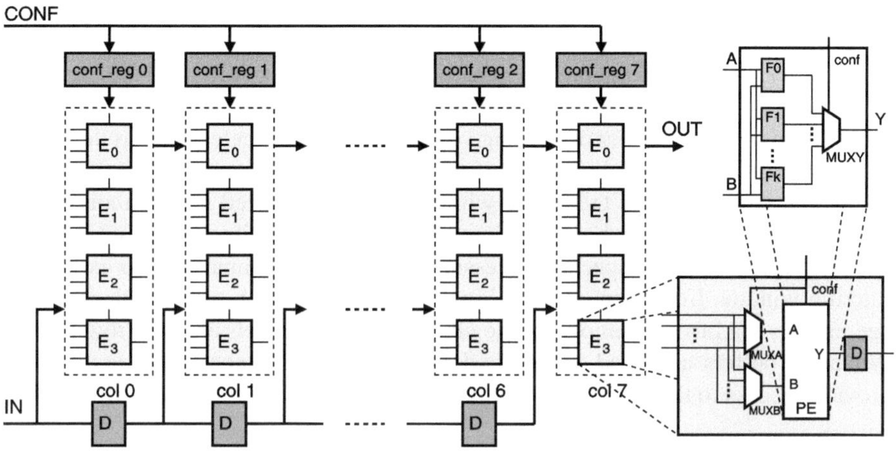

Fig. 7.4 VRC for symbolic regression problems.

The VRC consists of two-input programmable elements (PEs), denoted by E_i, placed in a grid of eight columns and four rows. The input of a PE may be connected either to the primary circuit input or to the output of a PE, which may be placed anywhere in the preceding column. Any PE can be programmed to implement one of 16 functions, given in Table 7.1. All of these functions operate with eight-bit operands and produce eight-bit result. The reconfiguration of the VRC is performed column by column. The computation is pipelined; one column of PEs represents one stage of the pipeline. Registers (denoted D) are inserted between the columns in order to synchronize the input pixels with the PE outputs. The configuration bit stream of the VRC, which is stored in a register array *conf_reg*, consists of 384 bits. A single PE is configured by 12 bits: four bits are used to select the connection of a single input and four bits are used to select one of the 16 functions. The selectors are implemented using multiplexers. The search algorithm operates directly with the configurations of the VRC; simply, a configuration is considered as a chromosome.

For the image filter design problem, the training data are stored in external SRAM memories. The fitness unit loads the training data and forwards them to the inputs of VRC. The outputs of the VRC, y_i, are compared with the required outputs, r_i (which are loaded from another external memory), and simultaneously stored into the external memory. The FU can be considered as an extension of the VRC pipeline

Table 7.1 Functions implemented by a programmable element

Code	Function	Description	Code	Function	Description
0	255	Constant	8	$x \gg 1$	Right shift by 1
1	x	Identity	9	$x \gg 2$	Right shift by 2
2	$255 - x$	Inversion	A	$swap(x,y)$	Swap nibbles
3	$x \vee y$	Bitwise OR	B	$x + y$	+ (addition)
4	$\bar{x} \vee y$	Bitwise $\bar{x}$ OR y	C	$x +^S y$	+ with saturation
5	$x \wedge y$	Bitwise AND	D	$(x+y) \gg 1$	Average
6	$\overline{x \wedge y}$	Bitwise NAND	E	$max(x,y)$	Maximum
7	$x \oplus y$	Bitwise XOR	F	$min(x,y)$	Minimum

because in each clock cycle, a temporary fitness value is updated by a new difference, $|y_i - r_i|$. Owing to the pipelined reconfiguration of the VRC and the execution of instructions, the evaluation of a candidate program (circuit) requires P clock cycles, where P is the number of training vectors. A similar accelerator was developed for the evolution of multiplierless multiple-constant multipliers [16].

As the 3×3 pixels of the image window w are not stored in neighbouring addresses of the SRAM, the hardware implementation of the fitness unit utilizes first-in-first-out raw buffers, special addressing circuits and comparators to extract the filtering window from memory (see Fig. 7.5). This approach assumes that one image pixel is read from memory in one clock cycle. The pixels are read row by row. When the buffers are filled (which is done with a fixed latency), this architecture provides access to the entire pixel neighbourhood with a latency of one clock cycle.

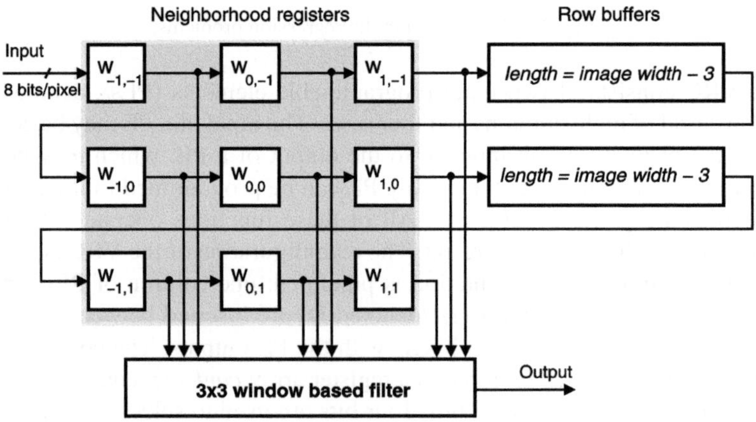

Fig. 7.5 Implementation of a 3×3 filter window in hardware.

7.3.3 VRC for Combinational-Circuit Evolution

The first accelerator for the evolution of combinational circuits was presented in [8]; however, it did not allow the size of the circuit to be optimized. The more advanced architecture which will be described here is similar to that of the VRC for symbolic regression. However, there are four main differences [14]:

- The PEs contain only elementary logic functions.
- A setting of $l = 2$ is supported in order to extend the space of possible designs.
- The size of the phenotype can be calculated and so optimized.
- A data-parallel operation of the PEs is introduced (the same as what was used in the parallel simulation in software described in Chap. 5).

The size of the data is denoted by the 'data width', dw. If the PEs operate at dw bits then the speed-up relative to the bit-level execution is dw times. In order to support $l = 2$, additional registers (D) are used to store the results of stage $i-2$ for stage i of the pipeline (see Fig. 7.6). The number of configuration bits for a single column is $2 * \log_2(n_i + 2n_r) + \log_2(n_f)$.

In contrast to the VRC for symbolic regression, the training data (truth table) is stored in BRAMs. For example, if $n_i = 16$, then 64 BRAMs are utilized. All possible input combinations are generated in the process of the fitness calculation. When the size of the circuit is not optimized, the maximum fitness value is $2^{n_i} n_o$.

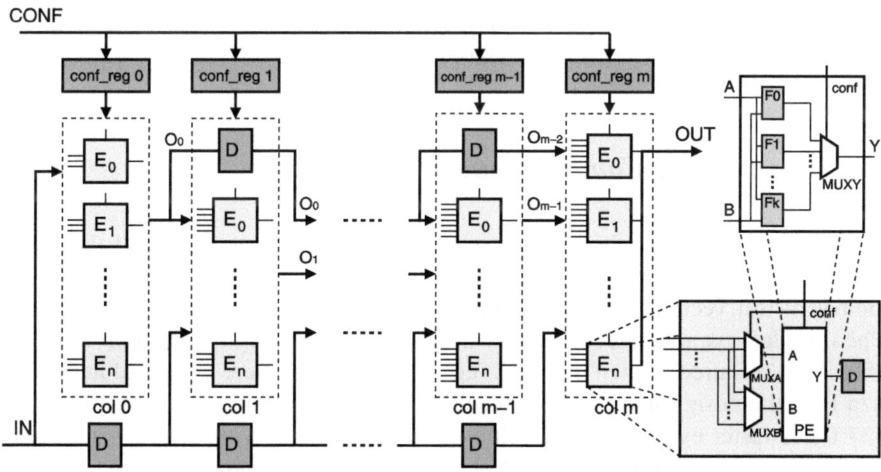

Fig. 7.6 VRC for evolution of combinational circuits.

Figure 7.7 explains the calculation of the size of a candidate circuit. Once a functionally perfect solution is found, the process of minimizing the number of gates is initiated.

The method assumes that each PE can implement a single wire. In order to make the implementation as easy as possible, the problem is formulated as a problem of maximizing the number of PEs that are configured as wires. The solution is based

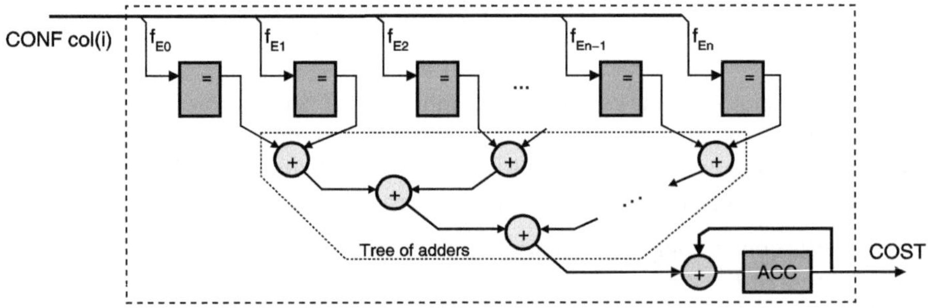

Fig. 7.7 Calculation of the size of a phenotype.

on analysis of the configuration of a single column of the VRC using comparators. Every comparator returns 1 if a particular PE operates as a wire. These 1s are added using a tree of adders. This calculation is performed when the column of PEs is configured. It costs no extra time. The size of the phenotype is stored in the eight least significant bits of the fitness value.

7.4 Performance Evaluation

The accelerators were described in the language VHDL, simulated using ModelSim and synthesized using the Mentor Graphics Precision RTL 2006a and Xilinx ISE 10.1 tools. The results [14], are summarized below.

7.4.1 Evolution of Combinational Circuits

Table 7.2 shows the results of the synthesis for various parameters of the VRC. The size of the VRC and the numbers of inputs and outputs were fixed, and the number of test vectors evaluated in parallel (i.e. dw) was increased from 1 to 12. When no data-parallel execution was used, the whole design used approximately 10% of the resources; when $dw = 12$ (i.e. 12 test vectors were evaluated in parallel by a PE), the design used approximately 90% of the resources. Using this set-up a 27 times faster evaluation was achieved in comparison with a highly optimized software implementation running on an Intel Xeon 3 GHz processor (and utilizing a parallel simulation with 32 bits), even if the VRC operated at 100 MHz.

Table 7.3 shows the results of the synthesis for various VRC sizes. The numbers of inputs, outputs and logic functions and the data width were fixed. The last row shows the number of configuration bits of the VRC.

In order to investigate the impact of the parameter l, two VRCs were created with $l = 1$ and $l = 2$. The FPGA implementations were evaluated using the task of multiplier evolution, a traditional benchmark problem for evolutionary circuit design (see Sect. 5.2.3). Table 7.4 summarizes the results of 10 independent experiments

Table 7.2 Results of synthesis for a VRC with 10×10 PEs, nine inputs, nine outputs and four logic functions per PE (XC2VP50-ff1517 Xilinx FPGA). DFF is the number of flip-flops and FG is the number of function generators

		# of vectors evaluated in parallel (dw)				
Resource	Available	1	2	4	8	12
BRAMs	232	14	16	20	28	36
	used	6.0%	6.9%	8.6%	12.1%	15.5%
DFFs	49788	2743	2993	3533	4709	5843
	used	5.5%	6.0%	7.1%	9.5%	11.7%
FGs	47232	4836	7813	14164	26734	41281
	used	10.2%	16.5%	30.0%	56.6%	87.4%

Table 7.3 Results of synthesis for various VRCs with nine inputs, nine outputs, four logic functions and $dw = 2$ (FPGA XC2VP50-ff1517)

		VRC size			
Resource	Available	10×10	12×12	14×14	16×16
DFFs	49788	1644	2336	3634	4664
	used	3.3%	4.7%	7.3%	9.4%
FGs	47232	6242	9012	26700	32352
	used	13.2%	19.1%	56.5%	68.5%
# of configuration bits		1200	2016	2744	3584

for each problem; the population always contained eight individuals. We can see that increasing the value of l has a positive effect on the average number of generations and the success rate. The results obtained are comparable to the best known results [17] (where the maximum value of l was allowed).

Table 7.4 Results for evolution of multipliers with the function set {wire, and, xor, $\bar{a}$ and b}

Parameters of evolution										
Multiplier	**2×2**		**2×3**		**3×3**		**3×4**		**4×4**	
l	1	2	1	2	1	2	1	2	1	2
VRC	8×8	8×8	10×10	10×10	10×10	10×10	10×10	10×10	16×16	16×16
n_i	4	4	5	5	6	6	7	7	8	8
Generations (max)	10k	10k	100k	100k	1M	1M	10M	10M	20M	20M
Results										
Success rate	91%	96%	92%	100%	72%	96%	18%	84%	0%	4%
Gates (min)	7	7	13	13	29	24	60	45	–	125
Gates (max)	19	13	20	21	45	47	67	68	–	156
Gates (avg)	9	8	15	15	34	33	61	57	–	138
Generations (avg)	1.8k	1.5k	20k	13k	22k	284k	4.84M	3.84M	–	14.2M

Table 7.5 compares the numbers of candidate circuits evaluated per second in a highly optimized software implementation and in our hardware accelerators. In the case of the software implementation, the time for a circuit evaluation depends on the size of the phenotype and the number of training vectors. On the other hand, in

the case of a hardware accelerator, this time depends only on the number of training vectors. Hence, an accelerator becomes more useful for larger VRCs and larger sets of training data.

Table 7.5 Number of evaluations per second. The VRC operated at 100 MHz ($dw = 4$); the measurements are indicated by 'HW' (hardware). The software (SW) implementation was executed on the Intel Xeon CPU 3.06 GHz ($dw = 32$)

	VRC size (SW)			VRC size (HW)			Evaluation
n_i	**10 × 10**	**12 × 12**	**16 × 16**	**10 × 10**	**12 × 12**	**16 × 16**	speed-up
6	400	296	222	6250	6250	6250	15–28
7	250	173	89	3125	3125	3125	12–35
8	154	95	51	1563	1563	1563	10–30
9	85	50	25	781	781	781	9–31

7.4.2 Symbolic Regression Problems

Similarly to the accelerator for logic circuit synthesis, the CGP accelerator for symbolic regression problems was implemented on a COMBO6X card equipped with a Virtex II Pro 2VP50ff1517 FPGA. The accelerator was intended for image filter evolution where CGP was applied with the settings $n_c = 8, n_r = 4, n_i = 9, n_o = 1, l = 1$. The results of synthesis are summarized in Table 7.6. While the PowerPC operated at 300 MHz, the logic supporting the PowerPC operated at 150 MHz. The remaining FPGA logic (including the VRC and FU) operated at 100 MHz.

The experimental results show that approximately 6000 candidate filters can be evaluated per second when the training set consists of one 128 × 128-pixel image which is 44 times faster than the same algorithm running on a Celeron 2.4 GHz processor [12]. This accelerator was utilized to discover the novel implementations of image filters that were presented in Sect. 6.2.6 [12, 13, 11].

We can observe that the VRC occupies 15% of the FPGA (i.e. 3540 slices). A typical filter evolved in the accelerator can subsequently be implemented using approximately 250 slices. It is clear that the overhead associated with the use of the VRC is significant.

Table 7.6 Results of synthesis for the image filter CGP accelerator

VRC	I/O blocks	BRAM	Slices	DFF
Available	852	232	23 616	49 788
4 × 8 CFBs	602	12	4 591	3 638
used	70%	5%	20%	7%
No VRC	602	12	1 240	2 479
used	70%	5%	5%	5%

7.5 Summary

We have shown that FPGA implementations of CGP can accelerate circuit evolution. The speed-up obtained (30–40) is significant, although only a single fitness unit was utilized and various restrictions were applied to the CGP settings (e.g. $l = 1$ or 2 and $n_r = 1$) in order to utilize a reasonable amount of FPGA resources. Recent work [15] has shown that by implementing four VRCs and corresponding fitness units in a single FPGA, the architecture can accelerate image filter evolution by 170 times in comparison with a highly optimized software solution running on a Celeron 2.4 GHz processor. Future work in this direction should lead to the utilization of described accelerators of this type in adaptive embedded systems.

7.6 Acknowledgements

This work was partially supported by the Grant Agency of the Czech Republic under Contract No. P103/10/1517 and by the research programme MSM 0021630528.

References

1. Glette, K., Torresen, J.: A flexible on-chip evolution system implemented on a Xilinx Virtex-II Pro device. In: Proc. International Conference on Evolvable Systems, *LNCS*, vol. 3637, pp. 66–75. Springer (2005)
2. Glette, K., Torresen, J., Yasunaga, M., Yamaguchi, Y.: On-Chip Evolution Using a Soft Processor Core Applied to Image Recognition. In: Proc. NASA/ESA Conference on Adaptive Hardware and Systems, pp. 373–380. IEEE Computer Society (2006)
3. Gwaltney, D., Dutton, K.: A VHDL Core for Intrinsic Evolution of Discrete Time Filters with Signal Feedback. In: Proc. NASA/DoD Conference on Evolvable Hardware, pp. 43–50. IEEE Computer Society (2005)
4. Hauck, S., DeHon, A.: Reconfigurable Computing: The Theory and Practice of FPGA-Based Computation. Morgan Kaufmann (2008)
5. Salomon, R., Widiger, H., Tockhorn, A.: Rapid Evolution of Time-Efficient Packet Classifiers. In: IEEE Congress on Evolutionary Computation, pp. 2793–2799. IEEE CIS (2006)
6. Sekanina, L.: Virtual Reconfigurable Circuits for Real-World Applications of Evolvable Hardware. In: Proc. International Conference on Evolvable Systems, *LNCS*, vol. 2606, pp. 186–197. Springer (2003)
7. Sekanina, L.: Evolvable components: From Theory to Hardware Implementations. Natural Computing. Springer (2004)
8. Sekanina, L., Friedl, S.: An Evolvable Combinational Unit for FPGAs. Computing and Informatics **23**(5), 461–486 (2004)
9. Sekanina, L., Martinek, T.: Evolving Image Operators Directly in Hardware, pp. 93–112. EURASIP Book Series on Signal Processing and Communications, Volume 8. Hindawi Publishing Corporation (2007)
10. Upegui, A., Sanchez, E.: Evolving Hardware with Self-Reconfigurable Connectivity in Xilinx FPGAs. In: Proc. NASA/ESA Conference on Adaptive Hardware and Systems, pp. 153–160. IEEE Computer Society (2006)

11. Vasicek, Z., Sekanina, L.: An Area-Efficient Alternative to Adaptive Median Filtering in FPGAs. In: Proc. International Conference on Field Programmable Logic and Applications, pp. 216–221. IEEE Computer Society (2007)
12. Vasicek, Z., Sekanina, L.: An Evolvable Hardware System in Xilinx Virtex II Pro FPGA. International Journal of Innovative Computing and Applications **1**(1), 63–73 (2007)
13. Vasicek, Z., Sekanina, L.: Evaluation of a New Platform For Image Filter Evolution. In: Proc. NASA/ESA Conference on Adaptive Hardware and Systems, pp. 577–584. IEEE Computer Society (2007)
14. Vasicek, Z., Sekanina, L.: Hardware Accelerators for Cartesian Genetic Programming. In: Proc. European Conference on Genetic programming, *LNCS*, vol. 4971, pp. 230–241 (2008)
15. Vasicek, Z., Sekanina, L.: Hardware Accelerator of Cartesian Genetic Programming with Multiple Fitness Units. Computing and Informatics **29**(7), 1–12 (2010)
16. Vasicek, Z., Zadnik, M., Sekanina, L., Tobola, J.: On Evolutionary Synthesis of Linear Transforms in FPGA. In: Proc. International Conference on Evolvable Systems, *LNCS*, vol. 5216, pp. 141–152. Springer (2008)
17. Vassilev, V., Job, D., Miller, J.F.: Towards the Automatic Design of More Efficient Digital Circuits. In: Proc. NASA/DoD Workshop on Evolvable Hardware, pp. 151–160. IEEE Computer Society (2000)
18. Wang, J., Chen, Q.S., Lee, C.H.: Design and implementation of a virtual reconfigurable architecture for different applications of intrinsic evolvable hardware. IET Computers and Digital Techniques **2**(5), 386–400 (2008)
19. Xilinx: Virtex-II Pro Platform FPGAs: Complete Data Sheet (2005). URL: http://www.xilinx.com/partinfo/ds031.pdf
20. Zhang, Y., Smith, S.L., Tyrrell, A.M.: Intrinsic Evolvable Hardware in Digital Filter Design. In: Applications of Evolutionary Computing, *LNCS*, vol. 3005, pp. 389–398. Springer (2004)

Chapter 8
Hardware Acceleration for CGP: Graphics Processing Units

Simon L. Harding and Wolfgang Banzhaf

8.1 Graphics Processing Units

Graphic Processing Units (GPUs) are fast, highly parallel units. In addition to processing 3D graphics, modern GPUs can be programmed for more general-purpose computation. A GPU consists of a large number of 'shader processors', and conceptually operates as a single instruction multiple data (SIMD) or multiple instruction multiple data (MIMD) stream processor. A modern GPU can have several hundred of these stream processors, which, combined with their relatively low cost, makes them an attractive platform for scientific computing. In the last two years, the genetic programming community has begun to exploit GPUs to accelerate the evaluation of individuals in a population [1, 4].

CGP was the first GP technique implemented in a general-purpose fashion on GPUs. By 'general purpose', we mean a technique that can be applied to a number of GP applications and not just a single, specialized task. Implementing CGP on GPUs has resulted in very significant performance increases.

In this chapter, we discuss several of our implementations of CGP on GPUs. To begin with, we start with an overview of the hardware and software used, before discussing applications and the speed-ups obtained.

8.2 The Architecture of Graphics Processing Units

Graphics processors are specialized stream processors used to render graphics. Typically, a GPU is able to perform graphics manipulations much faster than a general-purpose CPU, as graphics processors are specifically designed to handle certain primitive operations. Internally, a GPU contains a number of small processors that are used to perform calculations on 3D vertex information and on textures. These processors operate in parallel with each other, and work on different parts of the

J.F. Miller (ed.), *Cartesian Genetic Programming*, Natural Computing Series,
DOI 10.1007/978-3-642-17310-3_8, © Springer-Verlag Berlin Heidelberg 2011

problem. First, the vertex processors calculate a 3D view, and then the shader processors paint this model before it is displayed. Programming a GPU is typically done through a virtual-machine interface such as OpenGL or DirectX, which provides a common interface to the diverse GPUs available, thus making development easy. However, DirectX and OpenGL are optimized for graphics processing, and hence other APIs are required to use a GPU as a general purpose device. There are many such APIs, and Sect. 8.3 describes several of the ones that have been used to implement CGP.

For general-purpose computing, we wish here to make use of the parallelism provided by the shader processors (see Fig. 8.1). Each processor can perform multiple floating-point operations per clock cycle, meaning that the performance is determined by the clock speed, the number of pixel shaders and the width of the pixel shaders. Pixel shaders are programmed to perform a given set of instructions on each pixel in a texture. Depending on the GPU, the number of instructions may be limited. In order to use more than this number of operations, a program needs to be broken down into suitably sized units, which may impact performance. Newer GPUs support unlimited instructions, but some older cards support as few as 64 instructions. GPUs typically use floating-point arithmetic, the precision of which is often controllable, as less precise representations are faster to compute with. Again, the maximum precision is manufacturer-specific, but recent cards provide up to 128-bit precision.

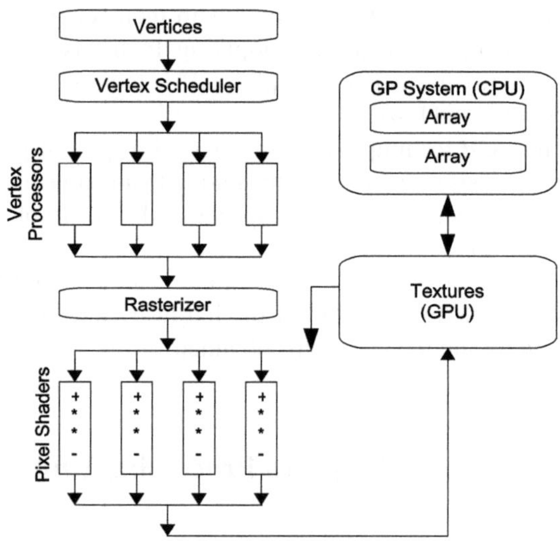

Fig. 8.1 Arrays, representing test cases, are converted to textures. These textures are then manipulated (in parallel) by small programs inside each of the pixel shaders. The result is another texture, which can be converted back to a normal array for CPU-based processing.

A variety of different graphics cards were used in our experiments. In particular, we used three cards from Nvidia: GeForce 7300 Go, GeForce 8600 and GeForce 9800.

8.3 Programming a GPU

We experimented with two different software APIs for programming GPUs: Microsoft's Accelerator and the API from Nvidia.

Accelerator is a .Net assembly from Microsoft Research that provides access to the GPU via the DirectX interface [15]. The system is completely abstracted from the GPU, and presents the end user with only arrays that can be operated on in parallel. Unfortunately, the system is available only for the Windows platform, owing to its reliance on DirectX. However, the assembly can be used from any .Net programming language. Conversion to textures and transfer to the GPU are handled transparently by the API, allowing the developer to concentrate on the implementation of the algorithm. The function set available for operating on parallel arrays is similar to those for other APIs. It includes element wise arithmetic operations, and square root, multiply–add and trigonometric operations. There are also conditional operations and functions for comparing two arrays. The API also provides reduction operators, such as the sum, product, minimum value and maximum value of an array. Further functions perform transformations such as shift and rotate on the elements of an array.

One benefit of this tool is that it uses lazy evaluation. Operations are not performed on the data until the evaluated result is requested. This enables a certain degree of real-time optimization, and reduces the computational load on the GPU. In particular, it enables optimization of common subexpressions, which reduces the creation of temporary shaders and textures. The movement of data to and from the GPU can also be optimized efficiently, which reduces the impact of the relatively slow transfer of data out of the GPU. The compilation to the shader model occurs at run time, and hence can automatically make use of the different features available on the various graphics cards supported.

CUDA is an extension of the C programming language developed by Nvidia. It allows low-level access to the GPU and therefore can make optimal use of the hardware. CUDA can be viewed as a set of language extensions that provide support for parallel programming using massive numbers of threads. The API provides all the tools needed to marshal data to and from the GPU and for the actual algorithm implementation. The programs look very much like C programs.

CUDA programs are typically written as pure C programs. However, many libraries exist that expose the programming interface of the CUDA driver to other programming environments. In the work described here, the CUDA.net library from GASS [3] was used. This library allows any .Net language to communicate with the CUDA driver.

Accessing the CUDA driver allows applications to access the GPU memory, load code libraries (modules) and execute functions. The libraries themselves still need to be implemented in C and compiled using the CUDA compiler. During this compilation process, the CUDA compiler can optimize the generated code.

In addition to the overhead of compiling shader programs, GPU programming also has a number of other bottlenecks. In particular, there is a small cost of transferring data to or from the GPU memory.

8.4 Parallel Implementation of GP

Implementations of GP on GPUs have largely fallen into two distinct evaluation methodologies: population-parallel and fitness-case-parallel. Both methods exploit the highly parallel architecture of the GPU. In the fitness-case-parallel approach, all the fitness cases are executed in parallel, with only one individual being evaluated at a time. This can be considered as a SIMD approach. In the population-parallel approach, multiple individuals are evaluated simultaneously. Both approaches have problems that impact performance.

The fitness-case-parallel approach requires that a different program is uploaded to the GPU for each evaluation. This introduces a large overhead. The programs need to be compiled (on the CPU side) and then loaded into the graphics card before any evaluation can occur. Previous work has shown that for smaller numbers of fitness cases (typically less than 1000), this overhead is larger than the increase in computational speed [2, 7, 6]. In these cases, there is no benefit from executing the program on the GPU. For some problems, there are fewer fitness cases than shader processors. In this scenario, some of the processors are idle. This problem will become worse as the number of shader processors increases with new GPU designs. Unfortunately, a large number of classic benchmark GP problems fit into this category. However, there are problems such as image processing that naturally provide large numbers of fitness cases (as each pixel can be considered as a different fitness case) [5, 8].

GPUs can be considered as processors (or in which each processor handles multiple copies of the same program, using different program counters [14]). When a population of programs is to be evaluated in parallel, the common approach so far has been to implement some form of interpreter on the GPU [14, 13, 16]. The interpreter is able to execute the evolved programs (typically machine-code-like programs represented as textures) in a pseudo-parallel manner. To see why this execution is pseduo-parallel, the implementation of the interpreter needs to be considered.

A population can be represented as a 2D array, with individuals represented in the columns and their instructions in the rows. The interpreter consists of a set of IF statements, where each condition corresponds to a given operator in the GP function set. The interpreter looks at the current instruction of each individual, and sees if it matches the current IF statement condition. If it does, that instruction is executed. This happens for each individual in the population in parallel, and for some individuals the current instruction may not match the current IF statement. These individuals must wait until the interpreter reaches their instruction type. Hence, with this approach, some of the GPU processors are not doing useful computation. If there are four functions in the function set, we can expect that on average at least 3/4 of the shader processors will be 'idle'. If the function set is larger, then more shaders will be 'idle'. Despite this idling, population-parallel approaches can still be considerably faster than CPU implementations.

The performance of population-parallel implementations is therefore impacted by the size of the function set used. Conversely, the fitness-case-parallel approaches are impacted by the number of fitness cases. Population-parallel approaches are also largely unsuited for GPU implementation when small population sizes are required.

In our work, we used a fitness-case-parallel version of the algorithm. Both Accelerator and CUDA compile the code before execution. Although this introduces a significant overhead, it does have some advantages, namely that the compiler can optimize the executed code.

With CGP, it is very commonly to observed that large numbers of nodes in the genotype are reused, and hence there are a lot of intermediate results that can be reused during execution. With an interpreter-based implementation, this would be very hard to implement. However, with compilation, we can rely on the compiler (and our own code generator) to spot where such work is redundant and remove it.

When running in a fitness-parallel fashion, the processing can be considered as vector mathematics. Each column in the data set is represented as a vector of values. The programs then operate on these vectors, in an elementwise fashion. Each shader program runs on one index of the vector, and hence each shader program is responsible for one row in the data set. Figure 8.2 illustrates this approach.

In the work demonstrated here, only the fitness evaluation was implemented on the GPU. The evolutionary algorithm (and other parts of the software) were all implemented on the CPU side, as shown in Fig. 8.1.

8.5 Initial GPU Results

In our initial papers [7, 6], we used an Nvidia 7300 Go GPU. The programming used the Accelerator package. At the time of writing, this GPU is still in use, but is a very dated and underpowered platform. The GP parser used here was written in C♯, and compiled using Visual Studio 2005. All benchmark tests were done using the Release build configuration, and were executed on a CLR 2.0 with Windows XP. The GPU was an Nvidia GeForce 7300 Go with 512 Mb video memory. The CPU used was an Intel Centrino T2400 (running at 1.83 Ghz), with 1.5 Gb of system memory.

In these experiments, GP trees were randomly generated with a given number of nodes. The expressions were evaluated on the CPU and then on the GPU, and each evaluation was timed. For each input-size/expression-length pair, 100 different randomly generated expressions were used, and the results were averaged to calcu-

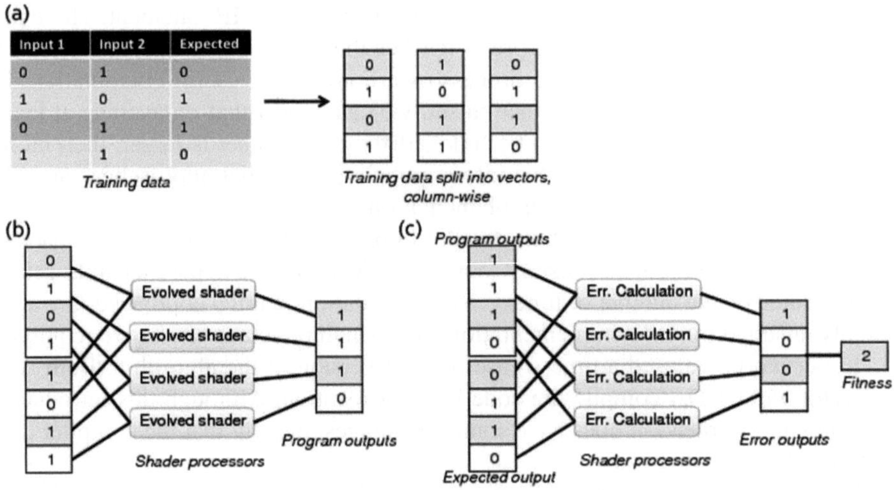

Fig. 8.2 (a) The data set is split into vectors. Each vector contains one column. (b) These vectors are passed to the evolved shader program. Each shader program runs on a single index, so it is able to see only one row of the data set. (c) The output of the evolved program is compared with the expected output, and a fitness value obtained. All these steps exploit the parallelism of the shader processor.

late acceleration factors. Therefore our results showed the average number of times by which the GPU was faster at evaluating a given tree size for a given number of fitness cases. Results less than 1 mean that the CPU was faster at evaluating the expression, and values above 1 indicate that the GPU performed better.

8.5.1 Floating-Point-Based Expressions

In the first experiment, we evaluated random GP trees containing varying numbers of nodes, and exposed them to test cases of varying sizes. The mathematical functions $+$, $-$, $*$ and $/$ were used. The same expression was tested on the CPU and the GPU, and the speed difference was recorded. The results are shown in Table 8.1. For small node counts and fitness cases, the CPU performance is superior because of the overhead of mapping the computation to the GPU. For larger problems, however, there is a massive speed increase for GPU execution.

Table 8.1 Results showing the number of times faster evaluating floating-point-based expressions is on the GPU compared with CPU implementation. An increase of less than 1 shows that the CPU is more efficient

	Test cases					
Expression length	64	256	1024	4096	16384	65536
10	0.04	0.16	0.6	2.39	8.94	28.34
100	0.4	1.38	5.55	23.03	84.23	271.69
500	1.82	7.04	27.84	101.13	407.34	1349.52
1000	3.47	13.78	52.55	204.35	803.28	2694.97
5000	10.02	26.35	87.46	349.73	1736.3	4642.4
10000	13.01	36.5	157.03	442.23	1678.45	7351.06

8.5.2 Binary

The second experiment compared the performance of the GPU at handling Boolean expressions. In the CPU version, we used the C♯ Boolean type – which is convenient, but not necessarily the most efficient representation. For the GPU, we tested two different approaches, one using the Boolean parallel array provided by Accelerator, and the other using floating-point numbers. The performance of these two representations is shown in Table 8.2. It is interesting to note that improvements are not guaranteed. As can be seen in the table, the speed-up can decrease as the expression size increases. We assume that this is due to the way in which large shader programs are handled by either Accelerator or the GPU. For example, the length of the shader program on the Nvidia GPU may be limited, and going beyond this length would require repeated passes of the data. This type of behaviour can be seen in many of the results presented here.

We limited the functions in the expressions to AND, OR and NOT, which are supported by the Boolean array type. Following some sample code provided with Accelerator, we mimicked Boolean behaviour using 0.0f as false and 1.0f as true. For two floats, AND can be viewed as the minimum of the two values. Similarly, OR can be viewed as the maximum of the two values. NOT can be performed as a multiply–add, where the first stage is to multiply by minus 1 then 1 is added.

8.5.3 Regression and Classifcation

Next we investigated the speed-up on both toy and real-world problems, rather than on arbitrary expressions. In the benchmark experiments, the expression lengths were uniform throughout the tests. However, in real GP, the lengths of the expressions vary throughout the run. As the GPU sometimes results in slower performance, we need to verify that on average there is an advantage.

Table 8.2 Results showing the number of times faster evaluating Boolean expressions is on the GPU, compared with CPU implementation. An increase of less than 1 shows that the CPU is more efficient. Booleans were implemented as floating-point numbers and as Booleans. Although representation as floating-point numbers provides faster valuation than that obtained with the CPU faster than the CPU for large input sizes, in general it appears preferential to use the Boolean representation. Using floating-point representation can provide speed increases, but the results are varied

Boolean implementation								
	Test cases							
Expression length	4	16	64	256	1024	4096	16384	65536
10	0.22	1.04	1.05	2.77	7.79	36.53	84.08	556.40
50	0.44	0.57	1.43	3.02	14.75	58.17	228.13	896.33
100	0.39	0.62	1.17	4.36	14.49	51.51	189.57	969.33
500	0.35	0.43	0.75	2.64	14.11	48.01	256.07	1048.16
1000	0.23	0.39	0.86	3.01	10.79	50.39	162.54	408.73
1500	0.40	0.55	1.15	4.19	13.69	53.49	113.43	848.29
Boolean implementation, using floating-point numbers								
	Test cases							
Expression length	4	16	64	256	1024	4096	16384	65536
10	0.024	0.028	0.028	0.072	0.282	0.99	3.92	14.66
50	0.035	0.049	0.115	0.311	1.174	4.56	17.72	70.48
100	0.061	0.088	0.201	0.616	2.020	8.78	34.69	132.84
500	0.002	0.003	0.005	0.017	0.064	0.25	0.99	3.50
1000	0.001	0.001	0.003	0.008	0.030	0.12	0.48	1.49
1500	0.000	0.001	0.002	0.005	0.019	0.07	0.29	1.00

8.5.3.1 Regression

We evolved functions that performed regression with respect to $x^6 - 2x^4 + x^2$ [11]. We measured the difference in the speed of evaluation using a number of test cases. In each instance, the test cases were uniformly distributed between -1 and $+1$. We also changed the maximum length of the CGP graph. Hence, the expression lengths could range anywhere from one node to the maximum size of the CGP graph. GP was run for 200 generations to allow convergence. The function set comprised $+$, $-$, $*$ and $/$. In C♯, division by zero of a float returns ‘Infinity’, which is consistent with the result from the Accelerator library.

The fitness was defined as the sum of the absolute errors for each test case and the output of the expression. This could also be calculated using the GPU. Each individual was evaluated with the CPU and then the GPU and the speed difference was recorded. Also, the outputs from the GPU and CPU were compared to ensure that they were evaluating the expression in the same manner. We did not find any instances where the two differed.

Table 8.3 shows results that are consistent with the results for the other tests described above. For smaller input sets and small expressions, it was more efficient to evaluate them on the CPU. However, for the larger test and expression sizes, the performance increase was dramatic.

Table 8.3 Results for the regression experiment. The results show the number of times faster evaluating evolved GP expressions is on the GPU, compared with CPU implementation. The maximum expression length is the number of nodes in the CGP graph

	Test cases			
Max. expressionlLength	10	100	1000	2000
10	0.02	0.08	0.7	1.22
100	0.07	0.33	2.79	5.16
1000	0.42	1.71	15.29	87.02
10000	0.4	1.79	16.25	95.37

8.5.3.2 Classification

In this experiment we attempted to solve the classification problem of distinguishing between two spirals, as described in [11]. This problem has two input values (the x and y coordinates of a point on a spiral) and has a single output indicating which spiral the point is found to be on. In [11], 194 test cases were used. In the present experiments, we extended the number of test cases to 388, 970 and 1940. We also extended the function set to include sine, cosine, $\sqrt{x}$, x^y and a comparator. The comparator looked at the first input value of a node, and if it was less than or equal to zero returned the second input, and 0 otherwise. The relative speed increases can be seen in Table 8.4. Again we see that the GPU is superior for larger numbers of test cases, with larger expression sizes.

Table 8.4 Results for the two-spirals classification experiment. The results show the number of times faster evaluating evolved GP expressions is on the GPU compared with CPU implementation. The maximum expression length is the number of nodes in the CGP graph

	Test cases			
Max. expression length	194	388	970	1940
10	0.15	0.23	0.51	1.01
100	0.38	0.67	1.63	3.01
1000	1.77	3.19	9.21	22.7
10000	1.69	3.21	8.94	22.38

8.5.3.3 Classification in Bioinformatics

In this experiment we investigated the behaviour on another classification problem, this time a protein classifier, as described in [12]. Here the task was to predict the location of a protein in a cell from the amino acids in that particular protein. We used the entire data set as the training set. The set consisted of 2427 entries, with 19 variables each and one output. We investigated the performance gain using several expression lengths, and the results can be seen in Table 8.5. Here, the large number of test cases used results in considerable improvements in evaluation time, even for small expressions.

Table 8.5 Results for the protein classification experiment. The results show the number of times faster evaluating evolved GP expressions is on the GPU compared with CPU implementation. The maximum expression length is the number of nodes in the CGP graph

	Test cases
Expression length	2427
10	3.44
100	6.67
1000	11.84
10000	14.21

8.6 Image Processing on the GPU

In a later set of experiments, we demonstrated the use of the GPU as a platform for the evolution of image filters, such as filters for noise reduction. An image filter is defined as a convolution kernel (or program) that takes a set of inputs (pixels in an image) and maps them to a new pixel value. This program is applied to each pixel in an image to produce another image. Such filters are common in image processing. For example, one commonly used filter to reduce noise in an image is a median filter, where the new state of a pixel is the median value of it and its neighbourhood.

Using CGP it is possible to evolve such filters. Further examples of this and other approaches can be seen in Chap. 6.

Running the filters on a GPU allows us to apply the kernel to every pixel (logically, but not physically) simultaneously. The parallel nature of the GPU allows a number of kernels to be calculated at the same time. This number is dependent on the number of shader processors available. The image filter consists of an evolved program that takes a pixel and its neighbourhood (a total of nine pixels) and computes the new value of the centre pixel. On a traditional processor, you would iterate over each pixel in turn and execute the evolved program each time. Using the parallelism of the GPU, many pixels (in effect all of them) can be operated on simultaneously. Hence, the evolved program is only evaluated once. Although the evolved program actually evaluates the entire image at once, we can break down the problem and consider what is required for each pixel. For each pixel, we need a program that takes it and its neighbourhood, and calculates a new pixel value. For this, the evolved program requires as many inputs as there are pixels in the neighbourhood, and a single output. In the evolved program, each function has two inputs and one output. These inputs are floating-point numbers that correspond to the grey-level values of the pixels. Figure 8.3 illustrates a program that takes a nine-pixel subimage, and computes a new pixel value.

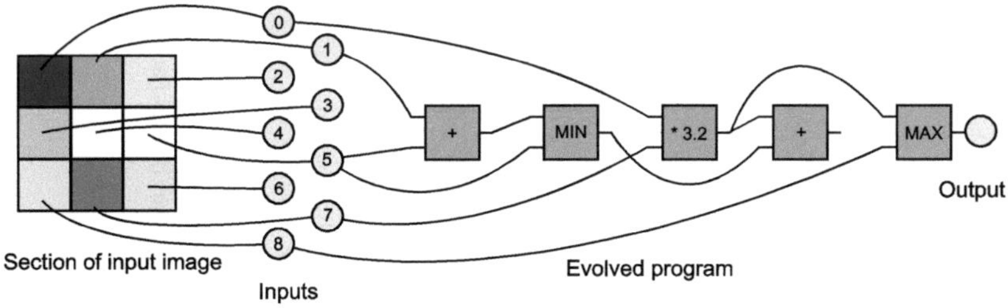

Fig. 8.3 In this example, the evolved program has nine inputs which correspond to a section of an image. The output of the program determines the new colour of the centre pixel. Note that one node has no connections to its output. This means that the node is redundant, and will not be used during the computation.

Mapping the image-filtering problem to the parallel architecture of the GPU is relatively straightforward.

It is important to appreciate that a GPU typically takes two arrays and produces a third by performing a parallel operation on them. The operation is elementwise, in the same way as for matrix operations. To clarify this, consider two arrays, $a = [1,2,3]$ and $b = [4,5,6]$. If we perform addition, we get $c = [5,6,9]$. With the SIMD architecture of the GPU, it is difficult to do an operation such as adding the first element of one array to the second of another. To do such an operation, the second array would need to be shifted to move the element in the second position to the first position.

For image filtering, we need to take a subimage from the main image, and use these pixels as inputs for a program (our convolution kernel), keeping in mind the vectorized operations of the GPU. To do this we take an image (e.g. the top left array in Fig. 8.4) and shift the array by one pixel in all eight possible directions. This produces a total of nine arrays (labelled (a) to (i) in Fig. 8.4).

Taking the same indexed element from each array will return the neighbourhood of a pixel. In Fig. 8.4, the neighbourhood is shaded grey and the dotted line indicates how these elements are aligned. The GPU runs many copies of the evolved program in parallel and, essentially, each program can can act on only one array index. By shifting the arrays in this way, we have lined up the data so that although each program can see only a given array position, by looking at the set of arrays (or, more specifically, a single index in each of the arrays in the set) it can have access to the given pixel and its neighbourhood. For example, if we add array (e) to array (i) the new value of the centre pixel will be 6, as the centre pixel in (e) has a value of 5 and the centre pixel in (i) has a value of 1.

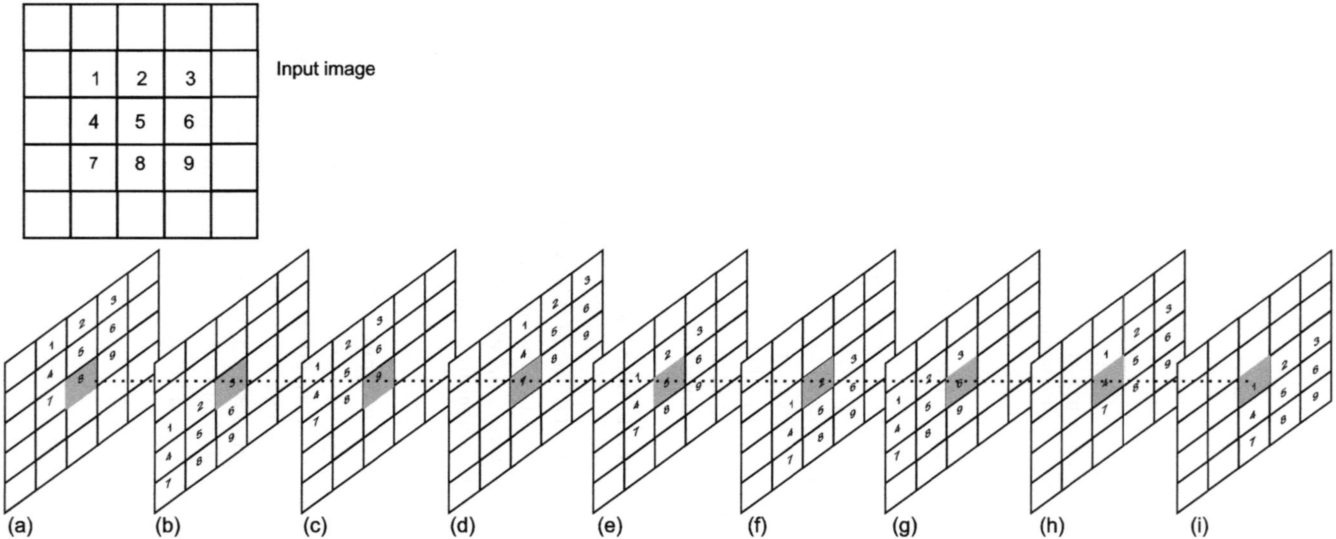

Fig. 8.4 Converting the input image to a set of shifted images allows the elementwise operations of the GPU to access a pixel and its neighbourhood. The evolved program treats each of these images as inputs. For example, should the evolved program want to sum the centre pixel and its top-left neighbour, it would add (e) to (i).

8.6.1 Evolving Image Filters Using Accelerator

In [5, 9], CGP was used with MS Accelerator to evolve a large number of different image filters. Chapter 6 contains more detail about the CGP component of this work, whereas here the focus is on the GPU side of the algorithm. In summary, the original input images (Fig. 8.5) were combined together to form a larger image. A filter was applied, using GIMP. We then used an evolutionary algorithm to find a mapping between the input and output images. The fitness function attempted to minimize the error between the desired output (the GIMP-processed image) and the output from the evolved filters.

GPU acceleration allows us to process images much more rapidly than on a CPU, and enables improvements to be made in the quality of the results collected. In [5] four different images were used for fitness evaluation, and none for validation. In [9], 16 different images (Fig. 8.5) were used; these were largely taken from the USC-SIPI image repository, with 12 used for fitness evaluation and four for validation. This allowed us to be confident that evolved filters would generalize well. As we employed a GPU for acceleration, we were able to test all the images at the same time and obtain both the fitness score and the validation score at the same time.

The fitness score of an individual was the average of the absolute difference (per pixel) between the target image and the image produced using CGP. The lower this error, the closer our evolved solution was to the desired output.

The fitness function was relatively complicated. The first step was to take the output from the evolved program, subtract it from our desired output and then find the absolute value of the result. The whole input image was processed at once (this input image contained 16 subimages). This provided an array containing the difference between the two images. Next, a mask was applied to remove the edges where the subimages meet. This was done because when the images were shifted, there would be overlapping data from different images that could introduce undesirable artefacts into the fitness computation. Edges were removed by multiplying the difference array by an array containing 0s and 1s. The 1s labelled where the fitness function would measure the difference.

To calculate the training error, the difference array was multiplied by another mask. This mask contained 1s for each pixel in the training images, and 0 for pixels that we did not wish to consider. The fitness value was the sum the contents of the array divided by the number of pixels.

The fitness function is illustrated in Fig. 8.6.

8.6.2 Results

Results for the quality of the image filters can be found in Chap. 6. Here, only the GPU acceleration is considered. Table 8.6 shows the performance of this implementation.

Fig. 8.5 The training and validation image set. All images were presented simultaneously to the GPU. The first column of images was used to compute the validation fitness, and the remaining 12 for the training fitness. Each image is 256 × 256 pixels, with the entire set of images containing 1024 × 1024 pixels.

It was found that using the GPU greatly decreased the evaluation time per individual. On our test system (Nvidia 8800 GTX, AMD Athlon 3500+, Microsoft Accelerator API), the average speed was approximately 145 million genetic programming operations per second (GPOps), and a peak performance of 324 Million GPOps was obtained. The processing rate was dependent on the length of the evolved programs. Some filters benefited more from the GPU implementation than others.

When the evolved programs were executed using the CPU, we obtained only 1.2 million GPOps, i.e. a factor of 100 times slower than for the GPU. However, using the reference driver incurs significant overhead and may not be an accurate reflection of the true speed of the CPU.

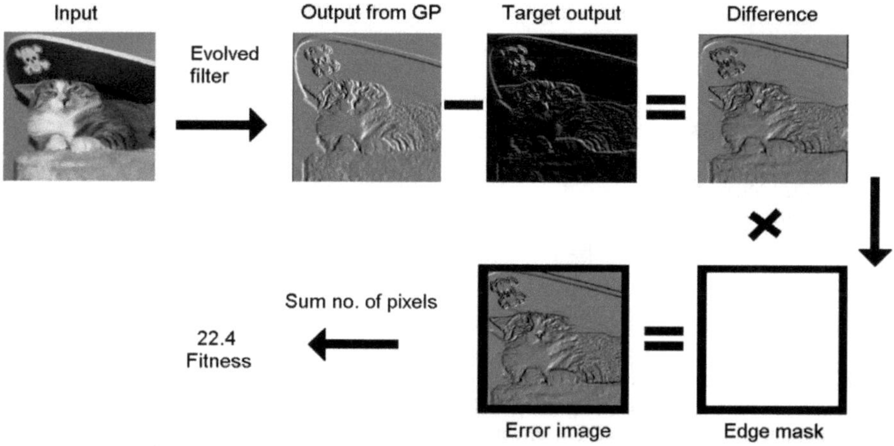

Fig. 8.6 Fitness calculation for images (simplified for a single image). The evolved program is run on the input image, and its per-pixel difference from the target image is calculated. This is then multiplied by an edge mask to remove artefacts created by the multiple subimages touching each other. Finally, the final error is calculated (the average pixel error).

Filter	Peak MGPOps	Avg MGPOps
Dilate	116	62
Dilate2	281	133
Emboss	254	129
Erode	230	79
Erode2	307	195
Motion	280	177
Neon	266	185
Sobel	292	194
Sobel2	280	166
Unsharp	324	139

Table 8.6 Maximum and average mega genetic programming operations per second (MGPOps) observed for each filter type

8.7 CUDA Implementation

The Accelerator implementations discussed so far suffer from many problems. Accelerator is very limiting (Windows only, bugs, too high-level), and does not appear to be under active development – although it is likely to resurface in future versions of DirectX.

The next implementation of CGP for GPUs that we describe is based on Nvidia's CUDA. However, doing this presented many challenges. To use a test-case-parallel approach, individual programs have to be compiled – and this introduces a significant time overhead. In [10], a new approach was introduced which aims to remove this overhead. This section discusses the most recent version of this approach.

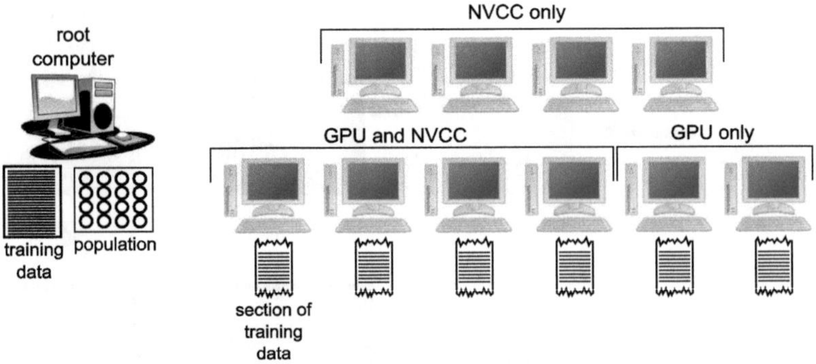

Fig. 8.7 Algorithm overview. The root computer administers the client computers. The root computer is responsible for maintaining the population and distributing sections of the data set to the GPU-equipped client computers. The client computers can be capable of executing programs (i.e. contain a compatible GPU) and/or be capable of compiling CUDA programs using NvCC. Different combinations of compiling and executing computers can be used, depending on the resources available.

8.7.1 Algorithm

The algorithm devised here is an implementation of a parallel evolutionary algorithm, where there is a master controller (called the 'root') and a number of slave computers (or 'clients'), as shown in Fig. 8.7. In summary, the algorithm works in the following manner:

1. A process is loaded onto all computers in the cluster.
2. The root process is started on the master computer.
3. Each computer announces its presence, and the root listens to these broadcasts and compiles a list of available machines.
4. After announcing their presence, each of the client machines starts a server process to listen for commands from the root.
5. After 2 minutes, the root stops listening for broadcasts and connects to each client machine.
6. Upon connecting to each machine, the root asks each client what its capabilities are (in terms of CUDA compiler and device capability).
7. The root distributes sections of the fitness set to each CUDA-capable computer.
8. The root initializes the first population.
9. The root begins the evolutionary loop.

    a. Convert population to CUDA C code.
    b. Parallel-compile population.
    c. Collate compiled code and resend to all compiling clients.
    d. Perform execution of individuals on all GPU-equipped clients.
    e. Collect (partial) fitness values for individuals from GPU-equipped clients.
    f. Generate next population, return to step 9a.

The algorithm here offloads both the compilation and the execution to a cluster of GPU-equipped machines. This significantly reduces the overhead and execution costs.

The cluster consisted of 28 computers, each with an Nvidia GeForce 8200 graphics card. This GPU is a very low end device. Each 8200 GPU has eight stream processors and access to 128 Mb of RAM. As the GPU was shared with the operating system, the free GPU memory on each machine was approximately 80 Mb. The root computer node used here was equipped with an Intel Q6700 quadcore, 4 Gb of RAM and Windows XP. The client computers used Gentoo Linux, AMD 4850e processors and 2 Gb of RAM. The software developed here uses a CUDA wrapper called Cuda.Net [3] and should work on any combination of Linux and Windows platforms.

8.7.2 Compilation and Code Generation

To mitigate the slow compilation step in the algorithm, a parallelized method was employed. Figure 8.8 shows how the compilation speed varies with the number of individuals in a population. The compilation time here includes rendering the genotypes to CUDA C code, saving the code and running the compiler. For a single machine, the compilation time increases quickly with the size of the population. For large populations, the compilation phase quickly becomes a significant bottleneck. Using distributed compilation it is possible to maintain a rapid compilation phase for typical population sizes.

CUDA allows library files (extension '.cubin') called 'modules' to be loaded into the GPU. Modules are libraries of functions that can be loaded, executed on the GPU and then unloaded. Module load times increase linearly with the number of functions (or individuals) they contain. However, there is a constant overhead to this, and for unloading the module, transferring the module across the network and other file-handling operations. Therefore it is preferable to minimize the number of module files that are created.

To compile a module, the user first creates a source file containing only CUDA functions. These should be declared as in the following example:

```
extern "C" __global__
    void Individual0(
           float* ffOutput,
           float* ffInput0,
           int width, int height)
```

When compiled, each of these functions is exposed as a function in the compiled library. Internally, the .cubin files are text files with a short header, followed by a set of code blocks representing the functions. The code blocks contain the binary code for the compiled function, in addition to information about memory and register usage.

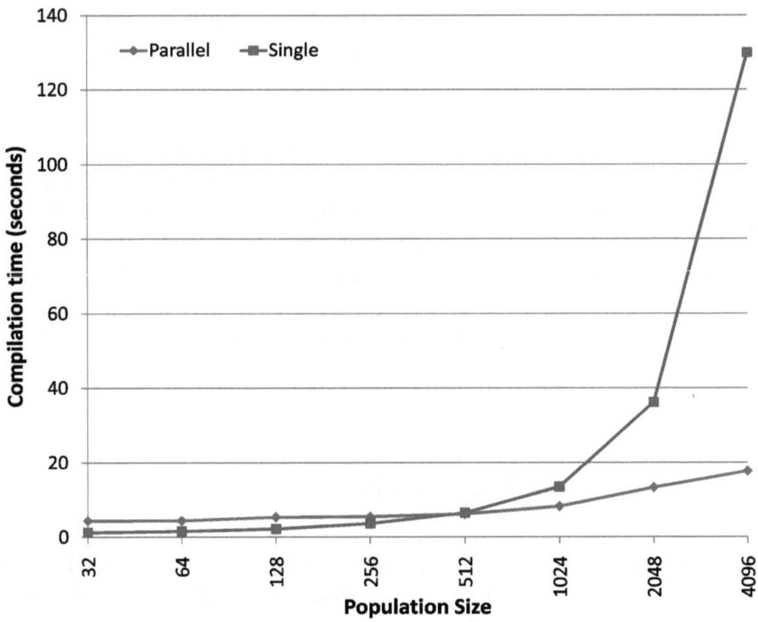

Fig. 8.8 How compilation speed varies with the number of individuals in a population, comparing single-machine compilation with compilation using 14 machines in the cluster.

It is possible to merge the contents of module files together by simply appending .cubin files together. This allows one to compile sections of a library independently of each other and then reduce these sections to a single module. However, only one header is permitted per .cubin file.

In this implementation, the following process was used to convert the genotypes into executable code.

1. Convert the population to CUDA. This process is performed on the root machine. At this stage, the population is divided up into sections, the number of which is equal to the number of machines available for compilation. Each section is then converted to CUDA using the code generator described in Sect. 8.7.2. This stage is performed in parallel across the cores on the root computer, where a separate thread is launched for each section of the population's code generator.
2. Send the uncompiled code to the distributed machines. In parallel, each section of uncompiled code is transmitted to a computer for compilation. The remote computer saves the code to the local disk, before compilation. The remote computer then loads the .cubin file from disk and returns it to the root computer.
3. The host computer merges the returned .cubin files together (in memory) and then sends the merged version back to all the computers in the grid for execu-

tion. Again, this happens in parallel, with the remote computers sending back their compiled files asynchronously.

Figure 8.8 shows the performance benefit obtained with this parallel compilation method. For typical population sizes, the parallel compilation means that each processor compiles only a small number of functions – and for such sizes the compiler is operating at its most efficient ratio of compilation time to function count. However, it should be noted that the compilation time does not appear to scale linearly with the number of compiling clients.

Tables 8.7 and 8.8 show the average times taken to compile populations of various sizes, with various program lengths. The results show a clear performance increase using the parallel compilation method.

Table 8.7 Time (in seconds) to perform the entire compilation process using 28 computers

Population size	Graph length			
	256	512	1024	2048
32	3.08	2.88	2.78	2.82
64	2.43	2.32	2.48	2.65
128	2.84	2.78	3.01	3.01
256	3.29	3.28	3.24	3.39
512	3.73	3.63	4.26	3.97
1024	4.96	5.89	5.41	6.35
2048	8.37	7.98	8.38	9.65

Table 8.8 Time (in seconds) to perform the entire compilation process using a single-threaded compilation method. Owing to memory allocation issues, some results are incomplete

Population size	Graph length			
	256	512	1024	2048
32	1.22	1.22	1.22	1.22
64	1.55	1.55	1.54	1.54
128	2.20	2.20	2.21	2.18
256	3.54	3.73	3.60	3.55
512	6.43	6.44	6.43	6.45
1024	13.33	13.56	13.43	13.45
2048	36.16	36.26	36.06	36.01

Code generation is the process of converting the genotypes in the population to CUDA-compatible code. During development, a number of issues were found that made this process more complicated than expected. One issue was that the CUDA compiler does not like long expressions. In initial tests, the evolved programs were written as a single expression. However, when the length of the expression was increased, the compilation time increased dramatically. This is presumably because the programs were difficult to optimize.

Another issue was that functions with many input variables can cause the compilation to fail, with the compiler complaining that it was unable to allocate sufficient registers. In the initial development, we passed all the inputs in the training set to each individual, regardless of whether the expression used them. This worked well for small numbers of inputs; however the training set that was used to test the system contained 41 columns. The solution to this problem was to pass the function only the inputs that it used. However, this requires each function to be executed with a different parameter configuration. Conveniently, the CUDA.Net interface does allow this, as function calls can be generated dynamically at run time. The other issue here is that all or many inputs may be needed to solve a problem. It is hoped that this is a compiler bug and that it will be resolved in future updates to CUDA.

```
extern "C" __global__ void Individual430(float* ffOutput,
    float* ffInput38, float* ffInput36, float* ffInput7,
    float* ffInput20, float* ffInput33, float* ffInput11,
    float* ffInput19, float* ffInput28)
{
    //set up indexes of where to read from
    unsigned int DataIndex = (blockIdx.x * blockDim.x) + threadIdx.x;
    //
    //Active nodes = 11
    float Temp788 = ((ffInput38[DataIndex]) +
        (ffInput36[DataIndex])) - (ffInput7[DataIndex]);
    float Temp1304 = ((-3395.2410) * ((ffInput20[DataIndex])
     * (ffInput33[DataIndex]))) * ((ffInput11[DataIndex]) *
     ((ffInput19[DataIndex]) - (ffInput28[DataIndex])));
    float Temp2196 = (Temp788) / ((ffInput38[DataIndex]) / (1175.2612));
    float r = ((1714.3846) + (Temp1304)) * (Temp2196);
    if (r<0)
        ffOutput[DataIndex] = 0;
    else
        ffOutput[DataIndex] = 1;
    //Operation Count = 11
    //Input Count = 8
}
```

Fig. 8.9 Example of the CUDA code generated for an individual.

The code in Fig. 8.9 illustrates a typical generated individual. As noted before, long expressions failed to compile within a reasonable time. The workaround used here was to limit the number of nodes that made up a single statement, and then use temporary variables to combine these smaller statements together. It is unclear what the best length is for statements, but it is likely to be a function of how many temporary variables (and hence registers) are used. The names of the temporary variables were related to the node index in the graph at which the statement length limit was introduced.

The final output of the individual was stored in a temporary variable (r). As this is a binary classification problem, it was convenient to add code here to threshold the actual output value to one of the two classes.

During code generation, it was also possible to remove redundant code. When a CGP graph is parsed recursively, unconnected nodes are automatically ignored and so do not need to be processed. It is also possible to detect code which serves no purpose, for example division where the numerator and denominator are the same, or multiplication by 0. This can be seen as another advantage of precompiling programs over simple interpreter-based solutions.

8.7.3 Fitness Function

The task here was to evolve an 'emboss' filter on a single image. A large image (3872 × 2592 pixels) was used. It should be noted that Accelerator could only cope with images of size 2048 × 2048 pixels.

The image was converted to grey scale, and then converted in a way similar to that described earlier, where each neighbourhood becomes a row in a data set. The desired output image was an 'embossed' version of the input. The data set contained 10,023,300 rows and 10 columns, which was approximately 400 Mb of data in memory. As each row was independent of every other row, the image could be split across the client computers in the same way as before.

The fitness was calculated as the sum of the differences between the image outputted by the evolved program and the target (embossed) image.

8.7.4 Results

Table 8.9 Average giga GP operations per second when the 'emboss' image filter was evolved using 28 computers

Population size	Graph length			
	256	512	1024	2048
32	3.3	3.6	4.5	5.4
64	5.2	6.8	7.3	8.4
128	7.2	8.4	9.9	13.8
256	11.4	13.2	15.3	18.6
512	14.2	16.6	18.3	21.8
1024	16.4	17.3	21.0	24.1
2048	16.4	18.3	21.6	26.2

Table 8.10 Peak giga GP operations per second when evolving the 'emboss' image filter was evolved using 28 computers

Population size	Graph length			
	256	512	1024	2048
32	7.2	7.5	8.9	10.0
64	8.0	9.9	13.8	16.2
128	12.3	15.7	17.9	22.5
256	19.5	21.7	26.2	29.7
512	22.1	24.5	28.5	32.6
1024	25.0	28.9	30.0	34.7
2048	24.1	27.3	32.2	34.2

Table 8.9 shows the average numbers of giga GP operations per second (GGPOps) and Table 8.10 shows the peak numbers of GGPOps for each of the population/graph size combinations used.

The results are considerably faster than those obtained using Accelerator on a single GPU, which had a peak of 0.25 GGPOps. The peak results here are over 100 times faster. The speed improvement is likely to be due to an increase in the number of shader processors available, although it is difficult to compare the results, given the differences in image size and fitness function.

8.8 Conclusions

GPUs are becoming increasingly common in all forms of scientific computing. For GP practitioners, we find that the nature of the problem of evaluating either individuals in parallel or test cases in parallel maps very well to the GPU architecture. As GPU technology improves, and particularly as the development tools improve, it is likely that more and more GP implementations will exploit this form of parallel processing.

More information about GPUs and genetic programming can be found at www.gpgpgpu.com.

8.9 Acknowledgements

WB and SH gratefully acknowledge funding from Atlantic Canada's HPC network ACENET, from the Canadian Foundation of Innovation, New Opportunities Grant No. 204503, and from NSERC under the Discovery Grant Program RGPIN 283304-07.

References

1. Banzhaf, W., Harding, S.L., Langdon, W.B., Wilson, G.: Accelerating Genetic Programming through Graphics Processing Units. In: R.L. Riolo, T. Soule, B. Worzel (eds.) Genetic Programming Theory and Practice VI, chap. 1, pp. 229–249. Springer (2008)
2. Chitty, D.M.: A data parallel approach to genetic programming using programmable graphics hardware. In: D. Thierens, H.G. Beyer, et al. (eds.) Proc. Genetic and Evolutionary Computation Conference, vol. 2, pp. 1566–1573. ACM Press (2007)
3. GASS Ltd.: CUDA.NET. http://www.gass-ltd.co.il/en/products/cuda.net/
4. Harding, S.L.: Genetic Programming on GPU Bibliography. http://www.gpgpgpu.com/
5. Harding, S.L.: Evolution of Image Filters on Graphics Processor Units Using Cartesian Genetic Programming. In: J. Wang (ed.) IEEE World Congress on Computational Intelligence, pp. 1921–1928. IEEE Press (2008)
6. Harding, S.L., Banzhaf, W.: Fast Genetic Programming and Artificial Developmental Systems on GPUs. In: International Symposium on High Performance Computing Systems and Applications, p. 2. IEEE Computer Society (2007)
7. Harding, S.L., Banzhaf, W.: Fast genetic programming on GPUs. In: Proc. European Conference on Genetic Programming, *LNCS*, vol. 4445, pp. 90–101. Springer (2007)
8. Harding, S.L., Banzhaf, W.: Genetic programming on GPUs for image processing. International Journal of High Performance Systems Architecture **1**(4), 231–240 (2008)
9. Harding, S.L., Banzhaf, W.: Genetic Programming on GPUs for Image Processing. In: J. Lanchares, F. Fernandez, J. Risco-Martin (eds.) Proc. International Workshop on Parallel and Bioinspired Algorithms, pp. 65–72. Complutense University of Madrid Press (2008)
10. Harding, S.L., Banzhaf, W.: Distributed Genetic Programming on GPUs using CUDA. In: I. Hidalgo, F. Fernandez, J. Lanchares (eds.) Proc. International Workshop on Parallel Architectures and Bioinspired Algorithms, pp. 1–10 (2009)
11. Koza, J.: Genetic Programming: On the Programming of Computers by Natural Selection. MIT Press (1992)
12. Langdon, W.B., Banzhaf, W.: Repeated Sequences in Linear Genetic Programming Genomes. Complex Systems **15**(4), 285–306 (2005)
13. Langdon, W.B., Banzhaf, W.: A SIMD Interpreter for Genetic Programming on GPU Graphics Cards. In: Proc. European Conference on Genetic Programming, *LNCS*, vol. 4971, pp. 73–85. Springer (2008)
14. Robilliard, D., Marion-Poty, V., Fonlupt, C.: Population Parallel GP on the G80 GPU. In: Proc. European Conference on Genetic Programming, *LNCS*, vol. 4971, pp. 98–109. Springer (2008)
15. Tarditi, D., Puri, S., Oglesby, J.: MSR-TR-2005-184 Accelerator: Using Data Parallelism to Program GPUs for General-Purpose Uses. Tech. rep., Microsoft Research (2006)
16. Wilson, G., Banzhaf, W.: Linear Genetic Programming GPGPU on Microsoft's Xbox 360. In: J. Wang (ed.) IEEE World Congress on Computational Intelligence. IEEE Press (2008)

Chapter 9
The CGP Developmental Network

Gul Muhammad Khan and Julian F. Miller

9.1 Introduction

Genetic programming has long been used for the creation of artificial morphogenetic systems. These are inspired by aspects of developmental biology. The essential idea is to encode the rules in an evolved genotype that will self-organize to produce a phenotype, rather than using a one-to-one relationship between a genotype and a phenotype. Indirect encoding schemes provide the ability to code much larger structures than do direct encoding schemes [18, 9, 2, 6].

The idea of constructing programs modelled on the brain is motivated by systems that exhibit intelligence, the ability to learn, self-replication and self-adaptation. The brain has many highly desirable features that are hard to replicate in conventional computer systems. It is developmental, in that it acquires increasingly sophisticated capabilities over time. The developing brain always retains its integrity as a learning system. The brain is adaptive and shows plasticity with respect to changes in its environment, so that new experiences and stimuli are incorporated into the neural system without altering existing capabilities. It shows tolerance to damage and the ability to self-repair and self-reorganize in such a way that it retains functionality. The brain is highly versatile in its ability to learn diverse tasks and to develop abstract symbolic models which enable a living system to operate effectively in complex environments.

It is apparent that although it is highly complex, the brain is made of essentially similar building blocks. These blocks are composed of highly interconnected networks of similar neurons. In the same way that the developmental engine of the biological cell contains the mechanism for building the human body, we believe that the key to the sophistication of the brain lies in the developmental power of the neuron. Thus obtaining a computational equivalent of the biological neuron is an important research objective.

One of the difficulties in attempting to create a dynamic computational model inspired by neuroscience is that the internal dynamics of biological neurons are ex-

J.F. Miller (ed.), *Cartesian Genetic Programming*, Natural Computing Series,
DOI 10.1007/978-3-642-17310-3_9, © Springer-Verlag Berlin Heidelberg 2011

tremely complicated and many of these processes may be unnecessary in a machine-learning technique. However, we take the view that the biology of neurons (i.e. their gross morphology and connectivity) *is* sufficiently well understood [1, 35] to allow us to identify essential subsystems that we must attempt to evolve in order to achieve a computational equivalent. Conventional models of neural networks do not consider the genetics of neurons and the development (as in biology) of a mature network during learning. Instead they are dominated by a static connectionist view of the brain. For a recent discussion of this issue see [25]. However, genetic programming (GP) offers the capability to represent neural programs and the transfer of genetic changes from generation to generation. Thus GP, in principle, provides us with a means to represent complex neuron 'engines' that can be evolved to exhibit properties similar to real neural systems, without the restrictions of a theoretical model of these systems. GP has been shown to be able to solve problems of this type [19] and often these solutions are not fragile and show unexpected emergent behaviours such as self-assembly and self-repairability [24, 23], which are natural properties of living systems.

In our work, we have used Cartesian genetic programming (CGP) [26] to construct this developmental network. Each neuron is considered to be a computational device, with each subprocessing part represented by a chromosome. The genotype of a neuron consists of a collection of chromosomes representing the subcomponents of the neuron. The chromosomes are evolved until the desired intelligent behaviour is obtained in the computational network.

Each neuron is provided with a structural morphology consisting of a soma, dendrites [30], axons with branches, dynamic synapses [8] and synaptic communication. Neurons are placed in a two-dimensional toroidal grid to give the branches a sense of virtual proximity. Branches are allowed to grow and shrink, and communicate with each other.

To achieve this, we have idealized seven neural components, which we have represented as CGP chromosomes encoding combinational digital circuits [14, 17]. These chromosomes encode distinct computational functions representing aspects of real neurons. This model allows neurons, dendrites and axon branches to grow, die and change while solving a computational problem. Also, the synaptic morphology can change and affect the information processing. While this model is undeniably quite complex and involves many variables and parameters, we feel this is justified by the evident enormous complexity of the brain. The computational network that forms when the seven chromosomes are run grows into a network of neurons, neurites and synapses that reflects its own internal dynamics and environmental interaction.

Learning in the brain occurs after evolution has finished (i.e. in the lifetime of the individual). In other words, evolution has created self-learning programs. These are programs which, when run, construct a system that is capable of learning from experience. We aimed to do the same. We evolved programs which, when run, continuously build and change a neural architecture in response to environmental interaction. We emphasize that no evolution takes place in the lifetime of the individual while it is learning.

The capabilities of the networks for learning have been tested on a well-known artificial intelligence problem known as Wumpus World [34]. The application of genetic programming to Wumpus World has already been considered [36, 37]. It is worth mentioning that although we tested the system in an artificial environment inspired by Wumpus World, the goal of the system was to evolve the capability of learning in general and not merely to solve Wumpus World.

In addition, we have examined and adapted the Wumpus World problem as a co-evolutionary competitive learning environment in which two agents (predator and prey) struggle to achieve the desired tasks and survive [10]. Each agent was independently controlled by an evolved CGP developmental network.

We have also evaluated the learning capability of our system in a game of checkers (or draughts). Although the history of research into computers playing games is full of highly effective methods (e.g. minimax, and the board evaluation function), it is arguable whether human beings use such methods. Typically, they consider relatively few potential board positions and evaluate the favourability of these positions in a highly intuitive and heuristic manner. They usually learn during a game. Indeed, this is how humans learn to be good at any game in general. So, the question arises: How is this possible? In our work, what we are interested in is how an *ability to learn* can arise and be encoded in a genotype that, *when executed*, gives rise to a neural architecture that can play a game well. Each agent (player) has a genotype that grows a computational neural structure during the course of the game. This allows our network to learn while it develops during its lifetime. The network begins as small randomly defined networks of neurons with dendrites and axo-synapses. The job of evolution is to come up with genotypes that encode *programs* that when *executed* develop into mature neural structures that learn through environmental interaction and continued development.

To test the capability of our approach for learning, we played two evolved players against each other using a form of coevolution. We have also investigated the evolution of an agent in competition with a minimax-based checker program (MCP). The results show that the agents' capability for playing checkers increases with evolution. The experimental results also demonstrate that the agents' capabilities improve with development without evolution.

9.2 Biology of Neurons

Neurons are the main cells responsible for information processing in the brain. They are different from other cells in the body in not only functionality but also in biophysical structure [12]. Neurons are the building blocks of the brain. They have different shapes and structures depending on their location in the brain, but the basic structure of neurons is always the same [20]. It has the following three main parts (as shown in Fig. 9.1):

- Dendrites (inputs). These receive information from other neurons and transfer it to the cell body. They have the form of a tree structure, with branches close to the cell body.
- Axons (outputs). These transfer information to other neurons by the propagation of a spike or action potential. Axons usually branch away from the cell body and make synapses (connections) to the dendrites and cell bodies of other neurons.
- Cell body (processing area or function). This is the main processing part of the neuron. It receives all the information from the dendrite branches connected to it in the form of electrical disturbances and converts it into action potentials, which are then transferred through the axon to other neurons. It also controls the development of neurons and branches.

9.3 The CGP Developmental Network

This section describes in detail the structure of the CGP developmental network (CGPDN). In addition, the rules and evolutionary strategy used to run the system are described. The CGPDN has two main aspects:

(a) Neurons with a number of dendrites, with each dendrite having a number of branches and an axon having a number of axon branches.
(b) A genotype, representing the genetic code of the neurons. Each genotype consists of seven chromosomes. Each chromosome is represented as a digital circuit.

The first aspect (a) defines mainly the neural components and their properties, and the second (b) is concerned with the internal behaviour of the neurons in the

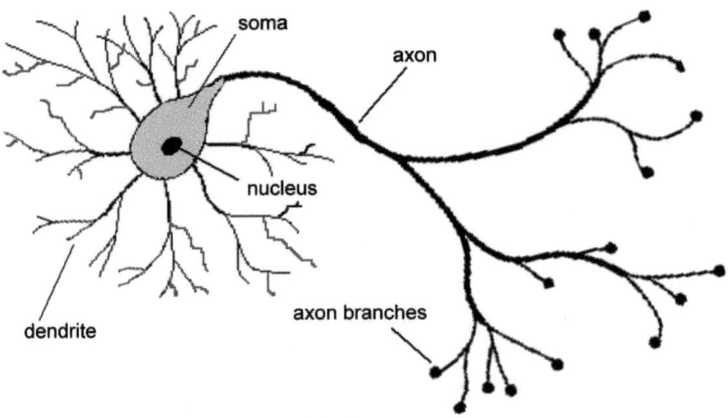

Fig. 9.1 Schematic of neuron showing its various parts: dendrites, soma, nucleus, axon and axon branches.

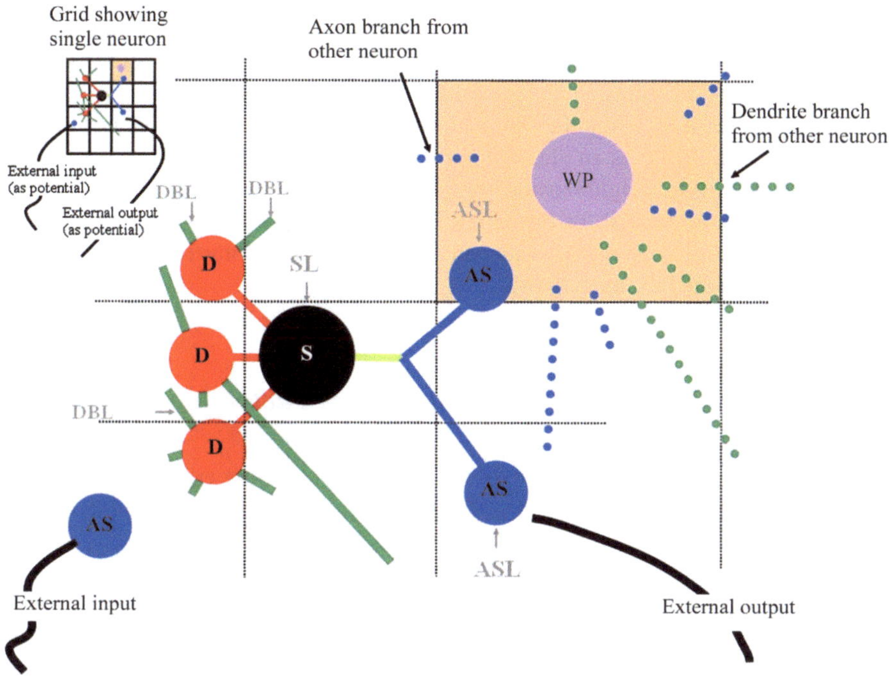

Fig. 9.2 On the top left, a grid is shown containing a single neuron. The rest of the figure is an exploded view of the neuron. The neuron consists of seven evolved computational functions. Three are 'electrical' and process a simulated potential in the dendrite (D), soma (S) and axo-synapse branch (AS). Three more are developmental in nature and are responsible for the life cycle of the neural components (shown in grey). They decide whether dendrite branches (DBL), the soma (SL) and axo-synaptic branches (ASL) should die, change or replicate. The remaining evolved computational function (WP) adjusts synaptic and dendritic weights and is used to decide the transfer of potential from a firing neuron to a neighbouring neuron.

network. The chromosomes represent the functionality of different parts of the neuron. During evolution, the second aspect (the genotype) is evolved towards the best functionality, whereas the first aspect (the neural components and their properties) changes only during the *lifetime* of the network, i.e. while it is performing the learning task.

In the CGPDN the neurons are placed randomly in a two-dimensional grid (the CGPDN grid), so that they are aware only of their spatial neighbours (as shown in Fig. 9.2) . The initial number of neurons is specified by the user. Each neuron is initially allocated a random number of dendrites, and dendrite branches, one axon, and a random number of axon branches. Neurons receive information through dendrite branches and transfer information through axon branches to neighbouring neurons. The dynamics of the network also changes during this process. Branches may grow or shrink, and move from one CGPDN grid point to another. They can produce new branches and can disappear. Neurons may die or produce new neurons. Axon branches transfer information only to dendrite branches in their proximity. This process is performed by passing the signals from all the neighbouring branches through a CGP program, acting as an electrochemical synapse, which updates the values of the potential only in neighbouring branches. An integer variable that mimics the electrical potential is used for the internal processing of neurons and communication between neurons. External inputs and outputs are also converted into potentials before being applied to the network.

9.3.1 Health, Resistance, Weight and State-Factor

Four variables are incorporated into the CGPDN, representing either the fundamental properties of the neurons (*health*, *resistance*, and *weight*) or as an aid to computational efficiency (*state-factor*).

Each dendrite branch and axo-synaptic connection has three variables assigned to it. These variables are called *health*, *resistance* and *weight*. The values of these variables are adjusted by the CGP programs (see below). The *health* variable is used to govern the replication and/or death of dendrites and axon branches. The *resistance* variable controls the growth and/or shrinkage of dendrites and axon branches. The biological basis for the *weight* variable has already been discussed. This variable is used in calculating the potentials in the network. Each soma has only two variables: *health* and *weight*. The use of these variables is summarized in Fig. 9.10 later in this chapter. *Health*, *weight* and *resistance* are represented as integers.

The variable *state-factor* is used as a parameter to reduce the computational burden by keeping some of the neurons and branches inactive for a number of cycles. When *state-factor* is zero, the neurons and branches are considered to be active and their corresponding program is run. The value of *state-factor* is affected by the CGP programs, as it is dependent on the outputs of the CGP electrical-processing chromosomes (see later). The bio-inspiration for *state-factor* is the fact that not all neurons and/or dendrite branches in the brain are actively involved in each process.

9.3.2 Cartesian Genetic Program (Chromosome)

The CGP function nodes used here consist of multiplexer-like operations [27]. Each function node has three inputs and implements one of four functions, as shown in Fig. 9.3. Here a, b and c are the inputs to the node. These functions are Boolean operations representing logical AND (represented by .) and logical OR (represented by +). The multiplexers require four genes each to describe the type of multiplexer function (underlined in Fig. 9.4) and its corresponding connections. All multiplexers operate in a bitwise fashion on 32-bit data (see Sect. 5.2.2 for further details). The

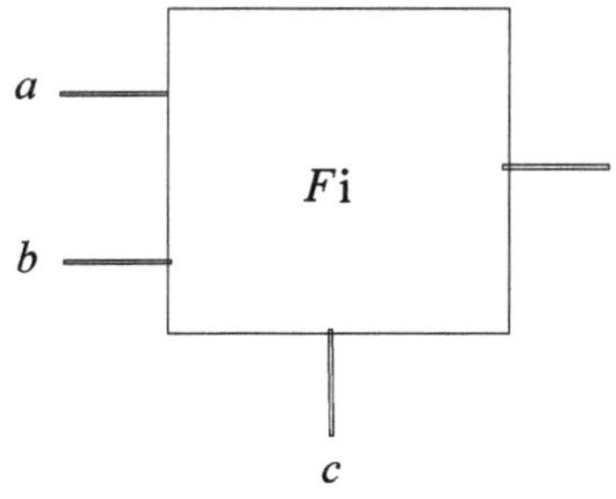

- $F1 = a.\overline{c} + b.c;$
- $F2 = \overline{a}.\overline{c} + b.c;$
- $F3 = a.\overline{c} + \overline{b}.c;$
- $F4 = \overline{a}.\overline{c} + \overline{b}.c;$

Fig. 9.3 Multiplexer diagram, showing inputs a, b and c, and the function F_i. The figure also lists all four possible functions that can be implemented by the multiplexer.

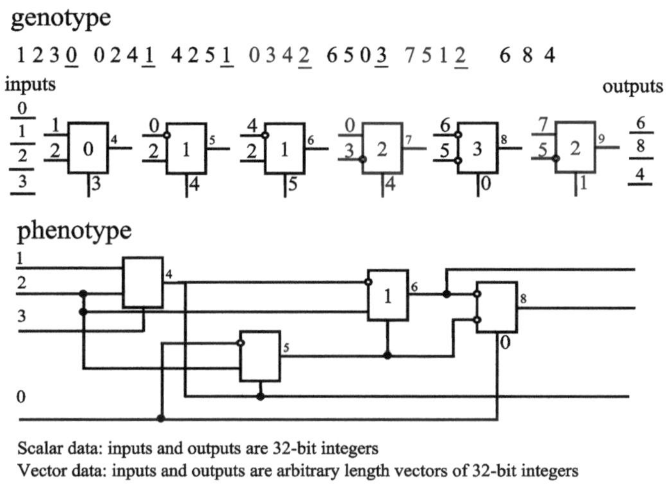

Fig. 9.4 Structure of CGP chromosome used for CGPDN, showing a genotype for a four-input, three-output function and its decoded phenotype. Inputs can be either integers or array of integers.

four functions in Fig. 9.4 are the possible input combinations of a three-input (two inputs and a control) multiplexer. Multiplexers can be considered as atomic in nature as it is well known that they can be used to represent any logic function [27, 4].

Figure 9.4 shows the genotype and the corresponding phenotype, obtained by connecting the nodes as specified in the genotype. The figure also shows the inputs and outputs of the CGP. Output is taken from the nodes as specified in the genotype (6, 8, 4). In the work described here, we did not specify the output in the genotype and instead used a fixed pseudo-random list of numbers to specify where the output should be taken from.

Here the number of rows used is one, as described earlier (see Chap 2). Inputs are applied to CGP chromosomes in two ways, either as a scalar or a vector. In

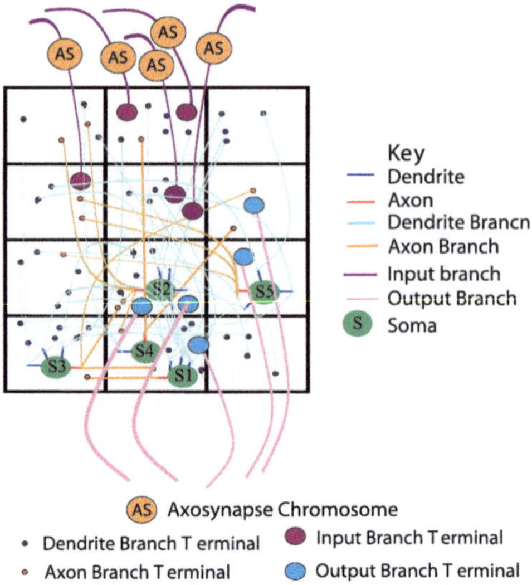

Fig. 9.5 Schematic illustration of a 3×4 CGPDN grid. The grid contains five neurons; each neuron has a number of dendrites with dendrite branches, and an axon with axon branches. Inputs are applied at five random locations in the grid using input axo-synapse branches by running axo-synaptic CGP programs. Outputs are taken from five random locations through output dendrite branches. The figure shows the exact locations of the neurons and branches used in most of our experiments in the initial network. Each grid square represents one location; the branches and soma are shown spaced for clarity. Each branch location is represented by where it terminates. Every location can have an arbitrary number of neurons and branches; there is no upper limit.

the former case, the inputs and outputs are integers, while in the latter case, inputs required by the chromosome are arranged in the form of an array, which is then divided into 10 CGP input vectors. If the total number of inputs cannot be divided into 10 equal parts, then they are padded with zeros. This allows us to process an arbitrary number of inputs with the CGP circuit chromosome simply by clocking through the elements of the vectors. In general, CGP cannot take a variable number of inputs.[1] The inputs are arranged in the form of vectors, and each vector can have an arbitrary number of elements. This method adds some noise, and this is more pronounced when the number of inputs is less than 10, as we pad the inputs with zeros when the number of inputs cannot be divided into 10 subvectors. But as the number of inputs increases, the effect of this noise is reduced.

[1] Another way of acquiring inputs is by introducing *functions* that read inputs. This is used in self-modifying CGP, which is discussed in Chap. 4; this method allows CGP to take a variable number of inputs at run time.

9.3.3 Inputs and Outputs

The inputs are applied to the CGPDN through axon branches by using axo-synaptic electrical-processing chromosomes. The axon branches are distributed across the network in a similar way to the axon branches of neurons as shown in Fig. 9.5. These branches can be regarded as 'input neurons'. They take an input from the environment and transfer it directly the axo-synapse input. When inputs are applied to the system, the program encoded in the axo-synaptic electrical branch chromosome is executed and the resulting signal is transferred to its neighbouring active dendrite branches.

Similarly, there are output neurons which read the signal from the network through output dendrite branches. These output dendrite branches are distributed across the network as shown in Fig. 9.5. The branches are updated by the axo-synaptic chromosomes of the neurons in the same way as for other dendrite branches. The output from the output neuron is taken without further processing after every five cycles. The number of inputs and outputs can change at run time (during development), a new input or output branch can be introduced into the network, or an existing branch can be removed. This allows CGPDN to handle arbitrary numbers of inputs and outputs at run time.

In the next section, we describe the complete neuron model along with its sub-processes.

9.4 CGP Model of Neuron

In the model, neural functionality is divided into three major categories.

- electrical processing;
- life cycle;
- weight processing.

These categories are explained in detail one by one in the subsections below.

9.4.1 Electrical Processing

The electrical-processing part is responsible for signal processing inside neurons and communication between neurons. It consists of the following three chromosomes (as shown in Fig. 9.6):

- electrical processing in a dendrite;
- electrical processing in soma; and
- electrical processing in an axo-synaptic branch.

The way they process electrical signals and transfer them to other neurons is shown in Fig. 9.7.

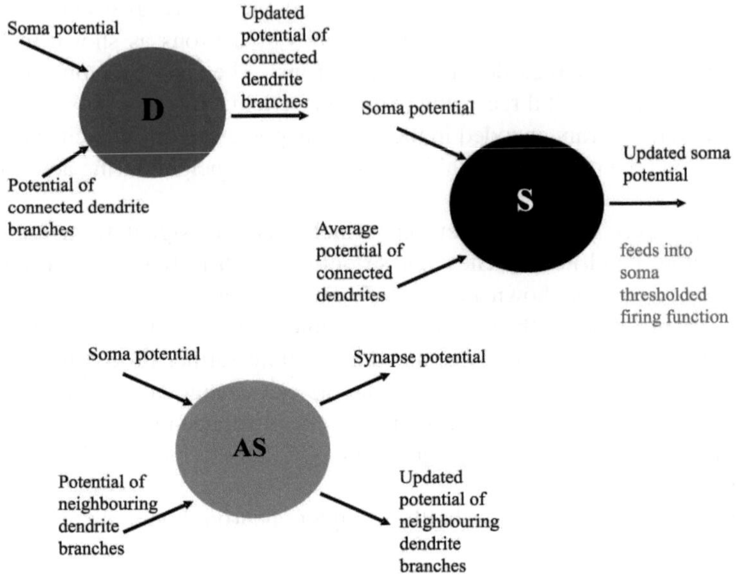

Fig. 9.6 Electrical processing in a neuron, showing the CGP programs for a dendrite branch, the soma and an axo-synaptic branch with their corresponding inputs and outputs.

9.4.1.1 Electrical Processing in Dendrite

This is a vector-processing chromosome. This chromosome handles the interaction between the potentials of different dendrite branches belonging to the same dendrite. Figure 9.6 shows the inputs and outputs.

The input consists of the potentials of all the active branches connected to the dendrite and the soma potential. The CGP program produces the new values of the dendrite branch potentials as output. Subsequently, the potential of each branch is processed by adding weighted values of *resistance*, *health* and *weight* using the following equation:

$$p' = p + \alpha_d h + \beta_d w - \gamma_d r. \tag{9.1}$$

P' and P are the updated potential and previous potential, respectively. The variables h, w and r are the *health*, *weight* and *resistance*, respectively, of the dendrite branch. α_d, β_d and γ_d are user-defined parameters having values between 0 and 1. In the work described here, they had the values of 0.02, 0.05 and 0.05, respectively.

The above equation shows that as the *health* and *weight* of the branch increase so does its potential and as the *resistance* increases, the potential decreases (the usual resistive behaviour). We allow increases in *health* to causes an increase in *potential* because it is reasonable to assume that healthy branches facilitate the flow of potential. Weights are responsible for amplification of potential. Thus high values of weights should cause an increase in potential.

The *state-factor* of each branche is adjusted based on the updated value of the branch potential. The branch is made more active (by reducing its *state-factor*) if there is an increase in potential after the dendrite program (D) has been run. This is done to encourage more sensitive branches to participate in processing by keeping them active. We set up a range of thresholds for assigning the *state-factor*. If any of the branches are active (has its *state-factor* equal to zero), its life cycle CGP program is run. The same process is repeated for all the dendrites and their corresponding branches. After all the dendrites have been processed, the average value of the potentials of all the dendrites is taken, which in turn is the average value of the potentials of the active dendrite branches attached to them. This potential and the soma potential are applied as inputs to the CGP soma electrical-processing chromosome as described below.

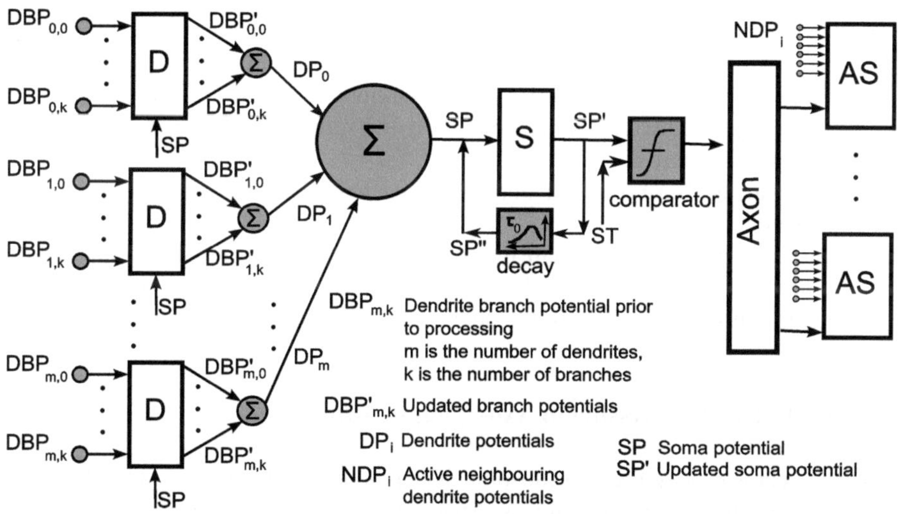

Fig. 9.7 Electrical processing in a neuron at different stages, from left to right branch potentials are processed by dendrites (D), then averaged. These average dendrite potentials are averaged at the soma, which adjusts the potential using the program encoded in the soma electrical chromosome (S), giving a final soma potential. This is fed in to a comparator which decides whether to fire an action potential. This is transferred using the program encoded in the axo-synapse electrical chromosome (AS).

9.4.1.2 Electrical Processing in Soma

This is a scalar-processing chromosome. This chromosome is responsible for determining the final value of the soma potential after receiving signals from all the dendrites, as shown in Fig. 9.6. The chromosome produces an updated value of the soma potential (P') as output which is further processed using a weighted summation of *health*(h) and *weight*(w) of the soma using the following equation:

$$P' = P + \alpha_s h + \beta_s w, \tag{9.2}$$

where α_s and β_s were assigned the values 0.02 and 0.05.

The processed potential of the soma is then compared with the threshold potential of the soma and a decision is made whether to fire an action potential or not. If the soma fires, it is kept inactive for a few cycles (a refractory period) by increasing its *state-factor*.

After this, the soma life cycle chromosome is run, and the firing potential is sent to the other neurons by running the axo-synapse electrical-processing chromosome. The threshold potential of the soma is also adjusted to a new value (the maximum) if the soma fires. If soma does not fire, the value of the processed potential is checked and the following actions are taken:

- If the processed potential is below one third of the maximum value, the *state-factor* is set to a higher value so that the soma is kept inactive for three cycles. This means that if the potential of soma is low so that it is unable to fire, the soma is kept inactive for more time. Only the somas which fire are encouraged and kept more active.
- If the value is above one third and below one half, then the soma is kept inactive for just one cycle.
- If the value is higher than half the maximum value, the soma is kept active for the next cycle and its life cycle is run. This means that if the soma has a high potential, it is kept active for the next cycle.

9.4.1.3 Electrical Processing in Axo-Synaptic Branch

This is a vector-processing chromosome. The potential of the soma is transferred to other neurons through axon branches. An axon branch and a synapse are considered as a single entity with combined properties. Axo-synapses transfer a signal only to the neighbouring active dendrite branches, as shown in Fig. 9.8. Branches sharing the same grid square form a neighbourhood. Figure 9.6 shows the inputs and outputs of the chromosome responsible for the electrical processing in each axo-synaptic branch.

The chromosome produces updated values of the neighbouring dendrite branch potentials and the axo-synaptic potential as output. The axo-synaptic potential is then processed as a weighted summation of *health*, *weight* and *resistance* of the axon branch using the following equation:

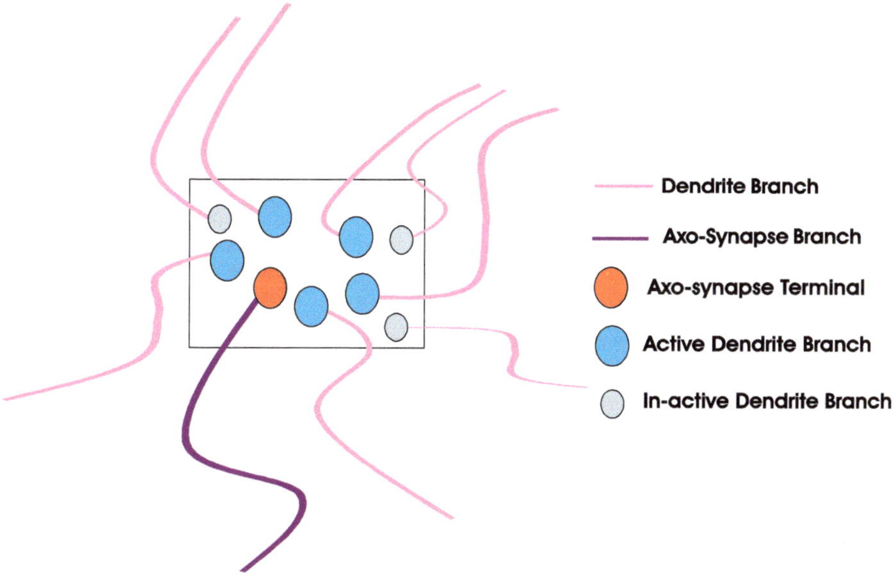

Fig. 9.8 Diagram showing one of the grid squares in which a signal is transferred from an axo-synapse to dendrite branches, showing inactive and active branches.

$$P' = P + \alpha_{as}h + \beta_{as}w - \gamma_{as}r, \tag{9.3}$$

where P' and P are the updated potential and previous potential respectively. h, w and r are the *health*, *weight* and *resistance*, respectively of the axon branch. α_{as}, β_{as} and γ_{as} are user-defined adjustment parameters having values between 0 and 1. In the work described here, these parameters hadvalues 0.02, 0.05 and 0.05, respectively.

The axo-synaptic branch weight-processing program (see Fig. 9.9) is run after the above process, and the processed axo-synaptic potential is assigned to the dendrite branch having the highest updated *weight*. The *state-factor* of each branch is adjusted based on the updated value of the branch potential. The branch is made more active if the potential increases after the execution of the program encoded in the axo-synaptic electrical chromosome (AS). We set up a range of thresholds for assigning the *state-factor*. If any of the branches are active (has its *state-factor* equal to zero), its life cycle CGP program is run, otherwise it continues processing the other axon branches. The axo-synaptic branch CGP is run in all the active axon branches one by one.

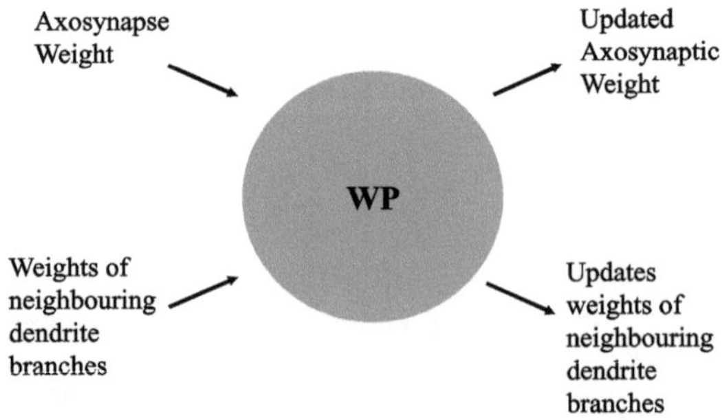

Fig. 9.9 Weight processing in an axo-synaptic branch, with its corresponding inputs and outputs.

9.4.2 Weight Processing

This is a vector-processing chromosome. Weight processing is responsible for updating the *weights* of branches. There is only one chromosome of this type. The *weights* of axon and dendrite branches affect their capability to modulate and transfer information (potential) efficiently. Weights affect almost all of the neural processes either by virtue of being an input to a chromosomal program or as a factor in the post-processing of signals.

Figure 9.9 shows the inputs and outputs to the weight-processing chromosome. The CGP program encoded in this chromosome takes as input the *weight* of the axo-synapse and the *neighbouring* (on the same CGPDN grid square) dendrite branches and produces their updated values as output. The synaptic potential produced at the axo-synapse is transferred to the dendrite branch having the highest weight after weight processing.

9.4.3 Life Cycle of Neuron

This part is responsible for increases or decreases in the numbers of neurons and neurite branches and also the growth and migration of neurite branches. It consists of three chromosomes:

- life cycle of dendrite branch;
- life cycle of soma;
- life cycle of axo-synaptic branch.

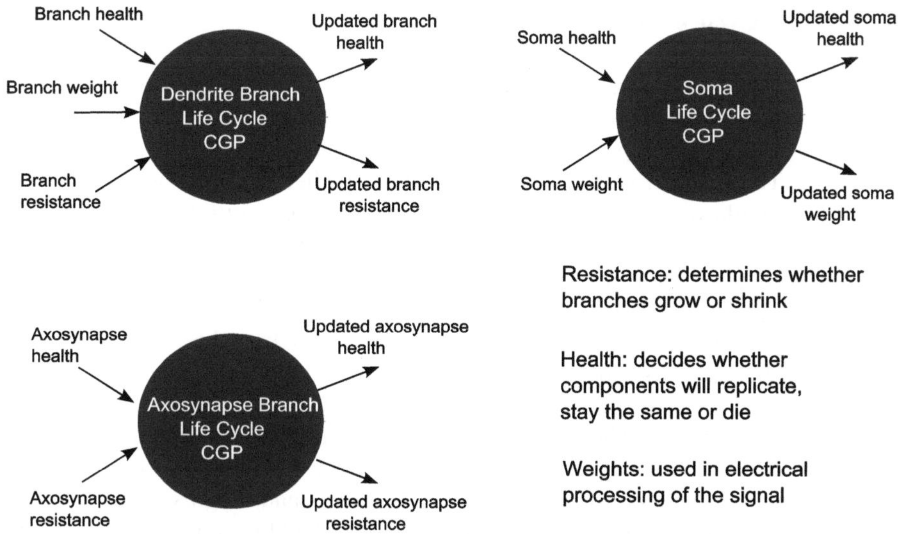

Fig. 9.10 Life cycle of neuron, showing CGP programs for life cycles in a dendrite branch, the soma and an axo-synapse branch, with their corresponding inputs and outputs.

9.4.3.1 Life Cycle of Dendrite Branch

This is a scalar-processing chromosome. Figure 9.10 shows the inputs and outputs of this chromosome. This process updates *resistance* and *health* of the branch. The variation in *resistance* of a dendrite branch is used to decide whether it will grow, shrink or stay at its current location. If the variation in *resistance* during the process is above a certain threshold (R_{db}), the branch is allowed to migrate to a different neighbouring location at random. This neighbouring location can be any one of the eight squares around the branch on a rectangular grid. The branch can move to any of the eight neighbouring squares at random. It can move only one square at a time. Changes in *resistance* can be negative (shrinkage) or positive (growth). In both cases, the absolute change in resistance is used to decide if the branch should move from its current grid square to another grid square. Growth and shrinkage do not occur within one grid square, as that shows only one location.

The updated value of the dendrite branch *health* decides whether the branch produces offspring, dies or remains as it was with an updated *health* value. If the updated *health* is above a certain threshold ($H_{db_{max}}$), the branch is allowed to produce offspring, and if it is below a certain threshold ($H_{db_{min}}$), the branch is removed from the dendrite. Producing offspring results in a new branch at the same CGPDN grid point, connected to the same dendrite.

The values of (R_{db}), ($H_{db_{max}}$) and ($H_{db_{min}}$) are specified by the user.

9.4.3.2 Life Cycle of Soma

This is a scalar-processing chromosome. Figure 9.10 shows the inputs and outputs of the soma life cycle chromosome. This chromosome is intended to evaluate the life cycle of a neuron. It produces updated values of *health* and *weight* of the soma as output. The updated value of *health* decides whether the soma should produce offspring, should die or continue as it is. If the updated *health* is above a certain threshold ($H_{s_{max}}$), the soma is allowed to produce offspring, and if it is below a certain threshold ($H_{s_{min}}$), the soma is removed from the network, along with its neurites. If the soma produces offspring, then a new neuron is introduced into the network, with a random number of dendrites and dendrite and axon branches, at a different location chosen randomly. The random number has an upper (B_{max}) and lower (B_{min}) limit. In all our experiments we chose B_{max} as 5 and B_{min} as 2 for the numbers of dendrites and branches. The new neuron is placed at a pseudo-random location (a square) on a grid. The soma and branches are provided with an initial value of *health* and pseudo-random values of *resistances*, *state-factors* and *weights*.

The values of $H_{s_{max}}$, $H_{s_{min}}$, B_{max} and B_{min} are specified by the user.

9.4.3.3 Life Cycle of Axo-synaptic Branch

This is a scalar-processing chromosome. The role of this chromosome is similar to that of the dendrite branch life cycle chromosome. Figure 9.10 shows the inputs and outputs of the axo-synaptic branch life cycle chromosome. It takes *health* and *resistance* of the axon branch as input and produces corresponding updated values as output. The updated values of *resistance* are used to decide whether the axon branch should grow, shrink or stay at its current location. As with the dendrite branches, if the variation in axon *resistance* is above a certain threshold (R_{as}), the axon branch is allowed to migrate to a different neighbouring location at random. The *health* of the axon branch decides whether the branch will die, produce offspring or merely continue with an updated value of *health*. If the updated *health* is above certain threshold ($H_{as_{max}}$), the branch is allowed to produce offspring and if it is below a certain threshold ($H_{as_{min}}$), the branch is removed from the axon. Producing offspring results in a new branch at the same CGPDN grid point, connected to the same axon.

9.5 Applications

We tested the capability of CGPDN for learning on the artificial intelligence problems of Wumpus World and the game of checkers. In both cases, we evolved and coevolved the CGPDN to study its learning behaviour. The next subsection gives an overview of Wumpus World, and Sect. 9.6 describes checkers.

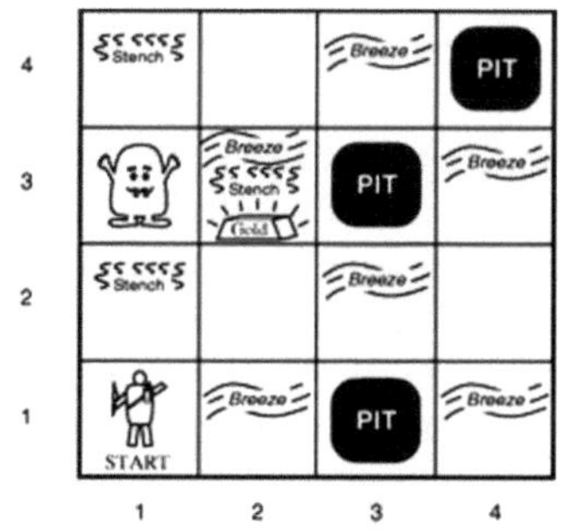

Fig. 9.11 A two-dimensional grid, showing the Wumpus World environment, having one Wumpus, one agent on the bottom left corner square, three pits and gold [34].

9.5.1 Wumpus World Problem

Wumpus World is a variant of Gregory Yob's 'Hunt the Wumpus' game [42]. It is an agent-based learning problem used as a testbed for various learning techniques in artificial intelligence [34]. Wumpus World was originally presented by Michael Genesereth [34]. It consists of a two-dimensional grid containing a number of pits, the Wumpus (a monster) and an agent [34], as shown in Fig. 9.11.

An agent always starts from a unique square (home) in a corner of the grid. The agent's task is to avoid the Wumpus and the pits, find gold and return it to the home square. The agent can perceive a breeze in the squares adjacent to the pits, a stench in the squares adjacent to the Wumpus and a glitter on the square containing the gold. Spector and Luke investigated the use of genetic programming to solve the Wumpus World problem [36, 37].

In the first case of the Wumpus World environment that we studied, we demonstrated the ability of the agent to learn during its lifetime and transfer the learned information genetically from generation to generation. The importance of the life cycle chromosomes was also investigated. In the second case, the network was tested in a coevolutionary environment with two agents learning to solve their tasks while interacting with each other [14]. We allowed the agents to have a quantity analogous to an energy level, which was increased or decreased as a consequence of their actions and encounters in the environment. The first agent increased its energy level (and, consequently ultimately its fitness) by avoiding deleterious encounters and achieving beneficial encounters. The opposing agent increased its energy level solely by direct antagonistic encounters with the first agent (which were beneficial to it). Paredis described this type of fitness as 'life-time fitness evaluation' and discussed how this 'arms race' provides a strong driving force toward complexity [31].

Nolfi and Floreano [29] adopted this approach when they coevolved two competing populations of predator and prey robots in order to explore how lifetime learning allows evolved individuals to achieve behavioural generality, i.e. the ability to produce effective behaviour in a variety of different circumstances. Since both of the populations change across generations, predators and prey face ever-changing and, potentially progressively more complex challenges.

Stanley and Miikkulainen's approach [39, 38] to achieving complexification is through the incremental elaboration of solutions by adding new neural structures. They started with a simple neural structure and, through a managed approach, developed it to produce a complex neural structure. In our model, the computational network is able to 'complexify' itself. This was inspired by the fact that the brain complexifies itself without any change in genetic code.

9.5.1.1 Experimental Set-up

In our experiments, we adapted the rules of Wumpus World slightly. Namely, when the agent encounters a pit or the Wumpus it is only weakened (thus reducing its life) but not killed. Also a glitter signal exists on squares adjacent to the square containing the gold. These changes were introduced to facilitate the capacity of the agent to learn. The CGPDN learns everything from scratch and builds its own memory (including the meaning of signals, pits and the Wumpus).

It is important to appreciate how difficult this problem is. The agent starts with a few neurons with random connections. Firstly, evolution must find a series of programs that build a computational network which is stable (does not lose all neurons or branches, etc.). Secondly, it must find a way of processing infrequent environmental signals. Thirdly, it must navigate in this environment using some form of memory. Fourthly, it must confer goal-driven behaviour on the agent.

The Wumpus World that we used is a two-dimensional grid (10×10), having 10 pits, one Wumpus and the gold. The locations of the Wumpus, gold and pits are chosen randomly. In the squares containing the Wumpus, gold, and pits, and in directly (not diagonally) adjacent squares, the agent can perceive stench, glitter and breeze, respectively. The agent is assigned a quantity called energy, which has an initial value of 200 units. If an agent is caught by the Wumpus, its energy level is reduced by 60%. If it encounters a pit, its energy level is reduced by 10 units. If it gets the gold, its energy level is increased by 50 units and, on arriving home, the agent ceases to exist. For each single move, the agent's energy level is reduced by 1 unit, so if the agent just oscillates in the environment and does not move around and acquire energy through solving tasks, it will run out of energy and die.

The fitness of an agent is calculated according to its ability to complete learning tasks. It is accumulated over the period that the agent's energy level is greater than zero (or before it returns home). The fitness value used in the evolutionary scheme is calculated as follows:

While (Energy ≥ 0):

1. For each move, the fitness is increased by one. This is done to encourage the agent to have a 'brain' that remains active and does not die.
2. If the agent returns home without the gold, its fitness is increased by 200.
3. If the agent obtains the gold, its fitness is increased by 1000.
4. If the agent returns home with the gold, its fitness is increased by 2000.

When the experiment starts, the agent takes its input from a Wumpus World grid square in the form of a potential representing the status of the square (home, gold, breeze, etc). This input is applied to the computational network of the agent through input axo-synapse branches distributed at five different locations in the CGPDN grid. This updates the potentials of the neighbouring dendrite branches in the CG-PDN grid connected to different neurons. The network is then run for five cycles (we refer to this as one *step*). During this process, the potentials of the output dendrite branches, which act as the output of the CGPDN are updated. After the step is complete, the updated potentials of all output dendrite branches are noted and averaged. The value of this average potential is used to decide the direction of movement for the agent. If there is more than one direction, the potential is divided into as many ranges as there are possible movements. For instance, if two possible directions of movement exist, then the agent will choose one direction if the potential is less than 128 and the other if it is greater than 128. The same process is then repeated for the next Wumpus World grid square. The agent is terminated if either its energy level becomes zero, all its neurons die, all the dendrite or axon branches die or the agent returns home.

Five randomly generated genotypes are produced, and the corresponding agent behaviour is assessed to obtain the fitness. The best agent genotype is selected as the parent for a new population.

The performance of the agent is determined by its capability to solve three kinds of tasks. The first task for the agent is to learn how to come back home, the second task is to find the gold, and the third and final task is to bring the gold back home. The agent has to perform all three tasks in one Wumpus World independently. However it obtains the largest fitness by completing the last task (No. 3 in Table 9.1). Each agent population's performance is tested in the same (one) Wumpus World where the gold, the pits and the Wumpus are located in the same places. During this process, the agent has to avoid pits and the Wumpus in order to increase its life span. The life span is the period of time during which the an agent's energy level is above zero before it returns home. An agent needs to sustain a neural network to solve its tasks. The performance of agents on these tasks was tested in independent evolutionary runs, with the same Wumpus World and different initial populations. Table 9.1 shows the performance of the agent in 20 independent evolutionary runs. From the table, it is evident that the agent solves the first and second tasks more quickly than the third task, which is much more difficult than the first two.

The network was tested further by starting with the same initial population and different Wumpus Worlds with the pits, Wumpus and gold at different locations. In each run, agents were evolved that could solve each task. The agent was able to solve all three tasks in numbers of evolutionary generations that were similar on average, to the numbers in Table 9.1, independent of the placement of the Wumpus, gold and pits in the Wumpus World environment.

The fitness function was devised in such a way that it initially forced the agent to sustain its network and so increase its life span (it was incremented with every move of the agent). At the start, the agent does not achieve any goals; the fitness is directly related to the time the agent spends in Wumpus World. For that time, it has

Table 9.1 The average, maximum, and minimum number of generations taken to perform various tasks in a fixed Wumpus World in 20 independent evolutionary runs

Task. No.	Task	Average No. of generations	Minimum No. of generations	Maximum No. of generations
1	Home empty handed	4	1	15
2	Find the gold	13	2	95
3	Home with gold	300	12	1108

to maintain its network, so in the initial generations, evolution tries to build a stable and sustainable computational network. Once this is achieved, evolution starts to produce agents that first learn how to come back home, then learn how to find the gold and, finally bring the gold back home.

At the start, the agent does not know anything about the gold, home, the Wumpus, the pits or the signals that indicate the presence of those objects. As the system is evolved, the genetic code in the agent allows a computational structure to be built that holds a memory of the meaning of these signals during the agent's life cycle. This knowledge is built from the initial genetic code when it is run on the initial randomly wired network. When different runs of the experiment were examined, it was found that in all cases, the agent learns to avoid the Wumpus and pits and tries to get the gold.

In the CGPDN, we evolved programs that build and continuously change a computational network within an external environment. Naturally, we were interested in whether these evolved programs could build a network that could lead to successful agent behaviour in different Wumpus Worlds (i.e. behaviour that is general and not specific to a particular Wumpus World). To test this, we took the programs for the best evolved agent for a particular Wumpus World and tested its performance on other Wumpus Worlds that had been generated at random. From various tests, we found that in cases when the gold was placed at the same location but the pits and the Wumpus were located at different locations, the agent was always able to get the gold and bring it home. In other cases, when the gold was not placed at the same location, sometimes the agent got the gold (30% of the time) but was unable to find its way home, and sometimes it could not find the gold (50% of the time) and returned home empty-handed. There were also cases when the agent could neither find the gold nor get back home (20% of the time), demonstrating that we have not yet obtained general problem-solving behaviour and there is still some noise in the network. Further observations during these experiments revealed that irrespective of the new environment, the agent always looked first for the gold at the place where the gold was when it was evolved. This is very interesting, as it shows that the evolved genetic codes are able to build a computational network that holds information (how to find the gold).

9.5.1.2 Development of Network over the Lifetime of the Agent

While solving the Wumpus World problem the CGPDN changes substantially in its structure, both in terms of the presence and absence of neurons and branches and in terms of whether they are active or inactive. Figure 9.12 shows the variation in the neural structures and the growth of the CGPDN at several different stages. As the network develops, dendrite branches vanish and new dendrite branches are introduced. After 10 steps, the network has gained three additional neurons. It is evident from Fig. 9.12 that the network changes quite randomly; however, it later adopts a more stable topology. It is interesting to observe that the network starts with neurons having two to four dendrites, and at the end of all the changes it ends up with similar numbers.

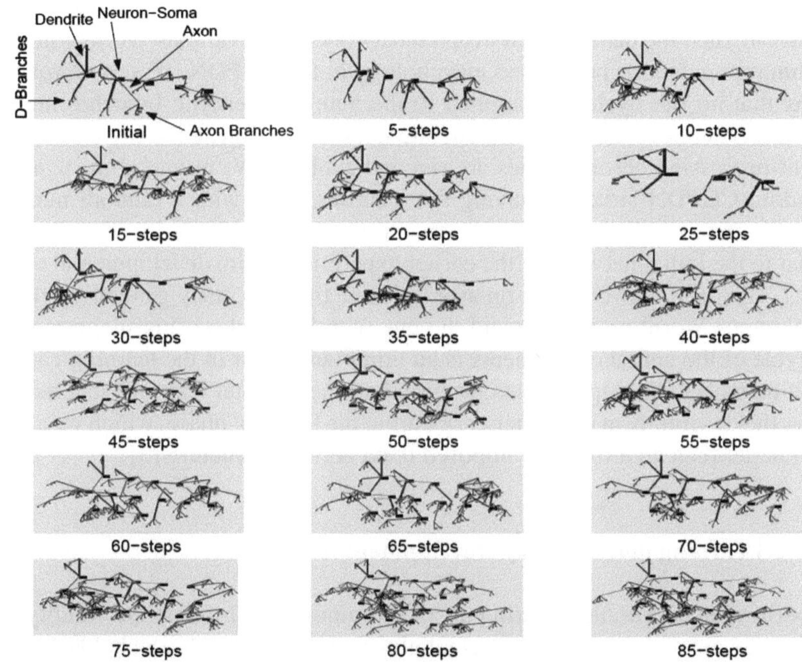

Fig. 9.12 Structural changes in the agent's CGPDN network at different stages of Wumpus World, shown against the number of completed steps. The network has five neurons at the start and 21 neurons after completing 80 steps. The black squares are somas, the thick lines are dendrites, and the long thin lines are axons. Note that this figure is not an exact indication of the locations of the network components in the 12-square CGPDN grid but only a schematic illustration to give an indication of the number of components in the network. Inputs and outputs are not shown as they are at a fixed locations, with inputs at locations 5, 2, 5, 2 and 4, and outputs at 8, 6, 11, 8 and 9.

Figure 9.12 illustrates some interesting network dynamics. The network dynamics appear random at the start, but soon appear to create a sustainable structure as the agent performs the task. The agent takes less time to get to its second goal (home) than to its first goal (gold). This may be because it has developed a sustainable network that allows it to achieve this quickly. This suggests that the agent may produce a map of its environment during its developmental stage, which is used to find the return path. In general, we observed that the return path to home is usually more direct than the path to the gold. If we imagine that the agent's movement is determined by the throw of a dice, it could take an infinite time to get the gold and bring it back home, as every empty square gives the same signal, so the agent might end up oscillating between two squares. As all the pits give the same signal, it is difficult to differentiate one pit from another.

9.5.1.3 Testing the Network without Life Cycle Programs

We investigated the behaviour of evolved networks on a Wumpus World with different initial populations but a fixed morphology in the CGPDN. 'Fixed morphology' means that no life cycle was applied to the neurons, dendrite branches and axon branches. Therefore the network structure does not change, and only electrical and weight-processing chromosomes are run and evolved. We provided the agent with a random CGPDN structure having five neurons, each with a random number of dendrites and axon branches. They were arranged in random order. This network is similar to the initial network of the earlier experiments with developmental chromosomes. The fixed network took on average three times as many generations to solve the Wumpus World problem as did the one with life cycles. This suggests that the life cycle of the neural components is an important factor in the learning capability of the networks. Chalup has also proposed an incremental learning scheme which allows the development of the network during the learning phase, which would work much better than an artificially imposed fixed network structure [3].

9.5.1.4 Learning and Memory Development in CGPDN

We performed additional experiments on Wumpus World to test the validity of the argument that the genotype of the agent obtained through evolution holds a memory of information about the Wumpus World environment in its genetic structure.

We examined the learning behaviour of 10 highly evolved agents in scenarios different from those in which they were trained. Instead of killing an agent after it had returned home with the gold, we always placed it on its home square immediately after it had obtained the gold, and allowed it to continue to live and move around in the environment. In some cases we found that the agent immediately took a path towards the gold (50% of the cases). Some of these agents moved around in the environment and then came back home, then went out again and visited different areas and came back home, and finally ended up oscillating near home (30% of the time). In other cases the agent left home but got stuck somewhere in a corner of the Wumpus World, oscillating between squares until it died (20% of the time). However, in most of the cases, the agent tried to get the gold again, taking a shorter path to the gold in several attempts. This suggests that a map-building ability (memory) is encoded in the genetic code of the CGPDN and also that the CGPDN possesses goal-driven behaviour so that it can get the gold again and again. Some of the agents ended up stuck in a corner, exhibiting oscillatory behaviour. This oscillatory behaviour may be caused by the agent being in the unusual situation of being in a corner. This may not be encountered often during the experiences of different generations.

9.5.2 Competitive Learning Scenario

Coevolutionary computation is largely used in a competitive environment. In competitive coevolution, the fitness of an individual is based on performance against an opponent. Thus the fitness shows the relative strengths of solutions, not the absolute strengths. These competing solutions create an 'arms race' of increasingly better solutions [5, 33, 40]. Feedback mechanisms between individuals based on their selection produce a strong force towards complexity [31]. Competitive coevolution is traditionally used to evolve interactive behaviours which are difficult to evolve with an absolute fitness function. To encourage interesting and sophisticated strategies in competitive coevolution, every player's network should play against a high-quality opposing player [38].

The performance of the CGPDN was also tested in a coevolutionary scenario, where *both* the agent and the Wumpus were provided with a CGPDN. The agent and Wumpus lived in a two-dimensional grid (10×10) containing 10 pits (as shown in Fig. 9.13). In this scenario, the Wumpus is also provided with a home square at the bottom right corner as also shown in Fig. 9.13. This time, we slightly modified the task for the agent. After the agent gets the gold, it is immediately placed back on its home square and its task is to get the gold again. The agent has to get the gold as many times as it can during its life time while avoiding pits and the Wumpus. The task of the Wumpus is to catch the agent as many times as it can. The job of the Wumpus is more difficult than that of the agent as the target for the Wumpus is mobile, while the target for the agent (the gold) is static. Both the Wumpus and the agent are placed back on their respective home squares each time the Wumpus catches the agent.

The task of learning in the coevolutionary scenario is much harder than in the single-agent World as each time the Wumpus catches the agent, it becomes more difficult for it to catch it again as the agent learns and tries to find a path to the gold that avoids the Wumpus. Pits affect only the energy level of the agent. The agent is weakened whenever it passes through squares containing pits. The Wumpus's home is diagonally opposite the agent's. Both the agent and the Wumpus perceive a breeze in squares adjacent directly to the pits and a smell in the squares directly (not diagonally) adjacent to each other. They also perceive a glitter when they pass close to the gold.

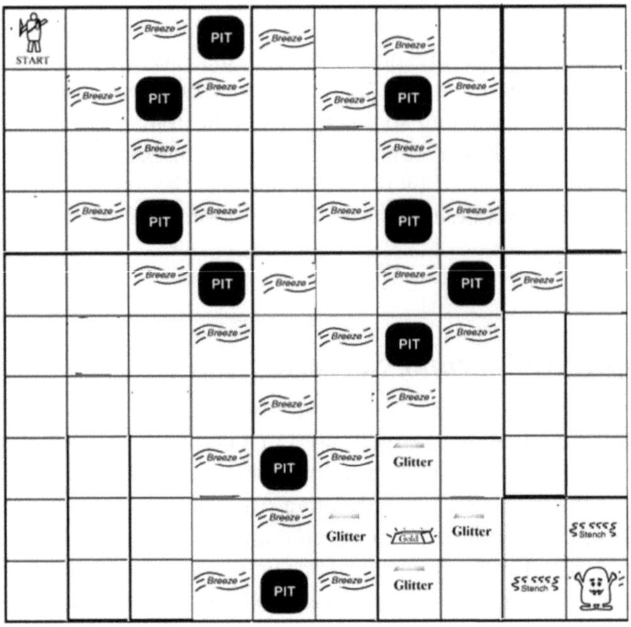

Fig. 9.13 A two dimensional grid 10 × 10, showing a Wumpus World environment having one Wumpus starting from the bottom right corner, one agent in the top left corner, 10 pits at various places and gold.

Both the agent and the Wumpus are assigned an initial energy level of 100 units. The agent and Wumpus have only one life cycle, i.e. they remain in the environment as long as their energy level is higher than zero. Both the agent and Wumpus move one square at a time. They do not have the option to stay in the same place. They can move in any of the possible directions to a square that is directly (not diagonally) adjacent. If the agent is caught by the Wumpus, its energy level is reduced by 60%. If it falls into a pit, its energy level is reduced by 10 units. If it gets the gold, its energy level is increased by 60%. The Wumpus is not affected by anything, except that its energy level is increased by 60% every time it catches the agent. For each single move, both the agent and the Wumpus have their energy level reduced by 1 unit.

The fitnesses of the agent and the Wumpus are accumulated over their lifetimes (while their energy level is non-zero) in the following manner:

1. For each move, the fitness of both the agent and the Wumpus is increased by one.
2. Every time the agent obtains the gold, its fitness is increased by 1000.
3. Every time the Wumpus catches the agent, its fitness is increased by 1000.

Each of five agent population members were tested against the best-performing Wumpus genotype from the previous generation. Similarly, each of the five Wumpus population members were tested against the best-performing agent genotype from the previous generation. The initial random network was the same for both the agent and the Wumpus. It is the genotypes in each generation which generate different networks and their associated functionality. The best agent and Wumpus genotypes were selected as the respective parents for the new population.

The idea behind choosing these experiments was meaningful in two ways. Firstly, we wanted to demonstrate that agents learn during their life time (i.e. post-evolution) while their energy level is above zero, thus demonstrating that evolution has conferred an ability to learn. Secondly, we intended to show that the genetic memory (learned experiences) obtained by an agent during its life time can be built by the genetic code into the mature network and can therefore be transferred to the next generation through the genetic code.

9.5.2.1 Results and Analysis

Figure 9.14 shows how the fitness of the agent and Wumpus change, in one particular evolutionary run, over 1250 generations. It is evident that there are increases and decreases in fitness at different stages for both agent and the Wumpus, with neither of them having the 'upper hand'.

The y-coordinate divided by 1000 gives the number of times the agent or the Wumpus achieves its goals. There are several different kinds of dynamic behaviour that the agent and Wumpus adopt because of the interactions between neurons resulting from their internal genetic codes. At the start, both the agent and the Wumpus know nothing about the gold, the pits, their interaction with each other and the signals that indicate the presence of these objects nearby. As they evolve, they develop their own memory of life experiences, and their ability to do this is encoded genetically. Each agent starts with a fixed randomly assigned neural structure that develops during the agent's lifetime enabling it to achieve its goal. The fitness graphs in Fig. 9.14 show that initially, the fitnesses of the agent and the Wumpus vary a great deal. However, as the number of generations increases, both the agent and the Wumpus become more skilful and the frequency of the variations in the fitness reduces. This suggests that the two CGPDN networks behave so that one of them benefits at the expense of the other. There are also some points along the fitness graph where the fitnesses of both the agent and the Wumpus go down. This occurs when they are both unable to achieve their goals. Often in the very next generation, however, they find a way to achieve their goals. It is interesting to observe that around generation 680, both the agent and the Wumpus become reasonably skilful. The agent obtains the gold twice (and on one occasion three times) while at the same time the Wumpus catches the agent twice (and later three times). This is followed by the agent and the Wumpus having fluctuating fortunes. Following that, at around generation 1100, both the agent and the Wumpus are more successful in achieving their goals, with the Wumpus being the more successful.

In further experimental analysis, we looked at the behaviour of the agent and the Wumpus in detail. The initial energy levels of the agent and the Wumpus were increased to 300 (from the value of 100 previously used during evolution). This allowed the agent to find the gold five times (see Fig. 9.15) while the Wumpus caught the agent once.

Figure 9.15 illustrates the movements of the agent (black arrows) and Wumpus (grey arrows) over six journeys; the Wumpus died during its fourth outing (panel D), and thus is not shown in panels E and F. The squares are numbered from 0 to 99 along rows from top left to bottom right; thus the gold is on square 86. In Fig. 9.15A, both the agent and the Wumpus begin with the same randomly assigned initial networks, which are converted into mature networks by running the seven CGP programs.

The agent takes a fairly direct route towards the gold, encountering two pits along the way. The Wumpus spends a great deal of time on squares 98 and 88 and moves towards the gold just after the agent has been replaced on its home square after obtaining the gold. Figure 9.15B illustrates the next journey of the agent, which

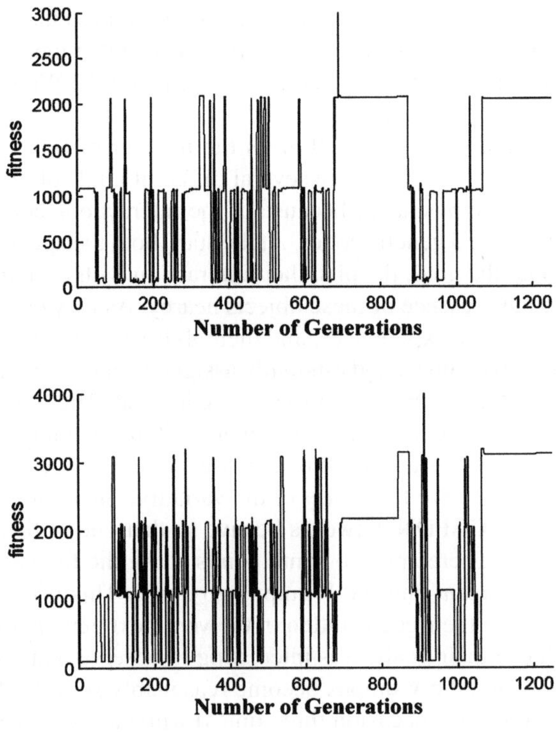

Fig. 9.14 Variation in fitness of the best agent (top) and the best Wumpus (bottom) with the number of evolutionary generations. The traces show oscillatory behaviour in the fitness of both at the start. After 680 generations, both the agent and the Wumpus reach a relatively stable behaviour, with the agent getting the gold twice and the Wumpus catching the agent two or three times. At generation 692, the agent retrieves the gold three times. However, at generation 900 large oscillations in fitness recur.

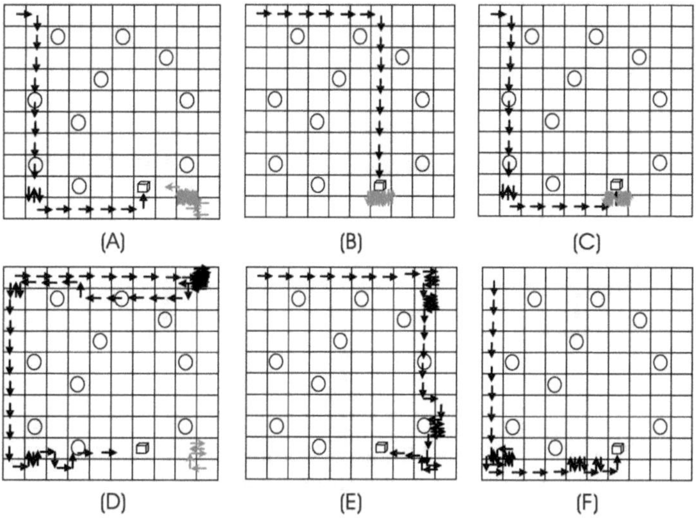

Fig. 9.15 Several paths followed by the agent and the Wumpus in successive journeys of the coevolutionary task. The agent starts towards the gold from its home square in the upper left corner, and the Wumpus starts towards the agent from its home square in the lower right corner. The movement of the agent is shown by black arrows, and the movement of the Wumpus by grey arrows. The pits are indicated by circles, and the gold by a box. The Wumpus CGPDN dies during journey D; therefore it is not available in E and F.

takes a different route to the gold (in fact following the minimum path) and avoids all pits. Meanwhile, the Wumpus just lurks near the gold, spending all its time on squares 96 and 86. We generally found that the Wumpus exhibited this kind of behaviour. This is a good strategy for it, as the agent (if it is any good) will eventually come close to the gold and this gives the Wumpus a chance of catching it. On this occasion, the Wumpus is unlucky and jumps away from the gold at the same time that the agent obtains it. Recall that we do not allow the agents to remain stationary. Figure 9.15C illustrates their third journey. The agent follows the same path to the gold as it did at the start. This is surprising, since its network has changed during its lifetime and its experiences.

This strongly suggests that it has somehow encoded a map of its environment in its network. However, when it arrives near the gold (square 96), it is attacked by the Wumpus and is relocated to its home square. Its subsequent behaviour (in Fig. 9.15D) is interesting. It now follows a very different, meandering path to the gold. It spends some time alternating between squares 8 and 9 before turning back and arriving home again, only to set off down the left-hand side, in the direction of the gold. The behaviour of the Wumpus is odd. Having been replaced on its home square (99) after attacking the agent, it moves around briefly in the bottom right four squares before its CGP developmental network dies. This illustrates an inter-

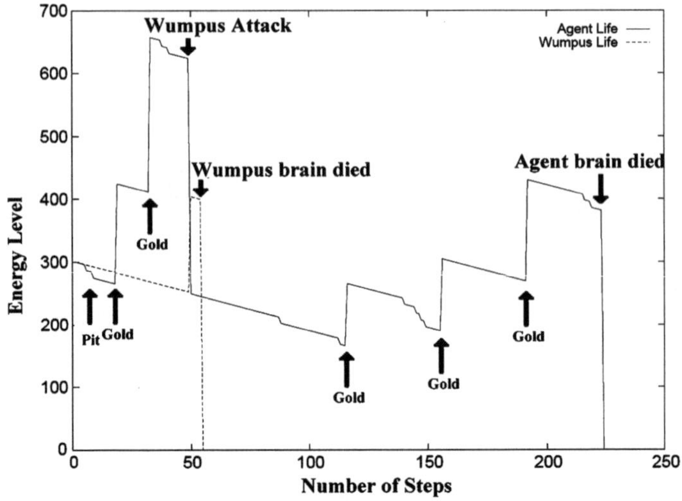

Fig. 9.16 Variation in the energy levels of the agent and the Wumpus during the consecutive journeys illustrated in Fig. 9.15. The energy shows a continual decrement of 1 per step, with larger changes when the agent and the Wumpus achieve their respective goals. When the agent finds the gold, its energy is increased by 60%. When the agent encounters a pit, its energy level is decreased by 10 and when caught by the moving Wumpus, its energy level is decreased by 60%. The Wumpus has its energy level increased by 60% when it catches the agent. In the scenario illustrated, the agent finds the gold five times. The Wumpus catches the agent once, around step 50. Shortly after this, the Wumpus CGPDN dies, indicated by the drop to zero in energy level.

esting but puzzling phenomenon that we observed with other evolved agents and Wumpuses. Often, their CGPDN dies when their energy level is a small number and becomes more active (with many branches and synapses) when they have a high value of their energy level. This is puzzling, since the energy level is not supplied as an input to the CGPDN. Beneficial encounters often result in more neurons and branching, whereas deleterious encounters often result in removal of neurons and branches from the network. In Figs. 9.15E and 9.15F, we see the subsequent behaviour of the agent, where it successfully obtains the gold again (panel E) and then dies before it reaches the gold (panel F).

These results suggest that the agent produces a memory map of the environment early in its evolutionary history. It is interesting to note that after the agent is attacked by the Wumpus, its CGPDN is strongly affected so that it follows a totally different path. However, when it does not find any gold, it returns to its home square and subsequently tries the same path that led to its attack by the Wumpus (with some minor variations), eventually finding the gold. In the subsequent trials, it appears to be looking for the shortest path to the gold. Even after the events shown in Fig. 9.15E, we found that when the agent was restored to its starting position, it again followed a short path to get to the gold but, unfortunately, as it reached the gold its network died.

We also examined the behaviour of this agent in a number of other situations. We removed all pits and found that the agent moved about in the environment apparently at random, and was unable to find the gold. This substantiates the notion that the agent uses environmental cues (i.e. pits) to navigate. We also moved the Wumpus to square 56 (and disabled its ability to move from this square). This lies directly on the path of the first agent in Fig. 9.15B. We found that the behaviour of the agent was identical to its previous behaviour, except that it was caught by the Wumpus (on square 56) and did not avoid that square when it encountered the smell signal on square 46. This shows that this agent's network-building program had not yet given it the ability to avoid the Wumpus. But this encounter affected its network and subsequently caused it to follow a totally different path to avoid the Wumpus. The agent and Wumpus used in these experiments came from generation 220, so they were not highly evolved. At this stage in evolution, the agent has not fully learned what to do in any situation, so the agent does not fully respond to the presence of the Wumpus and the degree to which it influences its energy level. We ran the experiment for longer (more than 2000 generations) but as the agent got better, so did the Wumpus, thus ending up with a stable point where the agent got the gold a few times and the Wumpus caught it a few times, as evident from Fig. 9.14.

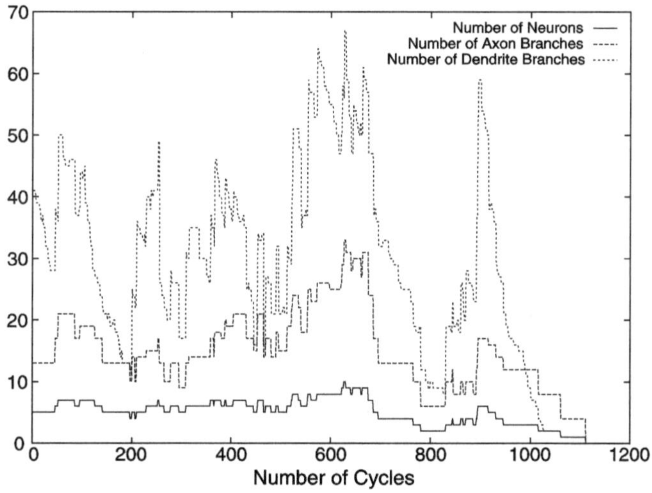

Fig. 9.17 Variation in the numbers of neurons, dendrite branches and axon branches (plotted against number of cycles) of the agent's CGPDN during the coevolutionary task illustrated in Fig. 9.15. Changes in the network structure appear to coincide with environmental events illustrated in Fig. 9.16. Note that each step in Fig. 9.16 requires five cycles to complete.

In earlier generations, both of the agents wander around in the Wumpus World environment. They do not achieve any goals (in fact, they do not know what their

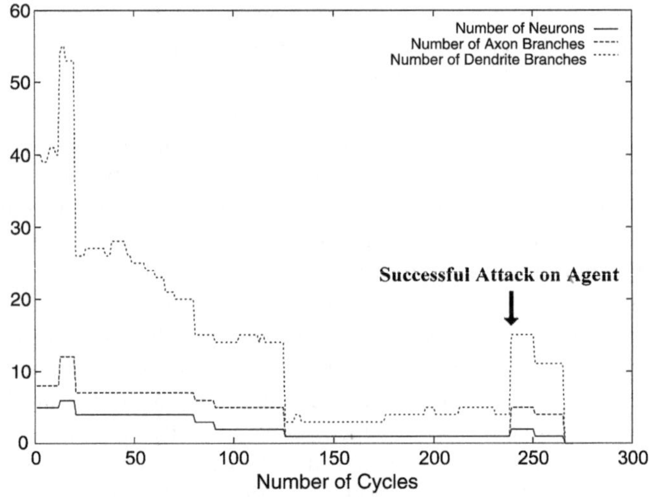

Fig. 9.18 Variation in the numbers of neurons, dendrite branches and axon branches (plotted against number of cycles) of the Wumpus CGPDN during the coevolutionary task illustrated in Fig. 9.15. There is a continuous drop in the numbers over time, resulting in the death of the Wumpus at cycle 270 (step 54 in Fig. 9.16).

goals are), thus causing their fitness values to be minimum, so they are not selected during the course of evolution. Once they happen to achieve a goal during evolution, their fitness increases and they are selected for the next generation. From this point onwards, these agents have a sense of their goal. Evolution makes one agent a Wumpus and the other an agent that looks for gold. The results shown in Figs. 9.15 and 9.16 are for well-evolved agents when they have a sense of their goals.

During an agent's and the Wumpus's lifetime, the structural development and activity of the CGPDN change considerably. We can illustrate this by looking at the variation in energy level and the changes in network morphology during the lives of the agent and the Wumpus whose behaviour was shown in Fig. 9.15. Figure 9.16 shows the variations in the energy levels of the agent and the Wumpus. The changes in energy level reflect their experiences. For the agent, increases show the number of times it found the gold. The decreases are of two kinds: smaller decreases after encountering a pit, and larger decreases when attacked by the Wumpus. Superimposed is a steady decrement in energy of 1 unit per step. The energy levels for the agent and the Wumpus increase only when they achieve their respective goals. When the Wumpus catches the agent, its energy level is increased by 60%. It can be seen how the energy level of the agent drops rapidly (by 60%) immediately after it is attacked by the Wumpus. Shortly after this, rather surprisingly, all the neurons in the Wumpus's CGPDN die.

Figure 9.17 shows the variation in the numbers of neurons, axon branches and dendrite branches during the agent's life. Figure 9.18 shows the corresponding numbers for the Wumpus.

We have also observed that in most of the cases, once the agent has been caught by the Wumpus, it is never able to get the gold again during its lifetime, because the interaction with the Wumpus has affected its network, leading to neuron death. In order to assess the validity of this argument, we tested the system by increasing the initial energy level of the agent, as demonstrated in Fig. 9.15. It was only through doing this that we found that the agent was able to get the gold after being caught by the Wumpus. However, in this case the Wumpus died soon after encountering the agent. It is evident that the network of the Wumpus was about to die (see Fig. 9.15) before it caught the agent, but the encounter appears to have increased its life span by triggering a brief increase in the numbers of neurons and branches.

We also evolved the agent and the Wumpus with various values of the initial energy level. If the energy is increased (by up to 300), it diminishes the influence of deleterious and beneficial environmental encounters. If it is decreased (to 50), the inimical encounters (pits and Wumpus) outweigh the beneficial effects (obtaining gold). We found that an initial life of 100 proved to be a good balance.

The CGPDN builds its network and any learned behavioural memory in the environment, so the system is, in a sense, self-learning. This environmental responsiveness, however, comes at a cost. Since the networks are time-dependent, it is possible for all neuron components to dwindle away.

9.6 Learning 'How to Play' Checkers

The ability of the system to learn how to play checkers was tested in two different scenarios: coevolution and evolution against a minimax-based checkers program. Figure 9.19 shows how the network was interfaced with the checkers board.

9.6.1 Coevolution of Two Agents Playing Checkers

Coevolution is a method of evolving two or more systems together such that they affect each other's evolution. Coevolutionary techniques have been applied to various games, including Othello [28], Go [21], chess [13], Kalah [11, 41] and checkers [7].

We used a form of coevolution where each agent was provided with a CGPDN as a central processing unit and played checkers against the other agent. Each agent's population consisted of five genotypes. Each of the five first-agent population members were tested against the best-performing second-agent genotype from the previous generation over a single game, as done by Stanley and Miikkulainen [38] and vice versa. The coevolution process was carried out as follows:

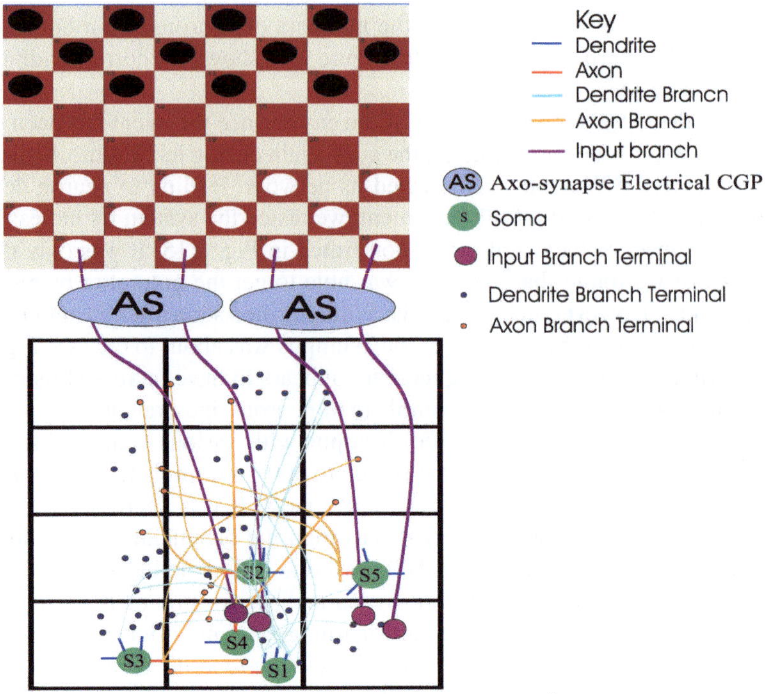

Fig. 9.19 Interfacing CGPDN with a checkers board. Board positions are applied in pairs to each grid square of the CGPDN. Four board positions are shown interfaced with the CGPDN.

1. A population of five genotypes is generated for both the agents.
2. In every generation, each of the first-agent genotypes is tested against the best genotype of the second agent, and vice versa.
3. The fitness of all the genotypes is calculated.
4. In each population, the agent with the highest fitness is selected as a parent for the next generation. Each parent is mutated to produce four offspring. The four offspring together with the parent comprise the new population for each agent.
5. The same process is repeated until a solution is found or a specified number of generations is reached.

In the first generation, both agents have the same initial random CGPDN structure. When the experiment starts, the agent playing black takes an input from the board. This input is applied to its CGPDN through input axo-synapses. The CG-PDN is then run for five cycles. During this process, it updates the potentials of the output dendrite branches. These updated potentials are averaged and used to decide the direction of movement for the corresponding piece. The potentials of the output branches for the pieces are also updated by the CGPDN. The updated values of these potentials are used to decide which piece to move, unless a jump is possible, which takes priority. For more than one jump, the piece with the highest potential makes the jump. The same process is repeated for the opponent, and the process is repeated and continues until the game stops. Every time a piece is removed, its branch is removed from the corresponding CGPDN grid square.

The game is stopped if the CGPDN of either an agent or its opponent dies (i.e. all its neurons or neurites die), if all of its or its opponent's pieces are taken, if the agent or its opponent cannot move anymore or if the allotted number of moves allowed for the game have been taken.

9.6.1.1 Results and Analysis

Learning to play checkers is difficult. To do this, the two agents must start with a few neurons with a random number of dendrites and branches and build a computational network that is capable of solving the task while maintaining a stable network (i.e. not losing all the neurons or branches). Secondly, it must find a way of processing the environmental signals and differentiating between them. Thirdly, it must understand the spatial layout of the board (the positions of the pieces). Fourthly, it must develop a memory or knowledge of the meaning of the signals from the board. Fifthly, it must develop a memory of previous moves and of whether they were beneficial or deleterious. Also, it must understand the benefits of making a king and of jumping over a piece and finally, and most importantly, it must do all these things while playing the game. Over the generations, the agents learn from each other about favourable moves. This learning is transferred through the *genes* from generation to generation.

To test whether more-evolved agents played the game better, we tested well-evolved agents against less evolved agents. It was necessary to do this to verify that later agents played better, because fitness is relative in coevolution as both players could be improving. We found that the well-evolved agent almost always beat the less well-evolved one. In some cases, the game ended in a draw but in those cases, the well-evolved agent ended up with more kings and pieces than the less well-evolved agent [15].

Above, we described the coevolution of two agents playing checkers and demonstrated how two developmental programs learned to play checkers through coevolution. The next subsection will describe the evolution of an agent playing against a minimax-based checkers program.

9.6.2 An Agent Plays Against a Minimax-Based Checkers Program

We also evolved a CGPDN checkers player against a minimax-based checkers program (MCP) rather than using coevolution. Unfortunately, we were unable to find any dll engine with an adjustable ply (level of play) search. The only suitable engine that we found was CHKKIT.DLL, which played at a high level and used a database of best moves. Each agent genotype played one game against the MCP with the agent starting each game from a random network. The best-playing genotype based on fitness was selected as the parent for the new populations and was promoted to the next generation unaltered, along with four offspring (mutational variants).

When the experiment started, the MCP made the first move and the updated board was then applied to the agent CGPDN. The CGPDN network was then run to decide on a move. The same process was repeated until the game was stopped.

9.6.2.1 Results and Analysis

During the course of evolution, when an agent was evolved against the MCP, the agent was never able to beat the MCP or even have a piece advantage within 20 moves. So it was difficult to assess from the variation in the fitness of the agent during the course of evolution whether it was learning or not. The MCP was playing at a higher level, and although the agent learned different moves during the course of evolution, the MCP still managed to beat the agent, thus causing its fitness to stay low. As the MCP produces a database of game moves and uses that in the calculation of its next move, it plays different games every time, even against the same opponent. Thus it is difficult for evolution to select the best genotype for the next generation. Although the genotype of the best agent is promoted unaltered to the next generation, it produces a different value of the fitness. As a result, it is difficult to obtain any improvement against the MCP.

To test whether any learning had taken place, we tested well-evolved agents against less well-evolved agents and found that the former almost always beat the latter. In some cases, the game ended in a draw but even in those cases, the well-evolved agent ended up with more kings and pieces than the less well-evolved agent [16]. Thus it is clear that the agents were improving their level of play.

9.7 Conclusions

We have described a neuron-inspired developmental approach to constructing a new kind of computational neural architecture which has the potential to learn through experience. We found that the neural structure controlling the agents grows and changes in response to their behaviour and their interactions with the environment, and allows them to learn and exhibit intelligent behaviour. The eventual aim is to see if it is possible to evolve a network that can learn from experience.

The model was inspired by principles taken from neuroscience and is more biologically plausible than previous artificial neural network models. The CGPDN is a developmental network, capable of self-configuring in a task environment. The genotype of the network is evolved so that, when executed, it produces a complete network of neurons, dendrites and dendrite and axon branches capable of learning.

We have tested the network both in the Wumpus World environment and in the game of checkers and found it to produce interesting learning capabilities. We have also tested the system for learning in both in evolutionary and coevolutionary scenarios.

In the case of Wumpus World, we found that a network, when tested on a different Wumpus World, preserves the stability of its network and its ability to avoid pits and the Wumpus. It is not possible, at present, to compare the effectiveness of this approach directly with that of other artificial neural networks, as others have not worked with the Wumpus World scenario.

We have also described a coevolutionary competitive learning environment in which the CGPDNs of two antagonistic agents grow and change in response to their behaviour, their interactions with each other, and the environment. We found that the agents can learn from their experiences and in some cases appear to build an internal map of their environment.

In the case of checkers, we have also tested the system both in a coevolutionary environment and by playing against a self-adaptive checkers program. Our results are encouraging and demonstrate that the system has the capability to learn.

In future work, we plan to evaluate this approach in richer and more complex environments. The eventual aim is to see if it is possible to evolve a general capability for learning. Also, we plan to examine whether a dynamic CGPDN is able to solve problems faster and more accurately merely by obtaining repeated experience of its task environment (post-evolution). Furthermore, we plan to investigate whether CGPDNs can be trained to solve a sequence of problems without forgetting how to solve earlier problems (conventional artificial neural networks suffer from a phenomenon known as 'catastrophic forgetting' [22, 32]).

References

1. Alberts, B.: Molecular Biology of the Cell, 3 edn. Garland Science (2002)
2. Belew, R.K.: Interposing an Ontogenetic Model Between Genetic Algorithms and Neural Networks. Advances in Neural Information Processing Systems (NIPS) pp. 99–106 (1992)
3. Chalup, S.K.: Issues of Neurodevelopment in Biological and Artificial Neural Networks. Proc. Biannual Conference on Artificial Neural Networks and Expert Systems pp. 40–45 (2001)
4. Chen, X., Hurst, S.L.: A Comparison of Universal-Logic-Module Realizations and Their Application in the Synthesis of Combinatorial and Sequential Logic Networks. IEEE Transactions on Computers **31**, 140–147 (1982)
5. Dawkins, R., Krebs, J.R.: Arms races between and within Species. In: Proc. Royal Society of London Series B, vol. 205, pp. 489–511 (1979)
6. Dellaert, F., Beer, R.D.: Toward an evolvable model of development for autonomous agent synthesis. In: R. Brooks, P. Maes (eds.) Artificial Life IV: Proceedings of the Fourth International Workshop on the Synthesis and Simulation of Living Systems. MIT Press (1994)
7. Fogel, D.B.: Blondie24: Playing at the Edge of AI. Morgan Kaufmann (2002)
8. Graham, B.: Multiple Forms of Activity-Dependent Plasticity Enhance Information Transfer at a Dynamic Synapse. In: J.R. Dorronsoro (ed.) Proc. International Conference on Artificial Neural Networks, *LNCS*, vol. 2415, pp. 45–50. Springer (2002)
9. Gruau, F.: Genetic Synthesis of Modular Neural Networks. In: Proc. International Conference on Genetic Algorithms, pp. 318–325 (1993)

10. Hillis, W.: Co-evolving parasites improve simulated evolution as an optimization procedure. Physica D **42**, 228–234 (1990)
11. Irving, G., Donkers, J., Uiterwijk, J.: Solving Kalah. International Computer Games Association Journal **23**(3), 139–147 (2000)
12. Kandel, E.R., Schwartz, J.H., Jessell, T.M. (eds.): Principles of Neural Science, 4 edn. McGraw-Hill (2000)
13. Kendall, G., Whitwell, G.: An Evolutionary Approach for the Tuning of a Chess Evaluation Function using Population Dynamics. In: Proc. IEEE Congress on Evolutionary Computation, pp. 995–1002 (2001)
14. Khan, G.M., Miller, J.F., Halliday, D.M.: Coevolution of Intelligent Agents using Cartesian Genetic Programming. In: Proc. Genetic and evolutionary Computation Conference, pp. 269–276 (2007)
15. Khan, G.M., Miller, J.F., Halliday, D.M.: Coevolution of Neuro-developmental Programs That Play Checkers. Proc. International Conference on Evolvable Systems **5216**, 352–362 (2008)
16. Khan, G.M., Miller, J.F., Halliday, D.M.: In Search of Intelligent Genes: The Cartesian Genetic Programming Computational Neuron (CGPCN). In: Proc. IEEE Congress on Evolutionary Computation, pp. 574–581. IEEE Press (2009)
17. Khan, G.M., Miller, J.F., Halliday, D.M.: Evolution of Cartesian Genetic Programs for Development of Learning Neural Architecture. Evolutionary Computation **19**(3) (2011). In press
18. Kitano, H.: Designing neural networks using genetic algorithms with graph generation system. Complex Systems **4**, 461–476 (1990)
19. Koza, J.R.: Genetic Programming: On the Programming of Computers by Means of Natural Selection. MIT Press (1992)
20. Kuffler, S.W., Nichols, J.G., Martin, A.R.: From Neuron to Brain, A Cellular Approach to the Function of the Nervous System, 2 edn. Sinauer Press (1984)
21. Lubberts, A., Miikkulainen, R.: Co-Evolving a Go-Playing Neural Network. In: Coevolution: Turning Adaptive Algorithms upon Themselves, Birds-of-a-Feather Workshop, Genetic and Evolutionary Computation Conference, p. 6 (2001)
22. McCloskey, M., Cohen, N.J.: Catastrophic Interference in Connectionist Networks: The Sequential Learning Problem. The Psychology of Learning and Motivation **24**, 109–165 (1989)
23. Miller, J.F.: Evolving Developmental Programs for Adaptation, Morphogenesis and Self-Repair. In: Proc. European Conference on Advances in Artificial Life, *LNAI*, vol. 2801, pp. 289–298 (2003)
24. Miller, J.F.: Evolving a self-repairing, self-regulating, French flag organism. In: Proc. Genetic and Evolutionary Computation Conference, vol. 3102, pp. 129–139 (2004)
25. Miller, J.F., Khan, G.M.: Where is the Brain inside the Brain? On Why Artificial Neural Networks should be Developmental. Memetic Computing **3**(3), YY–YY (2011). In press
26. Miller, J.F., Thomson, P.: Cartesian Genetic Programming. In: Proc. European Conference on Genetic Programming, *LNCS*, vol. 1802, pp. 121–132 (2000)
27. Miller, J.F., Vassilev, V.K., Job, D.: Principles in the Evolutionary Design of Digital Circuits – Part I. Genetic Programming and Evolvable Machines **1**, 7–35 (2000)
28. Moriarty, D., Miikkulainen, R.: Discovering Complex Othello Strategies Through Evolutionary Neural Networks. Connection Science **7**, 195–209 (1995)
29. Nolfi, S., Floreano, D.: Co-Evolving predator and prey robots: Do 'arms races' arise in artificial evolution? Artificial Life **4**, 311–335 (1998)
30. Panchev, C., Wermter, S., Chen, H.: Spike-Timing Dependent Competitive Learning of Integrate-and-Fire Neurons with Active Dendrites. In: J.R. Dorronsoro (ed.) Proc. International Conference on Artificial Neural Networks, *LNCS*, vol. 2415, pp. 896–901. Springer (2002)
31. Paredis, J.: Coevolutionary Computation. Artificial Life **2**(4), 355–375 (1995)
32. Ratcliff, R.: Connectionist Models of Recognition and Memory: Constraints Imposed by Learning and Forgetting Functions. Psychological Review **97**, 205–308 (1990)
33. Rosin, C.D., Belew, R.K.: New Methods for Competitive Evolution. Evolutionary Computation **5**, 1–29 (1997)

34. Russell, S., Norvig, P.: Artificial Intelligence: A Modern Approach. Prentice Hall (1995)
35. Shepherd, G.M.: The Synaptic Organization of the Brain. Oxford University Press (1990)
36. Spector, L.: Simultaneous Evolution of Programs and their Control Structures. In: P.J. Angeline, J.K.E. Kinnear (eds.) Advances in Genetic Programming 2, pp. 137–154. MIT Press (1996)
37. Spector, L., Luke, S.: Cultural Transmission of Information in Genetic Programming. In: J.R. Koza, D.E. Goldberg, D.B. Fogel, R.L. Riolo (eds.) Proc. Conference on Genetic Programming, pp. 209–214. MIT Press (1996)
38. Stanley, K., Miikkulainen, R.: Competitive coevolution through evolutionary complexification. Journal of Artificial Intelligence Research **21**, 63–100 (2004)
39. Stanley, K.O., Miikkulainen, R.: Evolving Neural Networks Through Augmenting Topologies. Evolutionary Computation **10**(2), 99–127 (2002)
40. Van Valin, L.: A new evolutionary law. Evolution Theory **1**, 1–30 (1973)
41. Wee-Chong, O., Yew-Jin, L.: An Investigation on Piece Differential Information in Co-Evolution on Games using Kalah. In: Proc. Congress on Evolutionary Computation, pp. 1632–1638. IEEE Press (2003)
42. Yob, G.: Hunt the Wumpus. Creative Computing, pp. 51–54 (1975)

Chapter 10
CGP, Creativity and Art

Steve DiPaola and Nathan Sorenson

10.1 Introduction

Over the last ten years, creative evolutionary systems have formed a new category of artificial intelligence that is characterized by the use of evolutionary computation to generate aesthetically pleasing structures in art, music and design. Within the field of computer visual art, these systems are often referred to as evolutionary art systems.

The creative evolutionary systems research discussed in this chapter is based on Ashmore and Miller's work [2, 1], which uses Cartesian genetic programming (CGP). CGP uses typical genetic programming (GP) evolutionary techniques (crossover, mutation and survival), but has many features, which we discuss in this chapter, that allow the GP system to favour creative solutions over optimized solutions. Portrait painting was chosen for this project as it relieves us of considering the creative space of all art paintings and instead allows us to consider a known portrait sitter–painter relationship well suited to exploring computer creativity. Our system uses the same approach as that discussed in Sect. 2.4.3; we generate portrait images using CGP, where the coordinates of each pixel in the image frame are the inputs to an evolved function that returns the final colour output. In our case, those outputs are in HSV (hue, saturation and value) space. Our main contribution is in adding an autonomous creative fitness function to the system, based on cognitive science research on human creativity.

According to Bentley and Corne, from their seminal book on the subject [4], a creative evolutionary system is designed to (1) aid our own creative process and (2) generate results for problems that traditionally require creative people to solve. Bentley and Corne go on to state that in achieving these goals, a creative evolutionary system may also appear to act 'creatively'– although this is still a source of debate. Unlike general evolutionary computation systems, creative evolutionary systems have been criticized because most of these systems use the presence of a human (often playing the role of a creative decision maker or fitness function) to guide the

J.F. Miller (ed.), *Cartesian Genetic Programming*, Natural Computing Series,
DOI 10.1007/978-3-642-17310-3_10, © Springer-Verlag Berlin Heidelberg 2011

direction of the evolutionary search. Our CGP-based portrait painter system specifically uses an automatic fitness function, albeit tailored to portrait painting where a portrait–sitter resemblance is encouraged, thereby attempting to work through the human-fitness-function dilemma and directly explore how computer algorithms can be autonomously creative.

10.2 Creativity and Art

Certainly, creativity is a broad and complex notion that does not permit simple characterization. Creativity can be seen as a quality pertaining to both historical movements and solitary events, as a defining characteristic of both societies and individuals, with understandable contention regarding the degree to which these instantiations are governed by the same phenomenon. Despite the inherent difficulties in constructing a comprehensive definition, much progress has been made in the characterization of creativity as an associative process. Indeed, Poincaré's famous metaphor of disparate ideas that 'rise in crowds' and 'collide until pairs interlock' [19] seems only to be affirmed by recent work in neuroimaging technologies: an increase in associative brain activity can be seen during moments of creative thought, as new neurological connections between association cortices are formed [13]. This dynamic associative process is certainly linked with the ability that creative individuals have to make surprising and unanticipated departures from existing modes of thought.

In fact, case-study research demonstrates that creators often work in highly structured domains with well-specified rules, against which they ultimately rebel [4, 3]. The traditional forms of the portrait, the sculpture or the symphony become a point of departure for the truly creative artist. This is not to say that Picasso, Michelangelo and Stravinsky were not masters of the traditional forms; indeed, it is precisely their mastery of existing techniques which made their innovations possible. In this sense, creativity is not simply the capacity to eschew traditional modes, but rather the ability to internalize and master them while still making associative connections that were previously not possible. This broad, associative imagination and narrow, focused mastery are two opposite impulses which characterize the creative thought process. As Feist states, 'It is not unbridled psychoticism that is most strongly associated with creativity, but psychoticism tempered by high ego strength or ego control. Paradoxically, creative people appear to be simultaneously very labile and mutable and yet can be rather controlled and stable' [12]. Many theorists recognize the existence of comparable categorizations of thought [6, 7, 8, 9]. Furthermore, many suggest that the ability to easily transition between these two modes is a defining characteristic of creativity [4, 10, 11]. This fluidity of mind is termed contextual focus [11], and requires both focused attention (typically linked with abstract thought and logical deduction) and a broad, expansive perspective (suited to the apprehension of unexpected correlations).

This dynamic contextual focus is central to our work. In order to explore the nature of computational creativity, we have developed a system to evolve artistic portrait paintings and implement a model of contextual focus to generate and evaluate works of art automatically. Our creative system simultaneously follows the precise and highly structured goal of representation, and the vague and associative notion of aesthetic quality.

10.3 Evolutionary Systems and Creativity

It is evident that the domain of artistic expression is particularly well suited to questions of machine creativity, as the standard techniques of domain-agnostic artificial-intelligence search do not apply. The process of creating art has no well-defined expected outcome; one cannot generate a creative artwork in the same way as one can search for an effective chess move or compute an optimal load-bearing bridge design. There is simply no readily identifiable 'problem' to be solved. The process used to formulate the problem definition is analogous to the artist's struggle to realize an underlying vision to guide a work of art. This lack of a clear problem specification is exactly the sort of issue that both necessitates creative thought and makes its presence most evident.

Systems such as Harold Cohen's AARON [6] and Karl Sims's genetic images [22] have popularized the notion that machines can autonomously produce output of aesthetic value. However, critics argue that the output is simply a function of the creativity of the system's designer, and not truly located within the machine. Indeed, how can the recognizable style of AARON be attributed to the machine and not its creator? Similarly, the evolution of the images produced by Sims's genetic system is directly guided by human interaction. Before one can claim to have embedded any degree of creativity within an automated system, it must be shown that the designers and users of that system are not ultimately responsible for the aesthetic decisions the system makes. We attempt to achieve autonomy by explicitly modelling the psychological process of contextual focus as a central component of our creative painter. Our system, therefore, exhibits the ability to evaluate its own designs in both a focused mode, with adherence to a specific, well-defined goal, and also a broad, imaginative mode, where more flexible judgment criteria are employed.

10.4 Evolutionary Art

Speaking broadly, creative evolutionary systems that combine with the aesthetic decisions of a human to judge fitness started well before computers. Standard historical selective-breeding practices, where a human selects the parents for each generation from a given evolved set of choices, is the basis for centuries of 'creatively' modified trees, roses, corn, dogs, cats, cows and so on. Current evolutionary art systems borrow from this time-tested approach. It was evolutionary biologist Richard Dawkins who first showed with his 'Biomorphs' program that accompanied his 1986 book 'The Blind Watchmaker' [8] that a computer can be combined with the aesthetic preference of a user to generate interesting results. Dawkins's work inspired artists such as William Latham and Stephen Todd [23], and Karl Sims [22].

Karl Sims's work went on to inspire many of the modern evolutionary artists today. In his 2D work [22], Sims used a very rich instruction set, containing image-processing functions as well as mathematical functions based on Lisp expression trees. As with most evolutionary art systems to follow, Sims's system evolved a number of images (16 in his case) and allowed the viewers to pick their favourites, thereby allowing the most 'aesthetically pleasing' images to survive and mutate to the next generation. Other well-known artists used similar techniques: Steve Rooke [21], also working in Lisp, is very well known for his artwork, which added evolvable fractals to the function set; and Penousal Machado [16], a researcher at the Artificial Intelligence Laboratory at the University of Coimbra, in contrast to Sims's complex function set, used a very simple function set which is believed to open up the possible search space.

These systems, as with most creative evolutionary systems, use a human (often the artist or viewer, using interactive control) to make the aesthetic decisions after each evolutionary generation. In contrast, the use of an automatic fitness function which is able to make qualitative judgements constitutes a uniquely challenging research problem. Because of this, automatic fitness functions in evolutionary art are less the norm than they are in the general field of evolutionary computing. However, a number of researchers are beginning to explore art-based automatic fitness strategies: Bentley and Corne [4], in the design space; Miller et al. [18] in the field of electronic-circuit construction; and Koza et al., using Koza's creative invention machine [15]. Automatic fitness functions for art, however, are especially difficult, and systems that use creative fitness functions in art are still quite naive. Ashmore and Miller [2] have attempted to use an automatic fitness function with CGP that gives preference to images that contain circular objects (detected with a Hough Transform) or exhibit a high degree of complexity. However, this function only initializes a population, and must defer to a human user's input for further evolution. Ashmore and Miller have also attempted to employ an automatic function for evolving towards a source image. We have based our system upon this notion of visual resemblance, with a more sophisticated similarity function and adapting their system to a portrait painter process.

10.5 Genetic Programming and Creativity

GP's successes in producing novel and, arguably, creative designs are well publicized, and implementing the creative process in an evolutionary algorithm such as CGP is conceptually pleasing, as the successive evolutionary stages of variation and selection map well to the engagement–reflection model of creative thought [5]. Engagement – the generation of possible ideas and solutions – manifests itself in the composition of simple building blocks defined by the GP's function set. Because these basic elements can be extremely basic and generic, the author of the algorithm need not inject too many pre-existing assumptions about anticipated solutions, which might restrict the output to a certain type and therefore hinder creativity. The reflection phase of creative practices is seen in the fitness function of the algorithm, and our model of contextual focus can be implemented as operations on the fitness function, modifying it to favour either precisely defined problems or broad and vague notions.

The classical GP approach, however, poses certain problems when tasked with modelling human creativity. Namely, GP excels at optimization problems that presume the existence of an optimal individual, which the search will then approximate. This is demonstrated by the fact that GP techniques typically use a single fitness function to evaluate every individual in every generation. The notion of an optimal individual is at odds with the process of contextually redefining and adjusting the goal of the search as it progresses. Indeed, problems that demand the use of creative intelligence do not have simple and stable evaluation criteria, and this is most certainly true of computer art, where the goal is not to produce an objectively optimal painting, but to explore variations and associations that are novel and unanticipated. Convergence to a particular individual solution halts exploration and stifles creativity. Indeed, regardless of the evaluation criteria used, if one individual in a population excels slightly better than the others, GP will tend to converge towards that value. By virtue of this process of optimization, diversity in the population is lost as all individuals assume the properties of the current leader. This loss of diversity has detrimental effects regarding the successful realization of the fluidity model of creativity: periods of narrow focus will damage the diversity of the population of solutions, essentially forgetting the individuals imagined during broad, associative phases. Returning to an associative phase from a narrow phase would essentially constitute starting from scratch each time the phases alternate. Clearly, this does not characterize the ease and fluidity that our model seeks to exhibit. Many solutions to the problem of maintaining diversity in evolutionary systems exist, and this is indeed a very well-researched subject. However, many solutions demand explicit organizational structures to be placed on them, such as sorting populations into different structural categories [20] or authoring a distance metric between individuals [14]. Such strategies rely on injecting a priori knowledge about the structure of the presumed solution into the system – something we wish to avoid.

10.5.1 Advantages of CGP in Creative Systems

We use CGP in our creative painting system, as it is particularly well suited to avoiding these concerns. Though CGP shares with GP the same general process of iterative selection and variation, it differs in its representation of the genotype. The encoding is not a simple tree as in GP, but rather is a graph of indexed nodes. Each node represents a single basic function from the original function set, and can have a number of inputs and outputs. When this representation undergoes mutation, the connectivity of the graph is altered, possibly causing some nodes to become disconnected from the final program output. As certain nodes do not connect to the output, the information they represent becomes redundant. Genetic information becomes latent, and this gives rise to the essential property of CGP: neutrality. The very same solution, or phenotype, can be the result of a wide variety of different graphs, or genotypes. Thus, a narrow mode of evaluation can indeed focus temporarily on converging towards a single individual without necessarily invoking a permanent loss of genetic diversity, as a subsequent broad, associative phase of exploration will still have access to all the latent genetic information not visibly expressed in the phenotypes of the population. Furthermore, not only does CGP preserve diversity, but it also allows us to encourage such latent diversity explicitly. For example, in one set of experiments we implemented the following rule: if the fittest individual of a population is identical to an individual in the previous generation for more than three iterations, the system chooses other genotypes that map to this same phenotype in favour of the current non-progressing genotype, thus promoting diversity in the latent genetic material.

10.6 Implementation

Our work is based on Ashmore and Miller's original application of CGP to genetic art [2]. Their basic approach, which we essentially follow, is outlined in Sect. 2.4.3. It consists of generating graphs with two inputs, namely the x and y coordinates of a pixel in the image plane, and three outputs, namely the hue, saturation and value (H, S, V) colour channels for that pixel. It should be noted that RGB outputs can be used and, in the case of Miller's initial work, either RGB or HSV could used. In our system, the HSV outputs provide better support for the artistic colour evaluation techniques used in our contextual-focus-based fitness function. Two of the three 'rules of art' described in Sect. 10.6.1 and Fig. 10.2 are easily described in terms of HSV; tone (non-colour graduation) can be inspected independently of colour through V, the value component, and the difference between warm and cool colour temperatures can be easily determined from H, the hue component.

The functions in the function set (see Table 13) also can also use a random constant `param` as an input, which can be altered by mutation. The functions are intentionally kept simple and neutral to avoid imposing unnecessary structure onto the ultimate results, thereby allowing a large search space. A graph of these functions constitutes an individual's genotype. When this compound function is evaluated for each pixel in an image plane, an image is produced, which is the individual's phenotype. It is this phenotype that is evaluated using our creative, contextually-focused fitness function.

Table 10.1 Functions 1–5 use simple arithmetic operators on the *x*,*y* coordinates of the image. Functions 6–13 contain logical or trigonometric functions that are able to express more geometric shapes and colour graduations

1. `x | y;`
2. `param & x;`
3. `(x + y) % 255;`
4. `if (x>y) x - y; else y - x;`
5. `255 - x;`
6. `abs (cos (x) * 255);`
7. `abs (tan (((x % 45) * pi)/180.0) * 255));`
8. `abs (tan (x) * 255) % 255);`
9. `sqrt ((x - param) * 2 + (y - param) * 2);` (thresholded at 255)
10. `x % (param + 1) + (255 - param);`
11. `(x + y)/2;`
12. `if (x > y) 255 * ((y + 1)/(x + 1)); else 255 * ((x + 1)/(y + 1));`
13. `abs (sqrt (x - param2 + y - param2) % 255);`

As with Ashmore and Miller's work, the genotype is stored as an array of integers of length $(n*4)+3$ where n is the number of nodes, as seen in Fig. 10.1. The last three integers in the chromosome are the output pointers for the hue, saturation and value colour channels. The number of nodes in the chromosome affects the complexity of the output image. A greater number of nodes results in more functions contributing towards the final image. In CGP, each node normally defines the inputs to the node and the function only. Our functions are limited by having outputs between 0 and 255 for H, S and V. To increase the flexibility, a further parameter has been added to each node, which may or may not be used by the specified function.

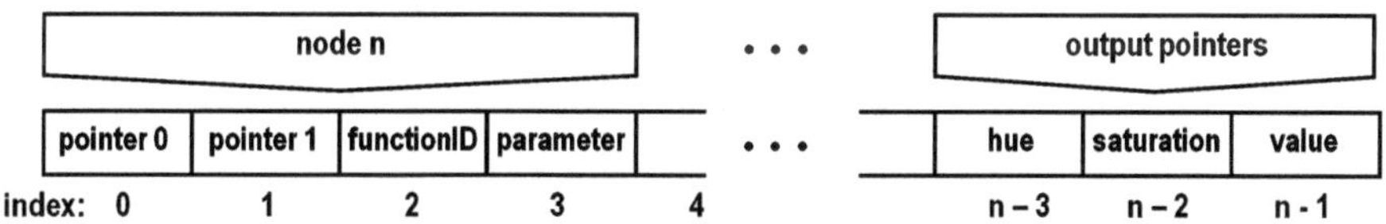

Fig. 10.1 Schematic illustration of the genotype.

Mutation occurs by calculating a random point along the chromosome, where a mutation rate specifies how many points are randomly chosen. Since along the chromosome some points can represent either a pointer, a function or a parameter, the mutation behaviour has specific constraints for each of these three types.

Crossover points are selected between whole nodes so data within the nodes is retained. To create an offspring, the nodes before the crossover point come from one parent while the nodes after the crossover point, including the output pointers, come from the other parent. A mechanism is in place that allows crossover to occur when the parents have genotypes of different lengths.

10.6.1 Fitness Function

Ashmore and Miller's original work consisted of initializing a population of interesting works of art, and allowing a human user to guide subsequent evolution by evaluating the images. Our goal is to remove the human from the system by providing an artistic evaluation function that produces painterly portraits and exhibits contextual focus in its search for an aesthetic image. The goal of portrait painting is not to perfectly reproduce the appearance of the subject (especially since the advent of photography), but rather to evoke a creative interpretation of the sitter. Therefore, the fitness function will, at times, emphasize the narrow and concrete goal of subject resemblance, while at other times it will defer to the fuzzy, associative and even contradictory 'rules' of abstract art, with a psychologically inspired model of contextual focus determining when to switch between them. CGP's phenotypic neutrality [24] ensures that the system does not destroy diversity when it seeks the narrowly defined goal of accurate resemblance; latent genetic material is still available for surprising associations later in the search process.

The fitness function determines resemblance by finding the mean squared error between an image in the population and the source image. In our case, we have taken an image of Charles Darwin as our subject. The abstract, painterly guidelines measure three different properties: the first is the composition of the face relative to the background; the second is the tonal similarity of the image, as matched to a sophisticated artistic colour space model emphasizing warm–cool colour temperature relationships according to analogous and complementary colour harmony rules; and the third is the presence of a dominant and a subdominant tone. These rules are drawn from the portrait painter knowledge domain, as detailed in [9].

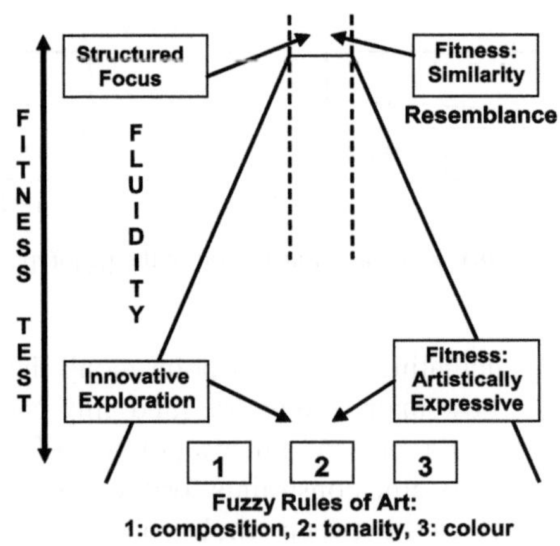

Fig. 10.2 Our fitness function mimics human creativity by moving between restrained focus (resemblance) to more unstructured associative focus (resemblance + more ambiguous art rules of composition, tonality and colour theory).

The fitness function, then, calculates four scores for each image (resemblance and the three painterly rules), using contextual focus to inform the way in which these values are combined as the search progresses, as seen in Fig. 10.2. Our model of contextual focus alternates between emphasizing the highly structured goal of resemblance and encouraging the spontaneous exploration of the aesthetic principles of artistic composition.

10.6.2 Contextual Focus

By default, the fitness function favours resemblance by rating paintings using a ratio of 80% resemblance to a 20% nonlinear combination of our three painterly rules. Several functional triggers can alter this ratio in different ways. For example, as long as a significant proportion of high-resemblance individuals exist in the population, in our case 80% ('resemblance patriarchs'), the system allows individuals with very high scoring under the painterly rules ('strange uncles'), to be accepted into the next population. These individuals with high painterly scores (weighted nonlinearly to allow for those with a very high score on just one rule) are saved separately, and mated with the current population; if the system remains in this default state of focused resemblance, further offspring continue to be tested with the default 80% resemblance and 20% painterly rule test. Therefore, though we pull out and save these 'strange uncles' to maintain artistic diversity, the focus of the genetic search is still towards resemblance.

The system, as a whole, will begin to favour the artistic rules when progress towards resemblance slows. When a plateau, or local minimum, is reached for a certain number of populations, the fitness function ratio begins to weight painterly rules higher than resemblance, on a sliding scale. Because artistic diversity has been explicitly encouraged in the focused resemblance phase, there is a great deal of artistically rich genetic information that is latent in the population. When the artistic rules are favoured over resemblance, this genetic information can manifest itself and a great deal of experimentation and exploration occurs.

Just as we saved artistically promising individuals during the focused stage, we are careful to isolate individuals with high resemblance during this artistic phase. These individuals are similarly allowed to pass on to the next generation when the condition of a certain proportion of aesthetically promising individuals is satisfied. Using this method, high-resemblance individuals always remain in the population. When the resemblance of these individuals shows a marked improvement beyond the previous plateau, the system returns again to the default focused resemblance mode.

10.7 Results

This system ran on one high-end PC for 50 days. Since the genes of each portrait can be saved, it is possible to recombine (marry) and re-evolve any of the art works in new variants (Fig. 10.3). As the fitness score increases, portraits look more like the sitter (Fig. 10.4). This gives us a somewhat known spread from very primitive (abstract) all the way to realistic portraits. So, in effect, our system demonstrates two ongoing processes: firstly, those portraits that pass on their resemblance strategies, making for more and more realistic portraits – the family of 'resemblance patriarchs' (Fig. 10.4) – and, secondly, the creative 'strange uncles', which are genetically related to the current resemblance patriarchs, but exhibit greater aesthetic creativity. This dual technique of evolving patriarchs and strange uncles models the contextual focus of creative individuals as discussed in Sect. 10.2, which is the paradoxical technique where creative people use the existence of some strong structural rules (such as the template of a sonnet, a tragedy or, in this case, a resemblance to the image of the sitter) as a resource or base to elaborate new variants beyond that structure (an abstracted variation of the image of the sitter). That is, novel ideas require a pre-existing system to serve as a reference point from which innovation can occur.

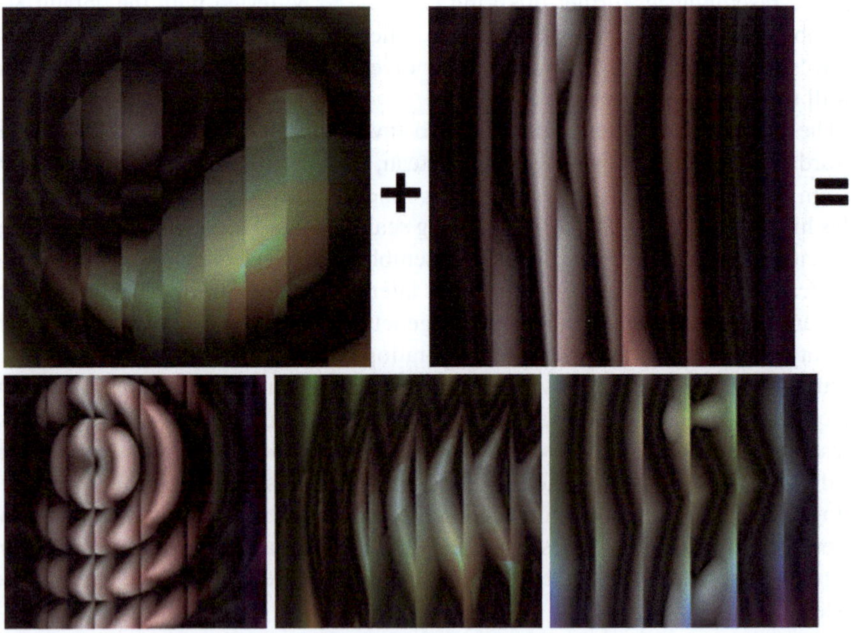

Fig. 10.3 Two portrait programs are mated together, showing merged strategies of the offspring.

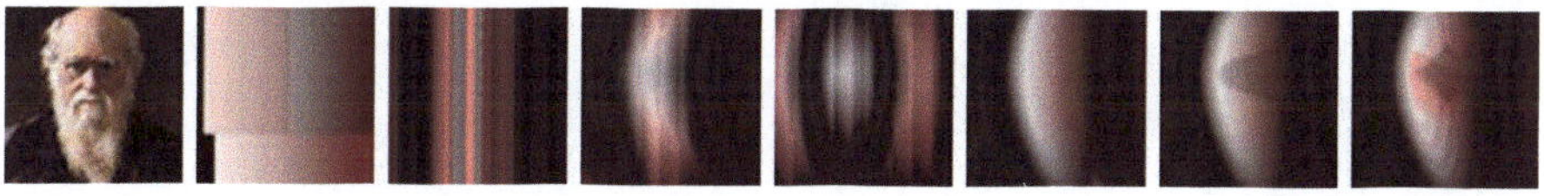

Fig. 10.4 Source Darwin portrait, which is part of the fitness function, followed by an evolved progression of portraits of best resemblance.

Another point worth highlighting is the difficulty of judging quality even in a population of equally poor paintings, which is currently an open research question regarding automated fitness functions [17]. Indeed, the initial population of images that the system produces will bear absolutely no resemblance to the portrait subject, yet it is critical that the population is sorted in a precise and meaningful way, in order to guide evolution towards an aesthetic goal. To overcome this problem, we ensure that our fitness function evaluates images effectively in any stage of evolution. The associative contextual mode of our fitness function contributes greatly to achieving this generality, as, owing to its broad nature, it is applicable to a similarly broad range of images. So, as opposed to direct resemblance, which only distinguishes between images meaningfully when there is a certain degree of resemblance already present, the abstract rules of colour and composition can be applied to virtually any image, regardless of how visually similar it is to its subject. Therefore, the system can switch its mental context at any point where it becomes difficult to distinguish between the images in its current population. Not only does this address the issue of ensuring effective evaluation at all stages of evolution, but it also models psychological creativity in a conceptually satisfying way: situations where there is no discernible way forward are precisely the times that call for creative exploration of alternatives.

However, it is ultimately those individuals that doggedly strive to resemble the Darwin image that move the system forward, as it is they that attain the highest resemblance scores and strategically move the system closer to the source image form in terms of resemblance. By allowing their related family members to be more innovatively artistic (via large local exploration) as safe variants from the patriarchs, we avoid the challenges to creativity that optimization presents, as discussed in Sect. 10.5. Figure 10.5 shows both types of individuals working synergistically, while Fig. 10.4 contains only the resemblance patriarchs. We should emphasize that our goal is not to reproduce the Darwin portrait, but to explore a family tree of related and living portraits that inherit creative painting strategies through an evolutionary process. Ultimately, it is our hope to extend this system so that it will be creative in a range of artistic and design-oriented spaces beyond artistic portrait painting.

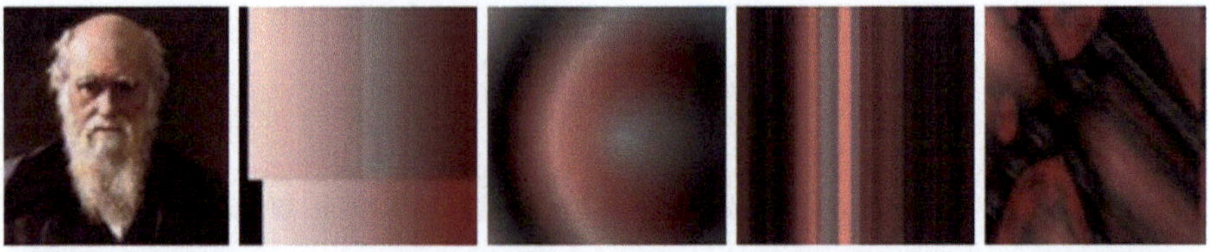

(a) In the first 100s of populations, colour and curves emerge, our first glimmer of a move toward approximating Darwin's image. 100s later we see bands resembling the vertical lighting of the portrait (#4), that then twist and curve

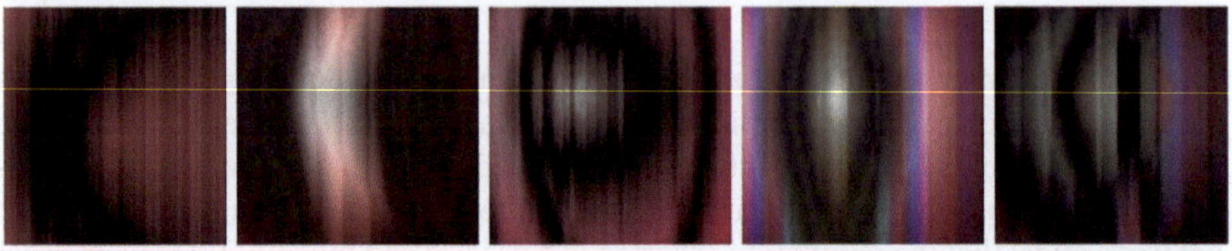

(b) Soon even thinner bands/twists strategies create the dominant form (#2) and from it the first 'head shapes' appear.

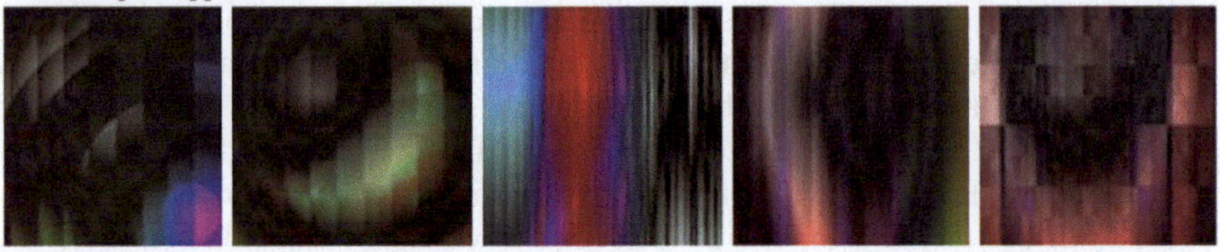

(c) The eventual expression of initially unexpressed modified nodes (via genetic drift) brings in a novel, colourful phase

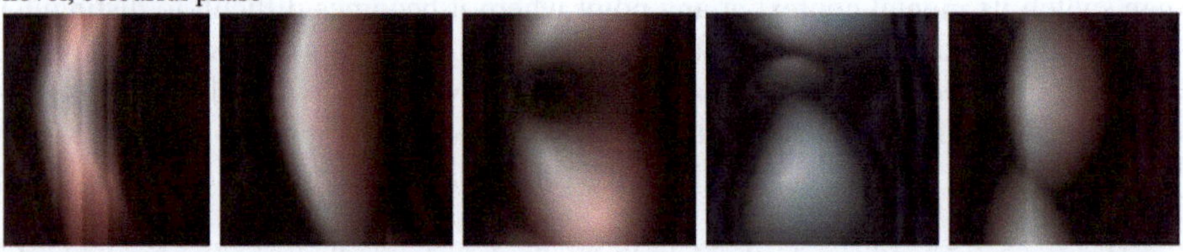

(d) A new best form #2 replaces #1. With this ramped dominant strategy, #2 below heralds in the soft blobby era.

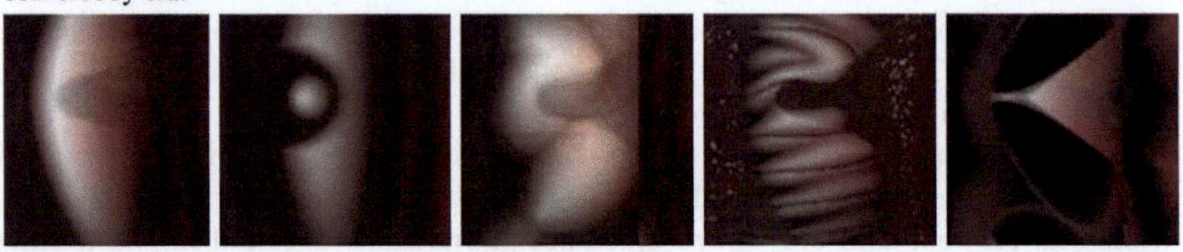

(e) The next major strategy to appear is the addition of the left 'racoon patch' eye area and the right eye.

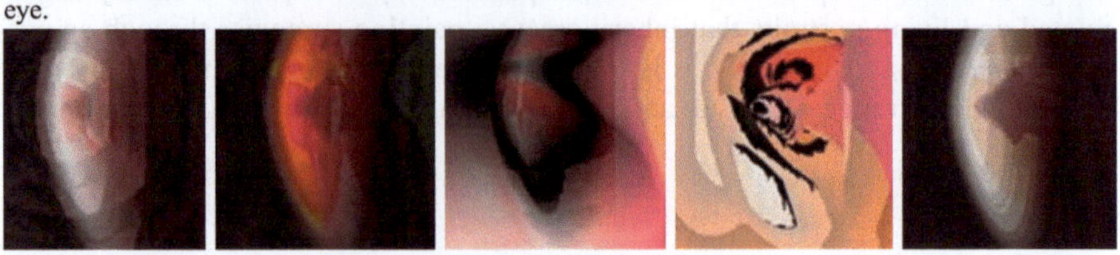

(f) A more painterly phase begins, combining head shape and texture. The last 2, show abstraction and resemblance.

Fig. 10.5 Portraits in chronological order, selected as examples of the process (from a larger sample at http://www.dipaola.org/evolve).

The images in Fig. 10.5 show selected portraits in chronological order. These represent a larger collection, and show both those that are best at resemblance and those that are artistically compelling. While the overall population improves at resembling Darwin's portrait, what is more interesting to us is the variety of recurring, emergent and merged creative strategies that evolve as the programs seek, in different ways, to become better abstract portraitists.

10.8 Conclusions and Future Directions

We have incorporated research on human creativity into a relatively new form of evolutionary computation, CGP, which has been successfully adapted to encourage the development of creative, painterly techniques. CGP exhibits genetic neutrality, which enables us to maintain the diversity needed to explore creative variations when faced with local minima. This technique has proved to be well suited to the development of our contextual-focus model of creativity, which requires the presence of such latent creative potential.

The domain of portrait painting was chosen because it leans heavily on resemblance (a closed and known issue for computer algorithms), but also has an open-ended creative element. As well as this, the portrait-sitter-to-painter relationship is well suited to exploring computer creativity. The system indeed evolves creative strategies to become a better abstract portraitist. We are continually refining the painterly portions of the automatic fitness function from lessons learned in past runs, and we are currently adding more creative, structural elements to this open-ended general system.

Key to this generality is increased understanding of how the potentiality of an idea changes and is affected by both the associative structure and the goals and desires of the mind it 'finds itself in'. To this end, future research will involve adding specific painterly and portrait knowledge with the goal of continuing to improve the automatic portrait painter system with human painterly knowledge. In addition, it is also possible (and this, possibly, will be the direction of our next version of the program) to evolve the associative aesthetic fitness function simultaneously with the rest of the system. This can alter the dimensionality of the search space, the parameterization and the representation of solutions, allowing the possibility of automation that is more creative.

Practically, to better approximate a human portraitist's technique, we are redesigning the functions in the function set to be reactions to the colour and position of the sitter image (the function set of the current system is blind to the sitter image, which is used only for evaluation). This way, any decision on a paint stroke output is a direct reaction to the recognition of an input (what the artist sees in the sitter scene). This would mean that, once a pleasing portrait image (individual) has been created, the program could use its same painterly strategies on any new sitter image, thereby creating a true portrait painter.

Furthermore, a successful portraitist program might even have 'one-man' shows and take commissions, allowing its human creator to play a background role as its talent agent. It could eventually even be bred with other successful portraitist programs similarly to what is done with racehorses, allowing experiments into cultural and collaborative creativity. This 'matching output stroke to input analysis' technique, with other modifications, would facilitate the realization of another goal: to have resolution-independent portraits, allowing small portrait sizes for speed during the evolving process, but larger sizes that reveal additional painterly and surface details for the final artwork – just as a human might make many creative sketches before the fully finished work.

We would like to explore the extent to which the techniques used here can be transported to other domains such as art and design, music, authoring, human–computer interfaces, entertainment, and gaming. The mechanisms will be kept general, since we believe it is the associative, domain-general (rather than specialized, domain-specific) aspect of a creative architecture (whether organic or artificial) that is its greatest asset. Finally, we foresee a possible research application as a test bed for simulating creative processes or as an educational tool for gaining hands-on understanding of evolutionary and creative processes.

10.9 Acknowledgements

We would like to thank Laurence Ashmore, Peter Bentley, Julian Miller and James Walker for their correspondence, as well as Ashmore and Miller for the initial Java-based system that we adapted for our creative experiments. We would especially like to thank Liane Gabora, whose original ideas on creativity formed a major of this chapter.

References

1. Ashmore, L.: http://www.emoware.org/evolutionary_art.asp (2004). Java evolutionary art software package
2. Ashmore, L., Miller, J.F.: Evolutionary Art with Cartesian Genetic Programming. Online technical report available from https://sites.google.com/site/julianfrancismiller/publications (2004)
3. Baker, E.: Evolving Line Drawings. In: S. Forrest (ed.) Proc. of the International Conference on Genetic Algorithms, p. 627. Morgan Kaufmann (1993)
4. Bentley, P., Corne, D. (eds.): Creative Evolutionary Systems. Morgan Kaufmann (2002)
5. Boden, M.: The Creative Mind: Myths and Mechanisms. Routledge (2003)
6. Cohen, H.: How to Draw Three People in a Botanical Garden. In: Proc. of AAAI, pp. 846–855. Morgan Kaufmann (1988)
7. Dartnell, T.: Artificial Intelligence and Creativity: An Introduction. Artificial Intelligence and the Simulation of Intelligence Quarterly **85** (1993)
8. Dawkins, R.: The Blind Watchmaker. W. W. Norton & Company Inc (1986)

9. DiPaola, S.: Painterly rendered portraits from photographs using a knowledge-based approach. In: Proc. of the SPIE Human Vision and Imaging. International Society for Optical Engineering (2007)
10. DiPaola, S., Gabora, L.: Incorporating characteristics of human creativity into an evolutionary art algorithm. Genetic Programming and Evolvable Machines **10**, 97–110 (2009)
11. Feinstein, F.: The Nature of Creative Development. Stanford University Press (2006)
12. Feist, G.J.: The influence of personality on artistic and scientific creativity, pp. 273–296. Cambridge University Press (1999)
13. Gabora, L.: Revenge of the 'Neurds': Characterizing Creative Thought in Terms of the Structure and Dynamics of Memory. Creativity Research Journal **22**, 1–13 (2010)
14. Goldberg, D.E.: Genetic Algorithms in Search, Optimization, and Machine Learning. Addison-Wesley (1989)
15. Koza, J.R., Keane, M., Streeter, M.J.: Evolving inventions. Scientific American **288**, 52–59 (2003)
16. Machado, P., Cardoso, A.: NEvAr - The Assessment of an Evolutionary Art Tool. In: G. Wiggins (ed.) Proc. of the AISB Symposium on Creative, Cultural Aspects and Applications of AI and Cognitive Science (2000)
17. McCormack, J.: Open problems in evolutionary music and art. In: F. Rothlauf, J. Branke, S. Cagnoni, D.W. Corne, R. Drechsler, Y. Jin, P. Machado, E. Marchiori, J. Romero, G.D. Smith, G. Squillero (eds.) Proc. of Applications of Evolutionary Computing, *LNCS*, vol. 3449, pp. 428–436. Springer (2005)
18. Miller, J.F., Kalganova, T., Lipnitskaya, N., Job, D.: The Genetic Algorithm as a Discovery Engine: Strange Circuits and New Principles. In: AISB Symposium: Proc. Workshop on Creative Evolutionary Systems, pp. 65–74 (1999)
19. Poincaré, J.H.: The Foundations of Science. Science Press (1913)
20. Potter, M.A., De Jong, K.A.: Cooperative coevolution: an architecture for evolving coadapted subcomponents. Evolutionary Computation **8**, 1–29 (2000)
21. Rooke, S.: Eons of genetically evolved algorithmic images. In: P.J. Bentley, D. Corne (eds.) Creative Evolutionary Systems, pp. 339–365. Morgan Kaufmann (2002)
22. Sims, K.: Artificial evolution for computer graphics. Computer Graphics **25**, 310–328 (1991)
23. Todd, S., Latham, W.: Evolutionary Art and Computers. Academic, New York (1994)
24. Yu, T., Miller, J.F.: Neutrality and the Evolvability of Boolean function landscape. In: J.F. Miller, M. Tomassini, P.L. Lanzi, C. Ryan, A.G.B. Tettamanzi, W.B. Langdon (eds.) Proc. European Conference on Genetic Programming, *LNCS*, vol. 2038, pp. 204–217. Springer (2001)

Chapter 11
Medical Applications of Cartesian Genetic Programming

Stephen L. Smith, James Alfred Walker, Julian F. Miller

11.1 Introduction

Evolutionary algorithms have been used in medical applications since they were first developed some 30 years ago. Although often regarded as a theoretical pursuit, research on and development of a wide range of real-world applications of genetic and evolutionary computation (GEC) has long been evident at conferences and in the scientific literature. Medicine and healthcare is no exception and this challenge, and worthy aim, has motivated many to apply GEC to a wide range of clinical problems.

This chapter presents examples of CGP applied to the diagnosis of three medical conditions: breast cancer, Parkinson's disease and Alzheimer's disease.

11.2 CGP Applied to the Diagnosis of Breast Cancer

Breast cancer is one of the leading causes of death in women in the western world. In 2007, some 46,000 new cases were detected in the UK alone, and 12,000 women died in 2008 as a result of the disease, making it the most common cancer in women and the second most common cause of death through cancer. The number of breast cancer-related deaths has fallen since screening programmes were introduced in 1988. Nonetheless, it is predicted that one in nine women will develop breast cancer at some point during their life [1]. The detection of breast cancer in the early stages of the disease significantly increases the survival rate of patients. The main method for screening patients is the mammogram, a high-resolution X-ray image of the breast [2]. The process of identifying and evaluating signs of cancer from mammograms is a very difficult and time-consuming task that requires skilled and experienced radiologists. This assessment is also, by its nature, highly subjective and susceptible to error, leading to cancers being missed and patients misdiagnosed. To achieve a more accurate and reliable diagnosis, computer aided detection (CAD)

J.F. Miller (ed.), *Cartesian Genetic Programming*, Natural Computing Series,
DOI 10.1007/978-3-642-17310-3_11, © Springer-Verlag Berlin Heidelberg 2011

systems have been investigated which provide an objective, quantitative evaluation. CAD systems have the potential to help in two main ways: (i) the detection of suspicious areas in the mammogram that require further investigation, and (ii) the classification of such areas as cancerous (malignant) or non-cancerous (benign) [3].

Two powerful indicators of cancer that are commonly used in evaluating mammograms are known as masses and microcalcifications. Masses (see Fig. 11.1) are the larger of the two indicators and can be either benign or malignant. Characteristics such as the border and the density of the mass, which is greater for malignant examples, can be used for classification. Traditionally, masses are more difficult to classify than microcalcifications. Microcalcifications are, essentially, small calcium deposits which occur as the result of secretions from ductal structures that have thickened and dried. They can have a great variety of mostly benign causes, but might also be an indication of malignancy. They are fairly common in mammograms and their appearance increases with age, so that they can be found in 8% of mammograms of women in their late 20s and in 86% of mammograms of woman in their late 70s [4]. Microcalcifications that indicate malignancy are usually less than 0.5 mm in size and are often grouped into clusters of five or more (see Fig. 11.2). Any calcification larger than 1 mm is almost always benign (see Fig. 11.3) [4].

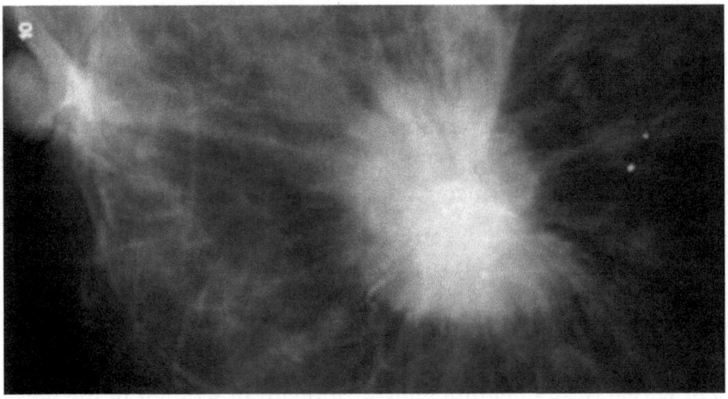

Fig. 11.1 An example of a speculated mass [5]. Image courtesy of the Radiology Assistant, www.radiologyassistant.nl

Over the past 20 years, there has been much research into the application of CAD to breast cancer, with numerous different approaches being exploited, and many of these involve image analysis of the digitized mammogram. A typical approach is to use a pattern recognition scheme that employs sensing, segmentation, feature extraction, feature selection and classification, to isolate and then characterize a feature of interest [6]. Each stage of this processing is a potentially complex operation requiring much investigation. Evolutionary algorithms and, in particular, genetic algorithms, have previously been used with some success in CAD of breast cancer, but

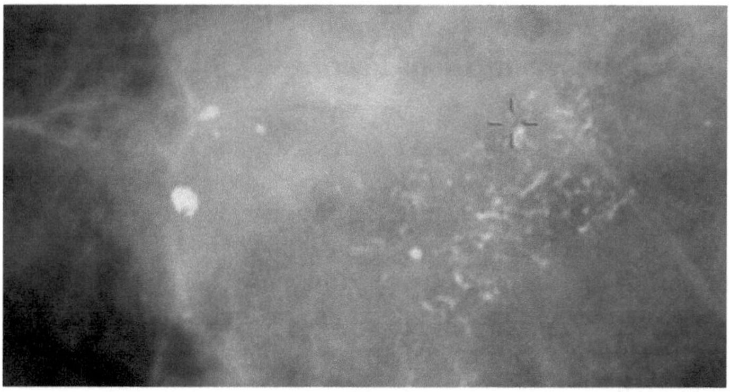

Fig. 11.2 Example of a cluster of malignant microcalcifications [5]. Image courtesy of the Radiology Assistant, www.radiologyassistant.nl

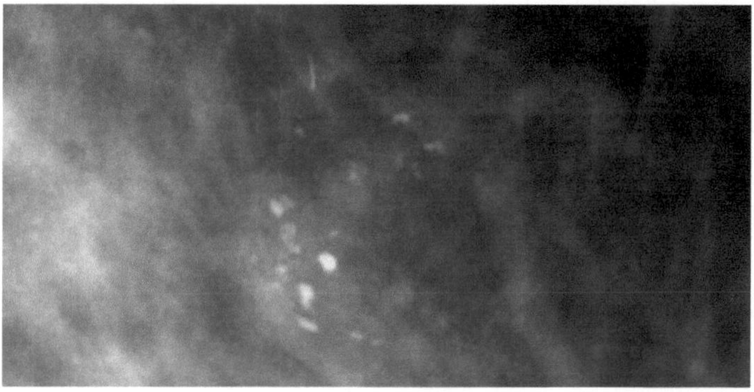

Fig. 11.3 Example of a cluster of benign microcalcifications [5]. Image courtesy of the Radiology Assistant, www.radiologyassistant.nl

typically only in the feature selection stage of the traditional image analysis scheme [7, 8, 9]. This limited use of evolutionary algorithms prompted Hope, Smith and Munday to investigate the use of CGP in the classification of mammograms [10]. This was undertaken in two distinct ways: first, by training the CGP network using a vector of conventional statistical features extracted from the mammogram, and, second, by presenting the raw pixel values of the mammogram directly to the inputs of the CGP network. In both cases, regions of interest (ROIs), comprising one or more microcalcifications, were manually segmented from a number of mammograms to form training and test data sets of benign and malignant images.

11.2.1 Classification of Mammograms Using a Vector of Conventional Statistical Features

The vector of conventional statistical features extracted from the pixel values in the ROI comprised the mean, the second moment and the third moment. The spatial grey-level dependence (SGLD) matrix was also calculated from the same ROI and the same features were extracted again, along with the following additional features: entropy, element difference moment and uniformity. Definitions and more information regarding these texture-based features can be found in [6]. This vector of conventional statistical features was then presented to the inputs of a CGP network as shown in Fig. 11.4.

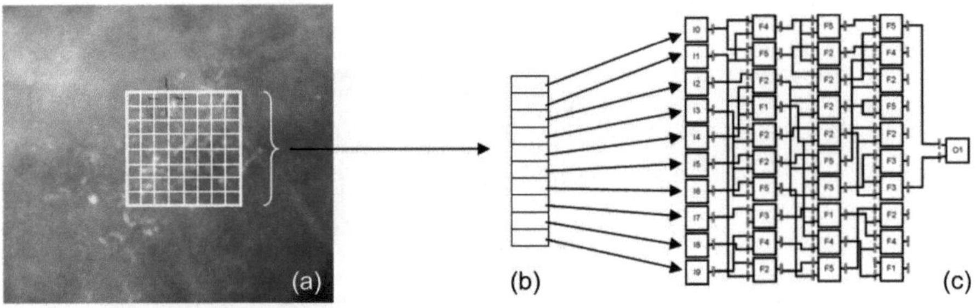

Fig. 11.4 CGP network for classification of mammograms using a vector of statistical features. (a) Region of interest in mammogram; (b) SGLD matrix; (c) CGP network.

A number of experiments were then undertaken to establish the effectiveness of the CGP network in classifying mammograms. The parameters used for configuring the CGP are given in Table 11.1, and the results obtained from leave-one-out cross-validation given in Table 11.2 (where 'FPR' means 'false-positive rate' and 'FNR' means 'false-negative rate'). Significant performance can be seen throughout these results, with high sensitivity and specificity. The overall classification accuracy is also good, at 70.0%. The analysis of the receiver operating characteristic (ROC) curve given in Fig. 11.5 shows an area under the ROC, of 0.69, which is comparable to values found in the literature [3].

11.2.2 Classification of Mammograms Using Raw Pixel Values

In the second approach that we investigated, the same ROI pixel values as were used in calculating the vector of features described above were presented directly to the inputs of the CGP network. This necessitates the number of inputs to the CGP

Table 11.1 CGP parameters for classification of mammograms using a vector of statistical features

Parameter	Value
Number of rows	5
Number of columns	16
Number of inputs per node	2
Number of generations	14,000
Mutation rate	0.1

Table 11.2 Results for classification of mammograms using a vector of statistical features

Sensitivity	Specificity	FPR	FNR	Training fitness	Test fitness
0.9	0.6	0.4	0.1	80%	70%

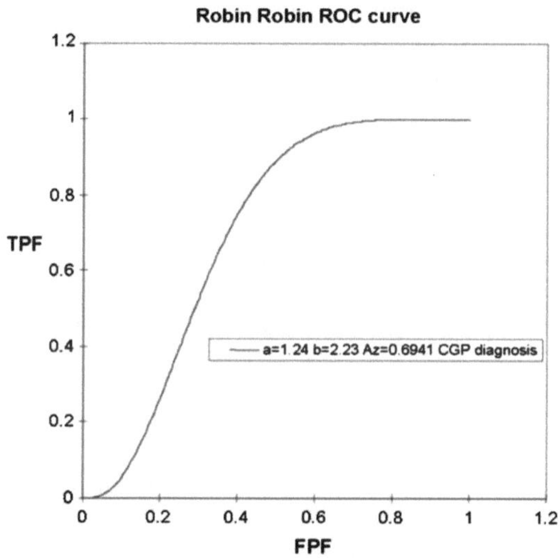

Fig. 11.5 Receiver operating characteristic (ROC) curve for classification of mammograms using a vector of statistical features [10].

network being increased to 64 and the number of rows to 16 to accommodate the extra processing required, as shown in Fig. 11.6.

A number of data sets were used to evaluate the ability of CGP using the parameters given in Table 11.3, and the results are presented in Table 11.4.

The training scores are lower than for the first experiment; high sensitivities are countered by low specificities, and so the overall classification is always higher than chance, but not significantly.

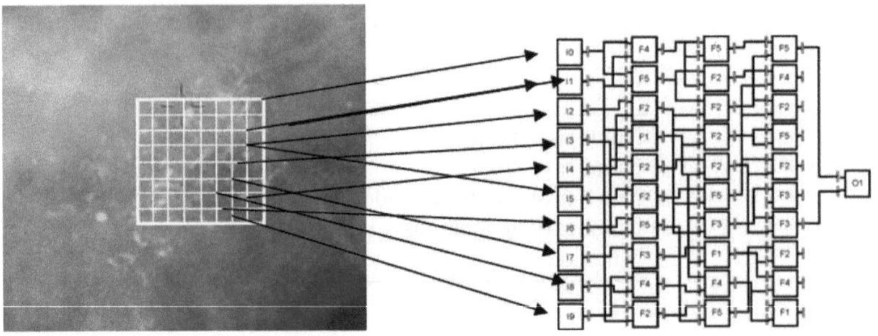

Fig. 11.6 CGP network for classification of mammograms using image raw pixel values. Region of interest in mammogram (left). CGP network (right).

Table 11.3 CGP parameters for classification of mammograms using image raw pixel values

Parameter	Value
Number of rows	16
Number of columns	16
Number of inputs per node	2
Number of generations	2000
Mutation rate	0.1

Table 11.4 Results for classification of mammograms using a vector of statistical features

Sensitivity	Specificity	FPR	FNR	Training fitness	Test fitness
0.9	0.44	0.55	0.1	57%	65%

An ROC analysis, shown in Fig. 11.7, indicates that the technique is more effective than the sensitivities and specificities would suggest. The area under the curve is close to 0.78, which is comparable to, if not better than, the performance of commercial classifiers; thus, some potential has been shown, but further work is clearly needed.

11.2.3 Classification of Mammograms Using Multi-chromosome CGP

Following Hope, Smith and Munday's work on training CGP using raw pixel values from ROIs in the mammogram [10], Völk et al. [11], with Walker et al. [12], investigated the use of multiple-chromosome CGP networks in the evaluation of mammograms. The premise for this approach was born out of the large physical

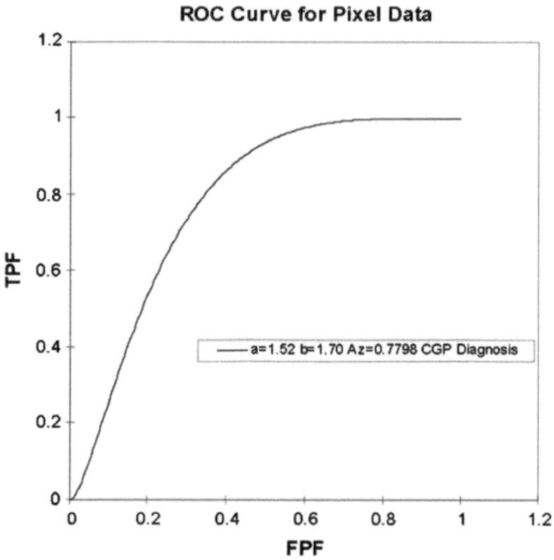

Fig. 11.7 ROC curve for classification of mammograms using image raw pixel values [10].

size of mammograms; current technology provides a pixel size of 50 μm^2, which equates to a full-field mammogram of some 4800×6000 pixels. It can therefore be appreciated that any processing using conventional CGP networks on full-field mammograms would take an unacceptably long time. Consequently, mammograms will have to be partitioned into smaller 'part' images to make processing feasible.

As with previous work described earlier in this chapter, regions of interest, featuring at least one microcalcification, were extracted from the mammogram and stored as a 128×128 pixel, eight-bit greyscale image. An example of such a region of interest, featuring two microcalcifications, is shown in Fig. 11.8. The images were then divided into 256 non-overlapping 8×8 pixel areas or parts. The 64 greyscale pixel values (0 to 255) for each of these parts formed the inputs to an individual chromosome, as shown in Fig. 11.9. Accordingly, 256 independent CGP chromosomes were employed, each of which used a large CGP network of 32 rows and 128 columns. This affords redundancy, which has been proven to be advantageous in the evolution of CGP networks, as reported in previous work by Miller and Smith [13].

The fitness of each network was based upon its ability to correctly classify mammograms in the training set as malignant or benign. This was achieved by considering each chromosome's output, which was thresholded about a predefined value, determined by experimentation.

As multiple networks were employed, the chromosomes could either be swapped or be replaced with another of the 256 chromosomes, according to a rearrangement mutation rate. The following rearrangement strategies were investigated over a series of independent evolutionary runs:

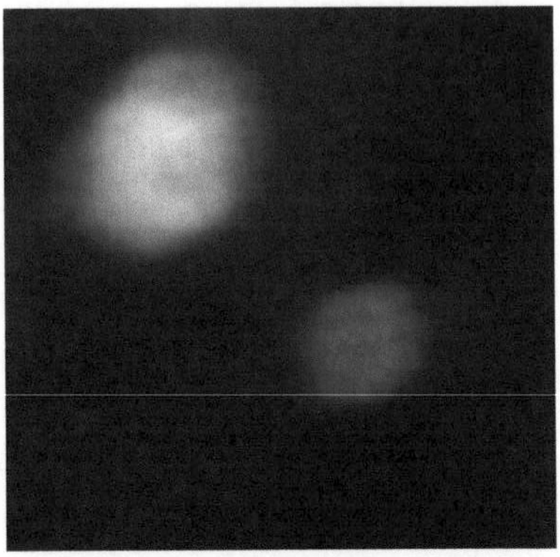

Fig. 11.8 Example mammogram region of interest, featuring two microcalcifications [12].

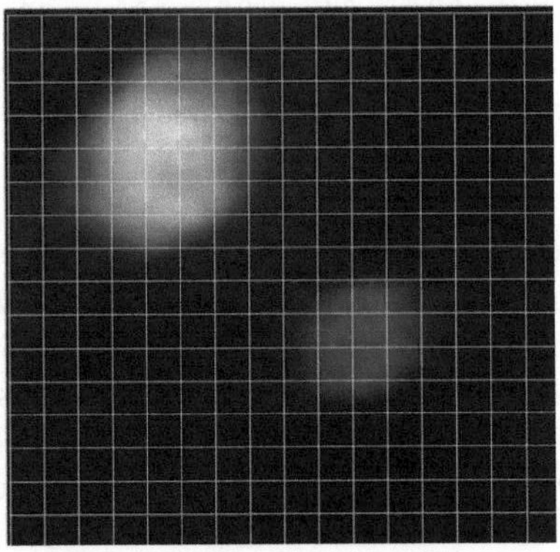

Fig. 11.9 Logical division of mammogram region of interest into 256 8×8 pixel parts, each of which is fed into a separate CGP network [12].

1. A random swap, in which any one chromosome may be swapped with another.
2. A neighbouring swap, in which a chromosome may only be swapped, at random, with one of its four direct spatial neighbours.
3. A copying operation, where a random chromosome is chosen to overwrite one other chromosome.

The neighbouring swap was implemented in order to target structures that continue from one part of the image to the next. Neighbouring parts might therefore have similar image properties, and are likely to respond equally well to the same chromosome.

If after a mutational rearrangement, the fitness of an individual's chromosome declined, compared with its fitness before the swap, then the rearrangement was disallowed. However, if the rearrangement made an improvement to the resulting fitness, then the exchange was preserved. Although there is a risk that the diversity of chromosomes might be reduced by deleting those that do not perform well and substituting them with fitter examples, this approach provides the genotypes with a greater opportunity for individual mutation, which in itself has the potential for restoring diversity to some extent. For example, if every chromosome is unique (and thus has no copies) then mutations can only be beneficial independently. If there are duplicated chromosomes, any mutation occurring in those would have to be, on average, beneficial to all of them. This means that one chromosome's fitness might be reduced if all other copies gained a higher fitness through the rearrangement.

As the majority of the single chromosomes classified only 50% of the test images correctly, a second classification strategy was employed, in which using the majority result from applying all the chromosomes to each part improved the classification accuracy. The results for this classification approach, for two different voting thresholds, are shown in Tables 11.5 and 11.6. From the results, it can be seen that the voting threshold has a pronounced impact on the results. For a voting threshold of one (more than one chromosome classifies the image section as malignant), the random-swap recombination strategy can classify 60% or 70% of the test images correctly, depending on whether reuse is used. However, with a voting threshold of two, the random-swap-with-reuse strategy does not perform well any more, but the neighbourhood-swap-with-reuse and the random-swap-without-reuse strategies both classify 70% of the test images correctly. Overall, it may be possible to deduce that the best-performing recombination strategy when chromosome voting is used is that of the random swap without reuse, as it appears to be more robust to changes of the voting threshold and never classifies any of the test images as false negatives, so a malignant test image is always detected.

The confusion matrices in the tables show the correlation between the actual (B and M) and predicted (b and m) numbers of benign and malignant classifications. The statistics provide details of the true-positive (TP), true-negative (TN), false-positive (FP) and false-negative (FN) rates, in addition to the precision (P) and the F1 measure (F1).

Table 11.5 Confusion matrices and statistics: test image classification using multi-chromosome voting with a voting threshold of 1 (VTH = 1)

	Confusion matrix				Statistics					
Recombination	Bb	Bm	Mb	Mm	TP	TN	FP	FN	P	F1
No swap, no reuse	0	5	0	5	1	0	1	0	0.5	0.67
No swap, reuse	0	5	0	5	1	0	1	0	0.5	0.67
Neighbouring swap, no reuse	0	5	0	5	1	0	1	0	0.5	0.67
Neighbouring swap, reuse	0	5	0	5	1	0	1	0	0.5	0.67
Random swap, no reuse	1	4	0	5	1	0.2	0.8	0	0.56	0.71
Random swap, reuse	2	3	0	5	1	0.4	0.6	0	0.63	0.77

Table 11.6 Confusion matrices and statistics: test image classification using multi-chromosome voting with a voting threshold of 2 (VTH = 2)

	Confusion matrix				Statistics					
Recombination	Bb	Bm	Mb	Mm	TP	TN	FP	FN	P	F1
No swap, no reuse	0	5	0	5	1	0	1	0	0.5	0.67
No swap, reuse	0	5	0	5	1	0	1	0	0.5	0.67
Neighbouring swap, no reuse	0	5	0	5	1	0	1	0	0.5	0.67
Neighbouring swap, reuse	3	2	1	4	0.8	0.6	0.4	0.2	0.67	0.73
Random swap, no reuse	2	3	0	5	1	0.4	0.6	0	0.63	0.77
Random swap, reuse	0	5	0	5	1	0	1	0	0.5	0.67

An example result is shown in Fig. 11.10. The two parts of the figure represent the same image part (8×8 pixel area): the number 1 signifies the multi-chromosome CGP algorithm's indication of malignant tissue in that area, and the number 0 indicates benign tissue. The radiologist's classification of malignancy is indicated by grey shading and by the white oval in the representation (Fig. 11.10a) and the mammogram ROI (Fig. 11.10b), respectively.

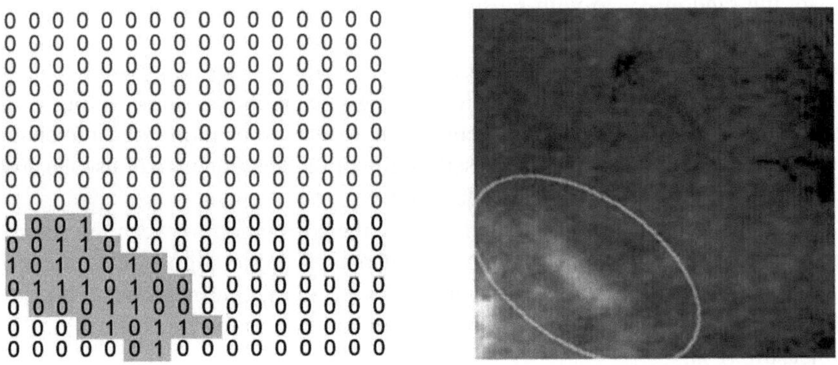

(a) (b)

Fig. 11.10 An example of a multi-chromosome CGP classification (a), for the corresponding region of interest in the mammogram (b). A '1' indicates a classification of malignant tissue by the multi-chromosome CGP and '0' indicates benign tissue, for the respective part of the region of interest. The grey shading and white outline depict the radiologist's classification of malignancy in (a) and (b), respectively [12].

11.2.4 Summary

In this section, we have seen how different representations of CGP have been used successfully to identify malignant microcalcifications in mammograms. The power and scalability of CGP are particularly well suited to this demanding pattern recognition problem, which deals with large, high-resolution images. Current work is investigating the application of these CGP representations to detect other malignant features within mammograms, such as spiculated masses.

11.3 CGP Applied to the Diagnosis of Parkinson's Disease

Parkinson's disease is a common, chronic, progressive neurodegenerative brain disease, afflicting about one person in 500 in the UK [14] and 10 million people worldwide [15]. The disease sets in insidiously, and in most patients progresses relentlessly, on average within ten years, to a state of total physical incapacitation. The symptoms usually start on one side of the body (hemiparkinsonism) but later spread to the other side. About 80% of patients suffer from idiopathic Parkinson's disease for which no cause is known. However, the diagnosis of idiopathic Parkinson's disease is based on clinical features, which can have poor sensitivity, with about 25% of patients diagnosed with the disease actually having other conditions [16]. Considerable research is being conducted to improve the diagnosis of the condition, but most studies to date are reliant on laboratory-based experimentation.

Smith et al. used CGP as part of a non-invasive computer-based assessment to diagnose two cardinal symptoms of Parkinson's disease, tremor and bradykinesia; this assessment can be conducted in the clinical environment and the doctor's surgery, using commonly available computing peripherals [17]. Parkinsonian tremor is probably the most well-recognized symptom of the disease, but is only present in 70% of cases [18]. It is an involuntary rhythmical movement, which is especially prominent at rest, but decreases, or even ceases, during active movement. The frequency of the tremor is commonly reported to lie between 3 to 8 Hz and is most evident, and is often first identified, in the fingers, hands or legs. Bradykinesia is the core disabling feature of Parkinson's disease and is defined as difficulty, slowness (bradykinesia proper) or virtual inability (akinesia) in initiating and executing movements, or in modifying ongoing motor activity. Poverty of spontaneous movement (hypokinesia), loss of normal associated movements, masked facial expression and sudden 'freezing' in the middle of a motor performance are all part of the disturbance.

The computer-based assessment comprised two parts: data acquisition and data processing. The data acquisition stage involved digitization of the patient's drawing activity while the patient was attempting a conventional figure-copying task. The study used a commercially available digitizing tablet, which has the advantage of utilizing a stylus with ballpoint refills that can be used in the same way as a conventional ballpoint pen on standard paper, thus reproducing a conventional 'pen and paper' environment (see Fig. 11.11).

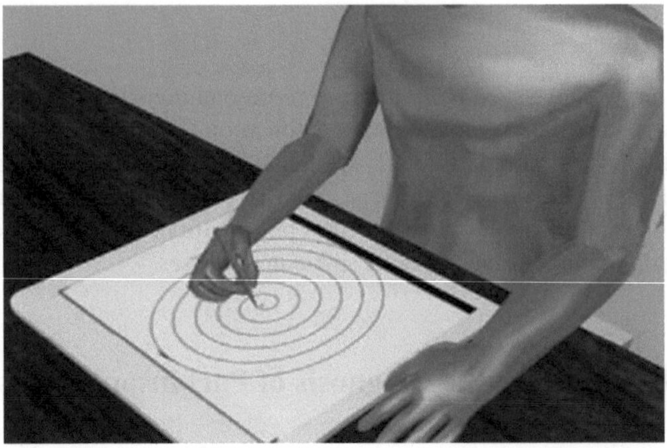

Fig. 11.11 Figure-copying task using a conventional digitizing tablet [17].

The patient's response to the task presented was a drawing activity that was digitized in realtime as a set of x–y coordinate pairs, providing information about pen position. The figure required to be copied by the patient was placed on the digitizing tablet in printed form and covered with a sheet of tracing paper, on which the patient traced a copy.

Although the term 'bradykinesia' is defined as the slowness of a performed movement [19], it is commonly used synonymously with 'akinesia' and 'hypokinesia', an expression of freezing and smaller movements, respectively. To quantify these symptoms, a measure of movement time and, particularly, the movement velocity of the patient's pen, was processed in the following way.

As the patient's drawing activity is digitized in realtime and at regular intervals, it is possible to determine the velocity of the pen at any instant. This can be achieved by calculating the distance between two coordinate positions and dividing this by the difference between the relative timestamps, as shown in the following equation:

$$v_{ij} = \frac{\sqrt{(x_j - x_i)^2 + (y_j - y_i)^2}}{t_j - y_i}, \tag{11.1}$$

where v_{ij} is the velocity of the pen between coordinates x_i, y_i and x_j, y_j, and t_i, t_j are the respective times at which the pen coordinates x_i, y_i and x_j, y_j were recorded

In order to train the CGP network to identify features associated with the symptoms of Parkinson's disease, the velocity profile needs to be converted into a form that will be suitable for input to the CGP network. From previous work by Smith at al., the relative velocities of the patient's pen, describing a two-part acceleration, are of particular interest in the detection of bradykinesia. Accordingly, the acceleration of the pen through the duration of the patient's velocity profile was calculated simply by differentiating with respect to time. The calculated acceleration or gradient values were then quantized and encoded according to the rules described in Table 11.7.

Table 11.7 Quantization and encoding of patient's data. The gradient is the acceleration of the patient's pen

Gradient range	Gradient encoding
gradient > 2	6
$1 \leq$ gradient ≤ 2	5
$0 <$ gradient < 1	4
gradient $= 0$	3
$-1 <$ gradient < 0	2
$-2 \leq$ gradient ≤ -1	1
gradient < -2	0

An implicit context representation of a Cartesian genetic program was used for the evolutionary algorithm in this application. A criticism of CGP (and GP in general) is that the location of genes within the chromosome has a direct or indirect influence on the resulting phenotype [20]. In other words, the order in which specific information regarding the definition of the GP is stored has a direct or indirect effect on the operation, performance and characteristics of the resulting program. Such effects can be considered undesirable, as they may mask or modify the role of the specific genes in the generation of the phenotype (or resulting program). Consequently, genetic programs are often referred to as possessing a direct or indirect context representation. An alternative representation for genetic programs in which genes do not express positional dependence has been proposed by Lones and Tyrrell [20, 21, 22], termed implicit context representation. Here, the order in which genes are used to describe the phenotype (or resulting program) is determined after their self-organized binding, based on their own characteristics and not their specific location within the genotype. The result is an implicit-context-representation version of traditional parse-tree-based GP termed 'enzyme genetic programming'. Smith et al. have implemented an implicit context representation of CGP, termed implicit-context-representation Cartesian genetic programming (IRCGP), specifically for the evolution of image-processing filters [23].

Implicit context representation employs an enzyme model comprising a shape, an activity and specificities (or binding sites) [24], as shown in Fig. 11.12. Along with inputs and outputs, the enzyme model can be considered as a program component, executing one of the functions listed in Table 11.8, from which a genetic program may be constructed. The shape describes how the enzyme is seen by other program components. Similarly, the binding sites determine the shape (and hence type) of the program component the enzyme wishes to bind to. Finally, the activity determines the logical function the enzyme is to perform. A typical IRCGP program will comprise a set number of inputs and outputs and a number of enzyme models or components. Initial values for each component's binding sites and logical function are assigned non-deterministically; the component's shape, however, is derived from a combination of its binding sites' shapes and logical function. Once initialized, the components are bound together to form a network, as shown in Fig. 11.13 The order in which components are bound is determined by the closeness of match between a component's binding site shape and another component's shape. The best-matching components are bound first, and the process is repeated until a network has formed in which no further binding is possible.

Over time, components may evolve through mutation. Mutation is applied to the component's binding sites and logical function with a predetermined probability. When this occurs, a new component shape is derived accordingly and may lead to different binding between components occurring. This, in turn, may result in a modified network.

The network of processing elements was arranged in 10 rows and three columns, as shown in Fig. 11.13. In addition, 10 input components and one output component can also be seen. The 10 input components were fed by 10 consecutive items of gradient data from the data described earlier. The value obtained at the output component was used to indicate whether a particular artefact (representing a Parkinsonian feature) is present.

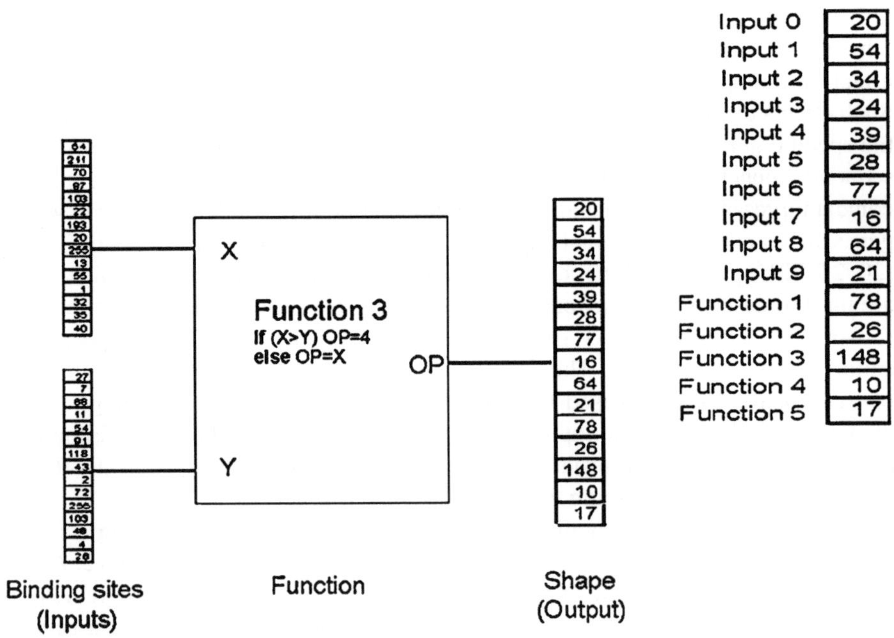

Fig. 11.12 Example of a processing element that forms part of an evolutionary network.

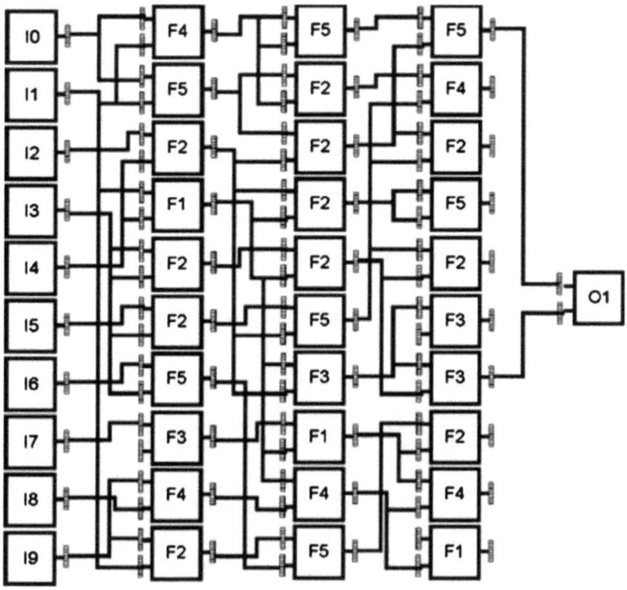

Fig. 11.13 Example evolutionary network.

Table 11.8 Functions available for processing elements. X is the first input value to the component. Y is the second input value to the component. OP is the output value of the component

Function index	Function definition
F1	**if** (X>Y+3) OP = 6 **else** OP=X
F2	**if** (X<Y+3) OP = 0 **else** OP=Y
F3	**if** (X>Y) OP = 4 **else** OP=X
F4	**if** (X<Y) OP = 2 **else** OP=Y
F5	OP = (X+Y)/2

The fitness function was based on the desire to identify some artefact representing a Parkinsonian symptom in the patient responses, but not in age-matched controls. The presence of the artefact was determined by a value returned by the output component of each individual network being greater than 3; a value less than or equal to 3 indicates the non-presence of the artefact. (The number 3 is the middle of the range of values possible at the output component.)

The fitness score comprised two parts, each dependent on whether a patient or age-matched control was being tested. For the response of a patient with Parkinson's disease, the fitness score was the number of artefacts detected; conversely, for an age-matched control's response, the fitness score was the number of non-artefacts detected (output values less than or equal to 3).

To achieve the aim of identifying a symptom of Parkinson's disease, it is only necessary to detect one such artifact in a patient's response. However, it is equally essential that no such artifacts are found in the responses of the age-matched controls. For this reason the fitness function used to evolve the network was weighted heavily in favour of non-detection of artifacts in the age-matched control responses. This was achieved by using an exponential function to bias the fitness scores accordingly.

For the results presented below, a network was evolved using a population size of 5 over 10,000 generations. A conventional elitist strategy was adopted, with a mutation rate of 6% for the function used by each component and 3% for each dimension of the binding sites' shapes.

Twelve patients with idiopathic Parkinson's disease (5/12, 42% female) were assessed as well as 10 controls (4/10, 40% female) who did not have Parkinson's disease or any other neurological disorder. Participants were enrolled from a Parkinson's disease specialist clinic and day hospital, after giving informed consent approved by the Liverpool Research Ethics Committee. The average age of Parkinson's disease patients enrolled in this study was 74.1 years (SD = 8.4 years), and the exclusion criteria included drug-induced parkinsonism, Parkinson-plus and multi-system atrophy syndromes, Alzheimer's disease, and significant cognitive impairment. The majority of the controls were relatives attending with patients at the day hospital, and their average age was 73.2 years (SD = 5.3 years). In order to assess the performance of the system under conditions normally found in out-patient clinics, patients were not given any specific instructions regarding medication, and were tested under their normal medication regime.

Once data had been collected from the patients and the age-matched controls, it was presented to the evolutionary algorithm in two ways. First of all, the patients' and age-matched controls' responses were arbitrarily split into a training set and a testing set of approximately equal sizes. Using the evolutionary algorithm described above, the evolution of a network that would discriminate between the responses of patients with Parkinson's disease and of the age-matched controls was attempted. After 101 generations, a network with a control fitness of 99.068% and a patient fitness of 0.168% was evolved. It should not be surprising that the age-matched control fitness was not 100%, as many of the patients would have been under the influence of medication when tested, compensating for some of the symptoms of Parkinson's disease. Subsequently, the evolved network was used to discriminate between 11 patients' responses and 19 control responses which were not included in the training stage, the results for which are shown in Fig. 11.14. For each response, the number of occurrences of the artifact identified by the evolved network is shown. Ideally, no artifacts should be present in the age-matched control responses and at least one occurrence of the artifact should be present in every patient's response. As can be seen, artifacts were located in every response, but, importantly, more were located in the patients' responses than in the age-matched controls. More specifically, the age-matched controls all have five or fewer occurrences of the artifact, whereas the patients' responses each have six or more occurrences of the artefact.

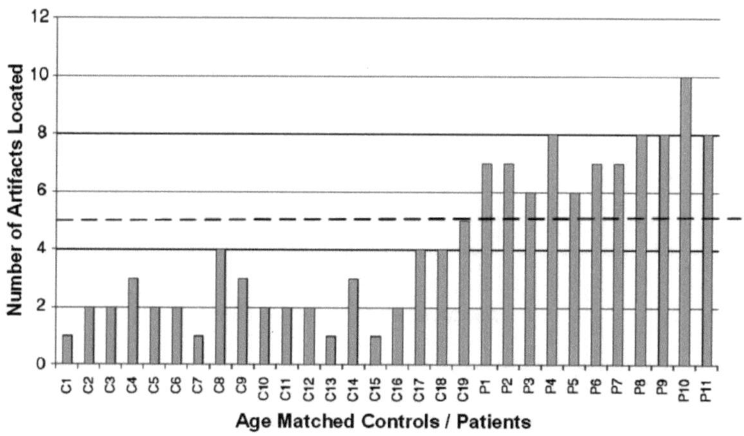

Fig. 11.14 Result of testing the evolved algorithm with patients (P1–P11) and age-matched controls (C1–C19) [17].

The second approach to processing the data was to apply leave-one-out cross-validation. In this case, all the data was included in the training set except for those responses relating to one patient. The network was evolved and the excluded patient data was then used as the test set. After the result was noted, this patient's responses were replaced in the training set and another patient's responses excluded, and so on until all patients' and age-matched controls' responses had been tested in this way. A graph showing the number of artefacts detected in each patient's and age-matched control's responses is shown in Fig. 11.15. The separation between patients and aged-matched controls is not as clear as for the arbitrary split of responses into training and testing sets, but the majority of patients have greater than three artefacts and the aged-matched controls this number or less.

This work demonstrates the potential of using a novel representation of CGP to discriminate between the responses of patients with Parkinson's disease and of age-matched controls to a simple figure-copying test. Although the discrimination between the two populations is not perfect, this is to be expected as the patients were receiving medication to suppress the very symptoms that were being measured.

11.4 CGP Applied to the Diagnosis of Alzheimer's Disease

The final case study of this chapter considers the application of CGP to the diagnosis of Alzheimer's disease, another demanding and important medical application.

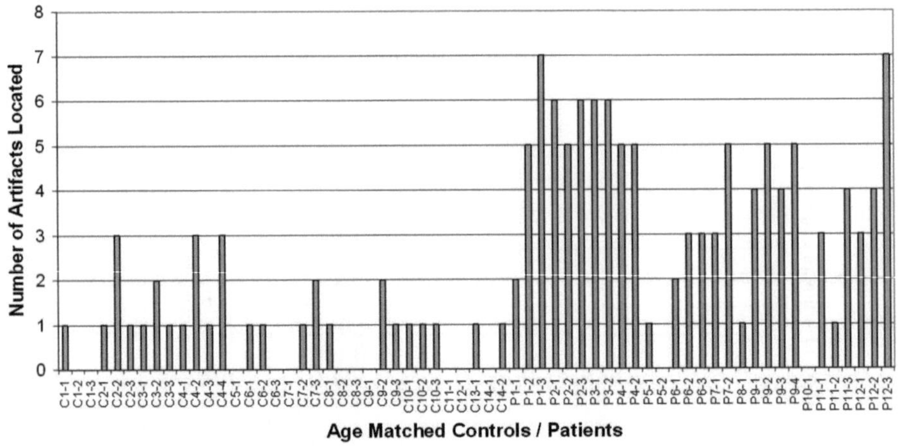

Fig. 11.15 Result of testing the evolved algorithm with patients (P1-1 to P12-3) and age-matched controls (C1-1 to C14-2) [17].

In their paper [25] Hazel and Smith used an infrastructure similar to that described above for the diagnosis of Parkinson's disease, using a commercially available digitizing tablet to record patients' responses to figure-copying tasks. However, in this case the pen movements were used to identify static structural components of the geometric figure drawn, rather than the dynamic acceleration/deceleration characteristics of the subject.

Alzheimer's disease is the most common form of dementia and accounts for 62% of dementia-type cases [26]. There are an estimated 24.3 million people in the world with dementia today and some 4.6 million new cases every year. The total number of affected people is expected to double every 20 years, with the largest prevalence, an estimated 60%, occurring in developing countries [27]. The disease occurs when amyloid β-protein forms miliary bodies (plaques) and dense bundles of fibrils (tangles) in the structure of the brain [28]. This leads to the disruption or death of neurons, causing symptoms such as confusion, disorientation, memory loss and communication problems [29]. As this is a progressive disease, the symptoms become increasingly more severe as further parts of the brain are damaged and, consequently, the patient's condition deteriorates. Whilst there is currently no cure for Alzheimer's disease, early diagnosis is essential, as it may permit treatment to slow down the progression of the disease and stabilize its symptoms. Medication can be prescribed to reverse non-neurodegeneration symptoms such as depression, and it is also important to allow the patient time to prepare for the future. Absolute diagnosis of Alzheimer's disease is possible only by examining brain tissue and is therefore impractical whilst the patient is alive. Owing to this difficulty, determining the presence of Alzheimer's disease is most often a diagnosis of exclusion, where the physician will try to find other causes of the symptoms, often by using laboratory tests and neuroimaging techniques. An important part of the diagnosis and monitoring of the disease is to perform a neurological examination to evaluate the extent of the impairment of the patient.

The most common method of diagnosis based on such examinations is the use of the NINCDS-ADRDA Alzheimer's Criteria [30], which examine eight cognitive domains: memory, language, perception, attention, constructive ability, orientation, problem solving and functional ability. Problems within these domains could suggest the onset of Alzheimer's disease, and the criteria lead to four possible outcomes: definite, probable, possible and unlikely Alzheimer's disease. Geometric shape-drawing tasks are often used as part of this assessment to evaluate visuo-spatial neglect. Several tests have been developed, such as the Clock Drawing Test, the Rey–Osterrieth Complex Figure Test and cube-drawing tests. Research into cube-drawing ability has shown not only that it is a useful tool in the detection of Alzheimer's disease but also that it is good at the detection of very mild Alzheimer's disease [31]. For cube-drawing assessments, detailed marking criteria are used to grade the cube and hence determine the level of impairment. One example of such a criterion is presented in [32]; this criterion is used to mark the development of cube-drawing ability in 7- to 10-year-olds and shows many similarities to the criteria used in [31] to mark drawings of elderly and Alzheimer's diseased patients. The scoring system, taken from [32], is as follows:

(a) A single square or rectangle of any orientation.
(b) A set of interconnected squares or rectangles numbering more or less than the number of visible faces in the cube (three) or a single trapezoid with some appropriate use of oblique lines.
(c) A set of three interconnected squares or rectangles not appropriately arranged to represent the visible arrangement of faces in the cube, or a set of interconnected squares or rectangles numbering more or less than the number of visible faces in the cube, including some appropriate use of oblique lines.
(d) A set of three interconnected squares or rectangles appropriately arranged to represent the visible arrangement of the faces of the cube, or an inappropriately arranged set of three outlines, including some appropriate use of oblique lines.
(e) Drawings that show only visible faces of the cube appropriately arranged (as noted previously) and that reveal crude attempts to show depth through use of oblique lines, curvature, or modification of angles.
(f) Drawings that approximate to oblique projection or linear perspective or drawings that approximate well to oblique projection or linear perspective, but are drawn to a horizontal ground line rather than to an oblique ground plane.
(g) Drawings that are close approximations to oblique projection or linear perspective but contain some inaccuracies in angular relations between lines.
(h) Accurate portrayals of a cube in oblique projection or linear perspective.

Figure 11.16 shows eight example drawings that have been classified based on this system.

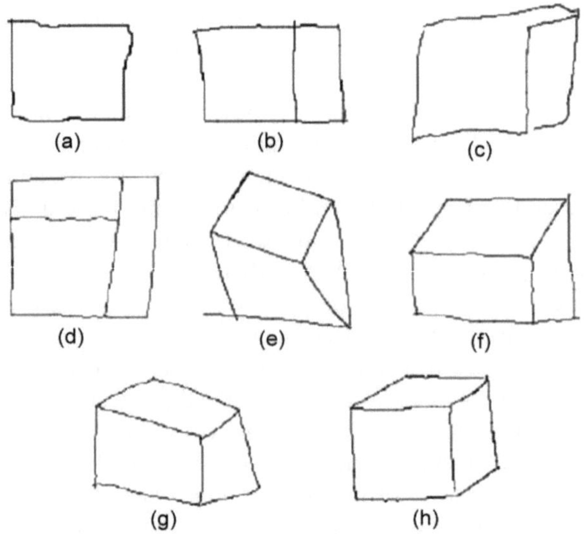

Fig. 11.16 Eight classifications of the data used in the study described here, based on the marking system set out by Bremner et al. [32].

Application of the assessment criteria by trained assessors can vary, and hence is arguably unreliable, so it is desirable to produce an assessment mechanism which will be able to classify cube drawings in a completely objective way.

There are considerable practical difficulties in obtaining data from Alzheimer's disease patients, owing to the availability of patients willing to participate and the lengthy, but necessary ethical approval procedures. However, as discussed earlier, there are close similarities between the development of visuo-spatial ability in 7- to 10-year-old children and the corresponding loss of visuo-spatial ability observed in Alzheimer's disease patients as the disease progresses; therefore, children can provide a much easier way of obtaining data. Drawings made by children can easily be digitized by using a commercial digitizing tablet, similar to that used in the Parkinson's disease work described in Sect. 11.3. Drawings were taken from children over an age range from 7 to 11 years. These children were divided into four year groups, and were given several attempts at drawing a cube. Once the data was collected, the cubes were manually classified, using the scheme described above into eight groups.

The preprocessing stage took the raw data, which was recorded from the graphics pad as a number of x and y coordinates, and classified it based on the angle between two points. Every twentieth point from this data was taken and the angles between consecutive pairs of points were calculated. A fairly large gap between points was chosen (i.e. 20) to reduce the effect of recorded noise that would have been introduced into the system if a very small increments between data points was used. An automated procedure then classified the resulting line into one of four groups: horizontal lines ranging from -10° and $+10^\circ$ (measuring from the horizontal) were classified as 1, lines between 25° and 45° were classified as 2, vertical lines between 80° and 100° were classified as 3 and all other angles were classified as 0. Figure 11.17 shows these ranges of angles.

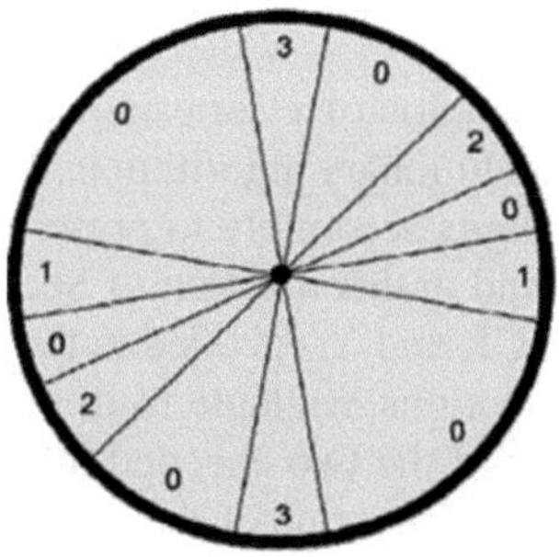

Fig. 11.17 Range of classifications used for drawings [25].

When the data was read into the algorithm there was an intermediate 'subwindow' which encoded the classification into a three-digit integer. The subwindow read in a number of classifications and calculated the total number of each classification. It then normalized this so that each number was between 0 and 9, and put them together so that the three-digit integer ABC referred to the relative amounts of horizontal, oblique and vertical components present in each section of line. This number could be between 000 (for when there were no classifications of 1, 2 or 3) and 900. Figure 11.18 shows three examples of the encoding.

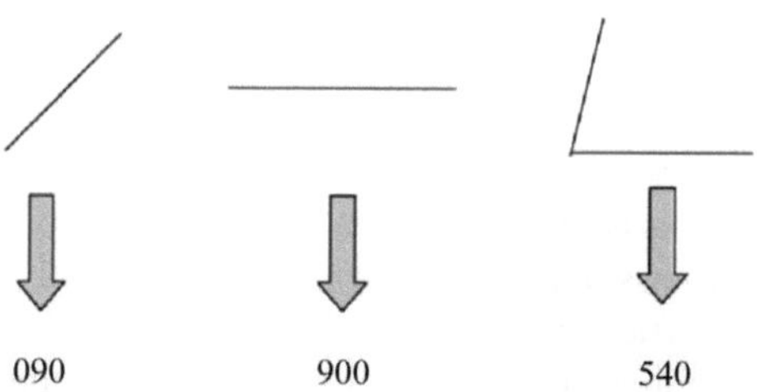

Fig. 11.18 Three examples of the encoding scheme [25].

The purpose of having an intermediate step between the preprocessing classification and the CGP is to preserve the context of a line, and to give the CGP a small overview of a particular section of the drawing. In this way, the relationships between the different-angled lines are considered by the algorithm, rather than just the presence or absence of different angles. For example, without the subwindow the actual numeric value of a number would have little significance, but with the encoding a larger number signifies that the component makes up a larger part of a section of the drawing. This method has the advantage of preserving the angular information in context but at the cost of reducing the amount of data the network uses.

An implicit context representation of a Cartesian genetic program as described in Sect. 11.3, was used for the evolutionary algorithm in this application. The functions always treated the different digits separately to preserve the encoding. There were two inputs to each function, and in this notation of Sect 11.3 the two inputs were X encoded in the form A1, B1, C1, and Y, encoded in the form A2, B2, C2.

The network had seven functions available to it, as shown in Table 11.9. Function 7 averages all three digits of the two encoded integer inputs (independently of each other) whilst the others compare one of the components of the two inputs (for example just the vertical component) and either returns one of the inputs or returns the average of both inputs.

Table 11.9 Function set

Function index	Function definition
F1	**if** (A1>A2) OP = A1 **else** OP=average(X+Y)
F2	**if** (B1>B2) OP = B1 **else** OP=average(X+Y)
F3	**if** (C1>C2) OP = C1 **else** OP=average(X+Y)
F4	**if** (A2>A1) OP = A2 **else** OP=average(X+Y)
F5	**if** (B2>B1) OP = B2 **else** OP=average(X+Y)
F6	**if** (C2>C1) OP = C2 **else** OP=average(X+Y)
F7	OP = average(X+Y)

Two fitness functions were used to differentiate between control and patient responses by making comparisons between the different components at the output of the network. The first fitness function, shown in Procedure 11.1, is based on the idea that the horizontal and vertical contents of patient and control files are different, with a patient response containing more vertical lines based on their neglect in drawing oblique components. The second fitness function (Procedure 11.2) directly compares the amount of oblique components to the amount of horizontal components. By making a direct comparison between two components (as opposed to making a comparison with a number), the effect of the difference in the size between different cube drawings is normalized.

For the results presented in this chapter a network was evolved using a population size of 4 over 5000 generations. A conventional elitist strategy was adopted, with a mutation rate of 6% for the function used by each component and 3% for each dimension of the binding sites' shapes.

Procedure 11.1 Fitness function 1

```
if (loop = 0) then
    if (A < C) then
        pat_fitness ← pat_fitness + 1
    end if
end if
if (loop = 1) then
    if (A ≥ C) then
        con_fitness ← con_fitness + 1
    end if
end if
```

Procedure 11.2 Fitness function 2

```
if (loop = 0) then
    if (3B < C) then
        pat_fitness ← pat_fitness + 1
    end if
end if
if (loop = 1) then
    if (3B ≥ A) then
        con_fitness ← con_fitness + 1
    end if
end if
```

The data consisted of 142 drawings from 27 children. They were first arbitrarily split into training and testing sets, and then split into patient and control groups so that cubes classified as 1–4 on the scale were grouped as patient responses and cubes classified as 5–8 were grouped as control responses. For these results, 22 patient and 51 control responses were used to train the algorithm, and 21 patient and 48 control responses were used to test it.

After 10 runs the highest-evolved chromosome from the first fitness function had a patient fitness of 75% and a control fitness of 94%. This chromosome was tested on the test set, and the percentages of artefacts found are shown in Fig. 11.19.

The second fitness function was then used to evolve a second network using the same training set, and after 10 runs the highest-evolved chromosome had a patient fitness of 69% and a control fitness of 79%. This chromosome was tested on the same test set, and the percentages of artefacts are shown in Fig. 11.20.

If a threshold of 50% is set, the first fitness function has 100% of the patient responses below the threshold and 85% of the control responses above the threshold. The second fitness function has 86% of the patient responses below the threshold and 98% of the control responses above the threshold.

The results from the two fitness functions can be averaged to help minimize anomalies, the results of which are shown in Fig. 11.21. For these combined results,

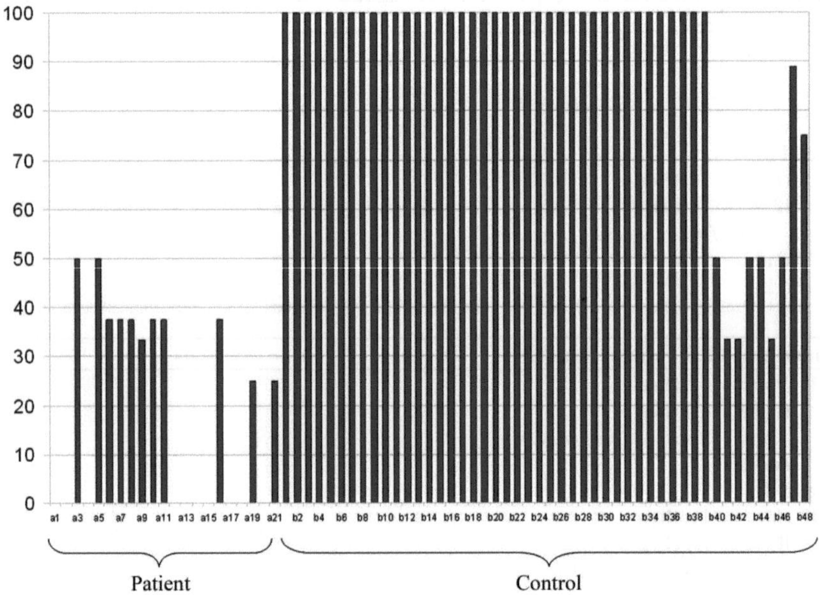

Fig. 11.19 Percentages of artefacts in test responses for fitness function 1 [25].

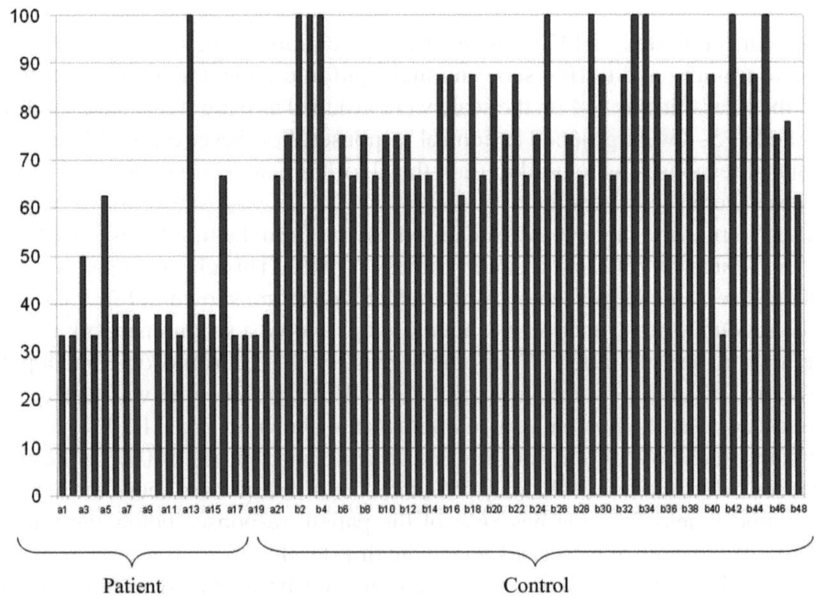

Fig. 11.20 Percentages of artefacts in test responses for fitness function 2 [25].

setting a threshold of 60% results in 100% of the patient responses being below the threshold and 98% of control responses being above.

The results show that a classification can be made using CGP by taking the very basic angular information, without the need for complex rule sets and image analysis. The accuracy might be improved further by extending the encoding scheme to include other properties in the data such as time taken, the rate of change of angles and more complex features such as tremor or hesitation. It could also possibly be developed by using a different more extensive classification scheme.

It is anticipated that the method could be extended to include multiple classifications, so that the algorithm could correctly and objectively group data into the eight groups shown in Fig. 11.16 to be used as a measure of the onset of Alzheimer's disease and as a diagnostic tool.

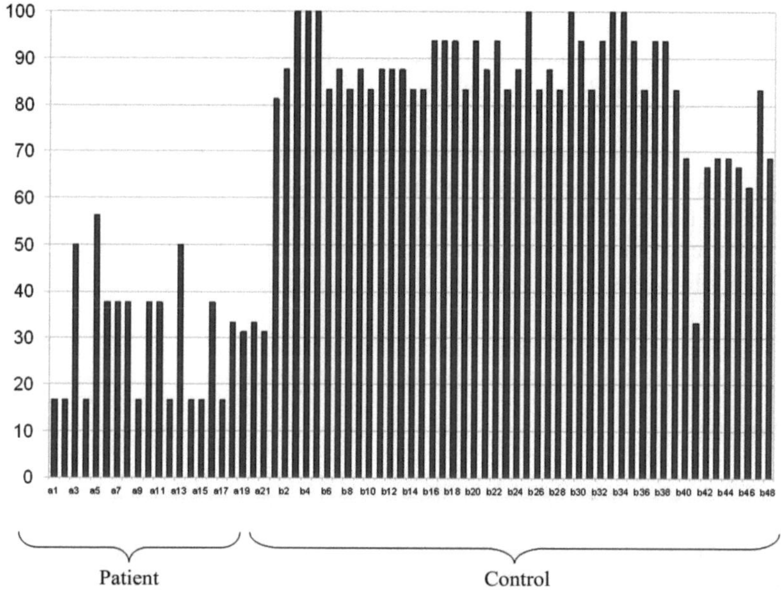

Fig. 11.21 Percentages of artefacts in test responses for the combined results of fitness functions 1 and 2 [25].

11.5 Conclusions

The work described in this chapter has demonstrated the effectiveness and flexibility of CGP in a range of medical applications. The flexible geometry of the network accommodates data from a range of sources, such as raw pixel values from images or a continuous stream of signal data. The use of a constrained directed graph provides benefits over conventional GP by providing a more predictable and 'well-behaved' behaviour, as well as performance benefits.

Medical applications are ripe for exploitation, providing numerous complex problems which traditional signal-processing algorithms have failed to solve. The novel approach of training evolutionary algorithms is one which suits the field of medicine well, by learning strategies and diagnostic procedures from health professionals providing objective and reliable results to assist in diagnosis and therapy.

There are, however, a number of challenges that need to be surmounted when one is working in the clinical environment. Gold-standard or ground-truth data sets are necessary to train CGP effectively to distinguish between different conditions, such as malignant or benign tumours, or Parkinson's or Alzheimer's disease, for example. Obtaining such data sets not only is expensive and time-consuming, but also requires 100% accurate diagnosis by conventional means, which is often very difficult as it commonly involves subjective assessment by health professionals. Obtaining ethical approval and statutory authorization is another consideration, which will require specialist resources and time to achieve.

At the end of this chapter, we are left with the exciting conclusion that CGP is a simple yet very powerful evolutionary algorithm which has been shown to be particularly effective in a range of medical applications and stands to make a significant impact in the future of healthcare. Clinical trials using some of the systems outlined above are already being undertaken in hospitals in the UK, and commercialization of the technology has begun with the filing of a patent.

References

1. *Cancer Research UK*. http://info.cancerresearchuk.org/cancerstats/types/breast/incidence/, 27-1-2011 ed.
2. V. Andolina, S. Lillé, and K. M. Willison, *Mammographic Imaging: A Practical Guide*. Lippincott Williams and Wilkins, 2001.
3. H. D. Cheng, X. Cai, X. Chen, L. Hu, and X. Lou, "Computer-aided detection and classification of microcalcifications in mammograms: a survey," *Pattern Recognition*, vol. 36, pp. 2967–2991, 2003.
4. D. B. Kopans, *Breast Imaging*. Lippincott Williams and Wilkins, 2006.
5. American College of Radiology, *Breast I Reporting and Data System Atlas (BI-RADS): Mammography*, 4 ed., 2003.
6. R. Gonzalez and R. Woods, *Digital Image Processing*. Prentice Hall, 2002.
7. J. C. Fu, S. K. Lee, S. T. C. Wong, J. Y. Yeh, A. H. Wang, and H. K. Wu, "Image segmentation feature selection and pattern classification for mammographic microcalcifications," *Computerized Medical Imaging and Graphics*, vol. 29, pp. 419–429, 2005.
8. M. Gavrielides, J. Lo, R. Vargas-Voracek, and C. Floyd, "Segmentation of suspicious clustered microcalcifications in mammograms," *Medical Physics*, vol. 27, pp. 13–22, 2000.
9. J. Kim and H. Park, "Statistical Textural Features for Detection of Microcalcifications in Digitized Mammograms," *IEEE Transactions Medical Imaging*, vol. 18, no. 3, pp. 231–238, 1999.

10. D. Hope, S. L. Smith, and E. Munday, "Evolutionary Algorithms in the Classification of Mammograms," in *IEEE Symposium on Computational Intelligence in Image and Signal Processing*, pp. 258–265, 2007.
11. K. Völk, J. F. Miller, and S. L. Smith, "Multiple Network CGP for the Classification of Mammograms," in *Applications of Evolutionary Computing*, vol. 5484 of *LNCS*, pp. 405–413, 2009.
12. J. A. Walker, K. Völk, S. L. Smith, and J. F. Miller, "Parallel evolution using multi-chromosome Cartesian genetic programming," *Genetic Programming and Evolvable Machines*, vol. 10, no. 4, pp. 417–445, 2009.
13. J. F. Miller and S. L. Smith, "Redundancy and Computational Efficiency in Cartesian Genetic Programming," *IEEE Transactions on Evolutionary Computation*, vol. 10, no. 2, pp. 167–174, 2006.
14. *Parkinson's UK*. http://www.parkinsons.org.uk/about_parkinsons/what_is_parkinsons.aspx, 28-1-11 ed.
15. Parkinson's Disease Foundation. http://www.pdf.org/en/parkinson_statistics, 28-1-11 ed.
16. C. Levine, K. R. Fahrbach, A. D. Siderowf, R. P. Estok, V. M. Ludensky, and S. D. Ross, *Diagnosis and Treatment of Parkinson's Disease: A Systematic Review of the literature*. No. 57, Agency for Healthcare Research and Quality, 2003.
17. S. L. Smith, P. Gaughan, D. M. Halliday, Q. Ju, N. M. Aly, and J. R. Playfer, "Diagnosis of Parkinson's Disease using Evolutionary Algorithms," *Genetic Programming and Evolvable Machines*, vol. 8, pp. 433–447, 2007.
18. The main symptoms of Parkinson's. http://www.parkinsons.org.uk/about_parkinsons/signs_and_symptoms/the_main_symptoms.aspx, 23-2-11 ed.
19. A. Berardelli, J. C. Rothwell, P. D. Thompson, and M. Hallett, "Pathophysiology of bradykinesia in Parkinson's disease," *Brain*, vol. 124, no. 11, pp. 2131–2146, 2001.
20. M. A. Lones, *Enzyme Genetic Programming*. PhD thesis, University of York, 2003.
21. M. A. Lones and A. M. Tyrrell, "Biomimetic representation with enzyme genetic programming," *Genetic Programming and Evolvable Machines*, vol. 3, no. 2, pp. 193–217, 2002.
22. M. A. Lones and A. M. Tyrrell, "Modelling biological evolvability: Implicit context and variation filtering in enzyme genetic programming," *BioSystems*, vol. 76, no. 2, pp. 229–238, 2004.
23. S. L. Smith, S. Leggett, and A. M. Tyrrell, "An implicit context representation for evolving image processing filters," in *Proc. Workshop on Evolutionary Computation in Image Analysis and Signal Processing*, vol. 3449 of *LNCS*, pp. 407–416, 2005.
24. M. A. Lones and A. M. Tyrrell, "Enzyme genetic programming," in *Proc. of the Congress on Evolutionary Computation* (J.-H. Kim, B.-T. Zhang, G. Fogel, and I. Kuscu, eds.), vol. 2, pp. 1183–1190, IEEE Press, 2001.
25. A. Hazell and S. L. Smith, "Towards an Objective Assessment of Alzheimer's Disease: The Application of a Novel Evolutionary Algorithm in the Analysis of Figure Copying Tasks," in *Proc. GECCO Workshop on Medical Applications of Genetic and Evolutionary Computation*, 2008.
26. M. Knapp and M. Prince, *Dementia UK: A report to the Alzheimer's Society on the prevalence and economic cost of dementia in the UK*. Alzheimer's Society, 2007.
27. C. Ferri, M. Prince, C. Brayne, H. Brodaty, L. Fratiglioni, M. Ganguli, K. Hall, K. Hasegawa, H. Hendrie, Y. Huang, A. Jorm, C. Mathers, P. R. Menezes, E. Rimmer, and M. Scazufca, "Global prevalence of dementia: A Delphi consensus study," *The Lancet*, vol. 266, no. 9503, pp. 2112–2117, 2006.
28. K. Blennow, M. Leon, and H. Zetterberg, "Alzheimer's disease," *The Lancet*, vol. 268, pp. 387–403, 2006.
29. W. Wolfson, "Unraveling the Tangled Brain of Alzheimer's," *Chemistry and Biology*, vol. 15, no. 2, pp. 89–90, 2008.
30. G. McKhann, D. Drachman, M. Folstein, R. Katzman, D. Price, and E. M. Stadlan, "Clinical diagnosis of Alzheimer's disease: Report of the NINCDS-ADRDA Work Group under the auspices of Department of Health and Human Services Task Force on Alzheimer's Disease," *Neurology*, vol. 34, pp. 939–944, 1984.

31. Y. Shimada, K. Meguro, M. Kasai, M. Shimada, S. Yamaguchi, and A. Yamadori, "Necker cube copying ability in normal elderly and Alzheimer's disease. A community-based study: The Tajiri project," *Psychogeriatrics*, vol. 6, pp. 4–9, 2006.
32. J. Bremner, R. Morse, S. Hughes, and G. Andreasen, "Relations between drawing cubes and copying line diagrams of cubes in 7-to 10-year old children," *Child Development*, vol. 71, no. 3, pp. 621–634, 2000.

Appendix A
Resources for Cartesian Genetic Programming

A.1 General Advice

The best and most reliable way to look for resources relating to CGP is to carry out internet searches using the names of well-known researchers who use CGP. Many of the authors of chapters in this book are just such people. Their web sites may change, but they will always be traceable somewhere on the internet! The first place to look is at the web sites associated with their institutions (listed at the beginning of this book).

A.2 Web Sites

There is a web site on CGP at

http://www.cartesiangp.co.uk

This contains links to many researchers who have used CGP. From these, you can obtain many of the publications on CGP.

A.3 Tutorial Material

Julian Miller and Simon Harding have given a tutorial on CGP over the last few years at the Genetic and Evolutionary Computation Conferences (and sometimes at other conferences too). The tutorials are freely available from Julian Miller's publications page

https://sites.google.com/site/julianfrancismiller/publications

J.F. Miller (ed.), *Cartesian Genetic Programming*, Natural Computing Series,
DOI 10.1007/978-3-642-17310-3, © Springer-Verlag Berlin Heidelberg 2011

A.4 Software

Software for CGP has been made available by various researchers.

A.4.1 CGP in C

Julian Miller has a C program implementation of CGP that is currently available from

https://sites.google.com/site/julianfrancismiller/professional

There is a collection of C programs there that can be compiled into programs that can handle three kinds of data with CGP, namely Boolean, integer and floating-point. Configuration files are included that will allow the user to run experiments on evolving Boolean circuits, mathematical expressions (Koza symbolic regression problems) and predicting prime numbers. Documentation is included about how to run the experiments and set parameters, etc. The C programs are well documented with instructions about how to adapt the programs for your particular application.

Zdenek Vasicek and Lukas Sekanina have developed a set of tools for Cartesian Genetic Programming (CGP) containing an input data reading module, CGP implementation and a CGP chromosome viewer (cgpviewer). Cgpviewer enables to display the CGP chromosome, simulate it and convert it to various output file formats (VHDL, bmp).

http://www.fit.vutbr.cz/research/view_product.php.en?id=61&notitle=1

A.4.2 CGP in Java

David Oranchak has implemented CGP in Java. Documentation is available at

http://oranchak.com/cgp/doc/

Java software is available at

http://oranchak.com/cgp/contrib-cgp-18.zip

He has implemented CGP using two representations. One is integer-based (i.e. the classic CGP described in Chap. 2), and the other is a real-numbered representation, which is mentioned in the same chapter in Sect. 2.6.2. His program can be run on the following example problems:

1. Symbolic regression.

2. Classification:
    a. iris data set;
    b. breast cancer.
3. Various parity problems.

There are also example runs of the program listed.

A.4.3 CGP in MATLAB

Jordan Pollack has kindly made available a MATLAB implementation of CGP that can handle symbolic regression problems. This is available at

https://sites.google.com/site/julianfrancismiller/publications

A.4.4 Evolutionary Art with Laurence Ashmore's CGP Program (in Java)

Laurence Ashmore has written an evolutionary art system using CGP. It is available from his web site

http://www.emoware.org/evolutionary_art.asp

Index

J.F. Miller (ed.), *Cartesian Genetic Programming*, Natural Computing Series,
DOI 10.1007/978-3-642-17310-3, © Springer-Verlag Berlin Heidelberg 2011